THE TELEOLOGIES IN HUSSERLIAN PHENOMENOLOGY

ANALECTA HUSSERLIANA

THE YEARBOOK OF PHENOMENOLOGICAL RESEARCH

VOLUME IX

Editor:

ANNA-TERESA TYMIENIECKA
The World Institute for Advanced Phenomenological Research and Learning

THE TELEOLOGIES IN HUSSERLIAN PHENOMENOLOGY

THE IRREDUCIBLE ELEMENT IN MAN

PART III

'Telos' as the Pivotal Factor of Contextual Phenomenology

Papers read at the VIth International Phenomenology Conference,
University of Arezzo/Siena, July 1–July 6, 1976

organized by

The International Husserl and Phenomenological Research Society,
Belmont, Massachusetts, U.S.A. and
Centro Italiano di Fenomenologia, Rome
Angela Ales Bello, Armando Rigobello, Mario Sancipriano

Edited by

ANNA-TERESA TYMIENIECKA

D. REIDEL PUBLISHING COMPANY

DORDRECHT : HOLLAND / BOSTON : U.S.A.

LONDON : ENGLAND

Library of Congress Cataloging in Publication Data

International Phenomenology Conference, 6th,
University of Arezzo/Siena, 1976.
The teleologies in Husserlian phenomenology.

(Analecta Husserliana ; v. 9)
Includes bibliographical references and index.
1. Husserl, Edmund, 1859-1938–Congresses.
2. Phenomenology–Congresses. 3. Teleology–Congresses.
I. Tymieniecka, Anna Teresa. II. Centro italiano di fenomenologia.
III. International Husserl and Phenomenological Research Society.
IV. Title. V. Series.
B3279.H94A129 vol. 9 193 79-16411
ISBN 90-277-0981-5

Published by D. Reidel Publishing Company,
P.O. Box 17, Dordrecht, Holland

Sold and distributed in the U.S.A., Canada, and Mexico
by D. Reidel Publishing Company, Inc.
Lincoln Building, 160 Old Derby Street, Hingham,
Mass. 02043, U.S.A.

Printed in The Netherlands

TABLE OF CONTENTS

INAUGURAL LECTURE

PART I: PROBLEMS OF TELEOLOGY IN THE SCIENCES OF NATURE AND IN THE HUMAN SCIENCES

PART II: THE TELIC PRINCIPLES

CLOSURE

COMPLEMENTARY SECTION: PHENOMENOLOGY IN ITALY

THE THEME

THE TELEOLOGIES IN HUSSERLIAN PHENOMENOLOGY

What issue has been more neglected in phenomenological scholarship than the problem of *Telos*?

In his last book, Husserl characterizes the historical process of Western humanity as the progressive elucidation of the *rationale* of man's existence within his *life-world*, with transcendental phenomenology as its *telos*. He interprets phenomenology, then, as the culminating stage of the philosophical foundation of all knowledge. And since the progress of knowledge is identified with the constitution of the forms of the *life-world*, the development of Western civilization can be seen as tending toward the final stage of transcendental phenomenology.

A deeper investigation of the nature of the constitutive progress of man and of his *life-world*, as conceived by Husserl, provides a glimpse of the *telos* of this very progress. The idea occurs at various stages and in different contexts in the development of Husserlian thought, but these instances are little more than hints dropped in passing. It is, therefore, little wonder that they have received scarcely any attention. They appear in a fuller light only in Husserl's personal notices in the latest publications of his posthumous work.

In point of fact, starting from ordinary givenness and its simplest forms, the constitutive genesis of man and his *life-world* unfolds progressively more and more complex structures. A categorial system of forms thus emerges, comparable to the hierarchical net of Plato's ideas, within which Husserl attributes a special role to the "form of all forms." It emerges as the *ultimate telos* of the constitutive process. In his recently published personal notices, the ultimate *telos* is identified with the idea of God: it is the "absolute reason," the *ethos* of humanity in constitutive progress.

Husserl's analysis did not encompass the complete range of the constitutive unfolding of man and of his *life-world*, which spreads in all directions. His personal reflections, which have only recently been published, lead one to wonder whether he did not reach a limit beyond which his investigation could not advance. Indeed, the ultimate telos seen as "absolute reason" is merely the prolongation of the rational thread of constitution. Yet, the thread of constitutive *rationale* does not reach the empirical conditions of its own origin and advance. Had he pursued his reflections further, he might have taken a

Tymieniecka (ed.), Analecta Husserliana, Vol. IX, ix–xiv.

new turn with respect to his previous eidetic and transcendental conclusions, and thus been obliged to begin, as it were, from a wholly new standpoint. His alleged utterance on his deathbed of a regret that his life is ending, at the very moment when "he could start all over again," expresses the very genius of Husserl. Indeed, we may wonder whether he would not be led to seek the ultimate telos different from that of "the form of all forms," and whether the framework laid down by Husserl would not have to be shattered, to accomplish this aim.

Here, the idea of "telos", not restricted to the constitutive genesis but as an *irreducible* principle of the total phenomenological project, must be made the focus of our attention and investigation.

The conception of telos in Husserlian phenomenology has so far appeared merely as the postulate of the processlike nature of the transcendental constitution of man and of his individual, personal, social, and cultural world. Consequently, when I invited scholars to collaborate on a common investigation of the "teleologies in Husserlian phenomenology" to be presented at our Arezzo conference, the plural (namely, "teleologies") of the proposal met with considerable surprise. In his reply, a noted historian of the phenomenological movement expressed incredulity if not misgivings. And yet, in general, the project elicited considerable interest and met with great enthusiasm, of which the present volume is but one manifestation.

We might say that the topic of our conference has remained hidden in the recesses of Husserlian reflection, concealed by the strata of other themes; and that, after a long germination, it has now come forth in a mature form. Yet, there were obstacles in the way; and we should, therefore, consider the conditions which favor such investigations as those presented at the Arezzo conference. Finally, we must ask how did the multiple teleologies manifest themselves in our common investigation?

1. To begin with the last question: the present book, which by its organization and coherence of thought and style might as well have been written by several authors working in common, is the result of the enthusiasm evoked by our original proposal. Each of our contributors, in fact, has found within his specific field of research, as approached from the phenomenological point of view, an intrinsic *telic principle*, a result the more remarkable when we consider that the fields of specific research extend from psychiatry through the methodology of science, historiography, anthropology, and natural science, to the spiritual life, the sacred, and eschatology. Within each of these seemingly separated fields phenomenological insight has laid bare a telic principle which appears as a fragment of a larger pattern. Indeed, the fragmentary

telic elements, partly completing each other, partly confirming each other's presence within a different realm of investigation, *seem to fall into one vast pattern*. Its arteries are still merely outlined. They emerge dimly but with a forceful, suggestive impact. The question then arises: "Are we not facing a vast network of purposeful organization at the heart of the HUMAN CONDITION?"

The problem of the telos appears, moreover, as *the point of convergence of disparate scholarly pursuits* if carried out within the phenomenological perspective.

2. Despite the fact that the idea of "telos" in Husserl's reflection has already been established, yet it is not the telos of the transcendental constitution that has inspired the present volume.

On the contrary, the idea of the "teleological" network inherent to the phenomenological program of the reconstruction of the human universe, arose from the subject matter itself in my research in cosmology, creative experience, and the sacred. The presence of the telos in the study of the sacred has since found confirmation in publications by both Angela Ales Bello and Mario Sancipriano, whose collaboration in our common venture is evident. At the same time, my researches into creative experience have played a critical role in fostering a congenial environment for the emergence of the idea of "teleology" in its widest significance.

Indeed, it is now time to retrieve the hitherto concealed, if not overlooked, significance of the problem. On the one hand, the analysis of the *creative function* of man has broken down the rigid system of intentionality to reveal the profound recesses of the *subliminal dimensions*. On the other hand, the renewed investigation of the Husserlian texts have pointed to the same *ante-predicative* realm by stating the presence of what M. Ricoeur calls "The surplus of meaning" within early Husserl's seemingly strictly defined notions.

Thus freed from the rigid methodological presumptions of the rational framework as well as from one-sided interpretations, phenomenological territory is ready for the telos to be investigated as the guiding insight or intuition of the phenomenologist Master.

3. As is well known, the fertility of the philosophical body of thought does not lie in its explicitly developed themes but in the profound insights and intuitions that have inspired their treatment. It is the richness of these insights of the human mind wrestling with the enigmatic nature of being that carries the quest of an original thinker like Husserl.

He attempts in his investigation, doctrine, or theory to find appropriate conceptual means for their transmission. Although these insights may fail to be developed and adequately conveyed by him, they present a challenge to other minds which, penetrating beyond his original reflection, will be led below its surface framework and seek deeper into its labyrinth. In the effort to reach the givenness through established theory, the interpreter is led again by these initial insights and intuitions. In order to retrieve them he may disclaim, as provisory, the established framework of thought.

We may then need to break it and attempt to project a new one more apt to thematize this hidden set of insights and intuitions, as the founding and leading principles of the whole philosophical edifice. In short, to acknowledge their proper place this edifice has to be revised, reinterpreted and built up anew.

Indeed, with the removal of the barriers set by the hegemony of intentionality and the expansion of the phenomenological investigation into the CONTEXTUAL ANALYSIS, which leads beyond the limits of intentionality and enlarges the criterion of the immediate givenness to the IRREDUCIBLE elements within man and the human condition, the TELOS emerges as the pivotal theme for a renewed phenomenological investigation as well as for a reinterpretation of the Husserlian endeavor approached from within.

4. It is with good reason that Paul Ricoeur presided over our Arezzo conference; also, it is not mere coincidence that it is to the Italian phenomenologists that we owe, in large measure, the fruitful accomplishment of the undertaking. In fact, quite naturally, we may call it our "Italian enterprise."

It was Paul Ricoeur who first discussed the telos of history in Husserl and who was concerned with eschatology; and it is the humanistic tradition and spirit which pervades Italian philosophy which makes the Italian phenomenologists particularly sensitive to this theme.

The intense interest which our convention awoke among Italian scholars has been characterized by Angela Ales Bello, in a report of the Conference published in *Sapienza*, as the "revival of phenomenology in Italy."

Although not well known abroad, phenomenology has taken deep root in Italian reflection, as we may judge from the following summary of Mario Sancipriano's account of the historical development of phenomenology in Italy.

The movement which originated in Germany at the beginning of this century with the advent of phenomenology was introduced to Italy in 1923 by Anthony Banfi in an original article dedicated to the *Logical Investigations* and was acknowledged, ten years later, by Annibale Pastore, as a new mode of philosophizing which was to result in a renewal of logic itself. In 1939 there appeared a comprehensive introductory volume by Sofia Vanni Rovighi, *The Philosophy of Edmund Husserl*; we then had, with the contributions of other scholars, the *first wave* of phenomenological studies in Italy; but World War II halted its progress: not even in the immediate postwar era was there a noticeable revival. In 1955, however, after the congress at Gallarate (Varese) dedicated to phenomenology, there was a definite renewal of interest in it: a new current of studies developed, tending toward a *spiritualistic* and from a certain aspect, *realistic* interpretation of Husserlian phenomenology. In this same line of inquiry a rereading of Bergson, from a phenomenological viewpoint, was proposed in 1957.

At the same time, Marxist students eager to discover the foundation of intersubjective relations gathered round the periodical *Aut-aut* directed by Enzo Paci. This provided impetus for the *second wave* of phenomenological studies, about the second half of the century. Critical investigations continued and various volumes of homage to Husserl were produced.

In 1959, phenomenology was presented as a "method"; but the publication of Husserl's later manuscripts actually came to confirm phenomenology even further as a doctrine with its own speculative meaning. We then had in Italy numerous works conducted with rigor, historiographically, in order to reconstruct Husserlian phenomenology in its objective contents, independently from any ideology.

Today, we wish to propose again an objective reconstruction, with a "return to Husserl," to the central problems which interested him, and to his painstaking analyses. It was this same program – of the reconstruction of the authentic Husserlian philosophy and of return to its original sources – that the Study Group of the "International Husserl and Phenomenological Research Society", *Centro Italiano di Fenomenologia* of Rome (which has promoted the present Congress), was constituted in 1975 by A.-T. Tymieniecka.

In a Western cultural world such as this, which has been accused of having destroyed reason, it may be useful to return to a philosophy which first reaffirms, in the *method*, the purity of Logic, and second rediscovers, in the *content*, the dynamism of human thought toward its rational end.

It is, accordingly, only natural that we have convened in the city of Arezzo, home of Petrarca and of Michelangelo, to debate in the spirit of a new Renaissance the topic of *telos* – a humanistic topic par excellence!

ANNA-TERESA TYMIENIECKA

ACKNOWLEDGMENTS

Now that the present volume containing the papers read at the 6th International Phenomenology Conference of *The International Husserl and Phenomenological Research Society* and of the affiliated *Centro Italiano di Fenomenologia*, Rome, is published, it is the privilege of the Editor, who prepared the program of the Conference, to express her warmest thanks to all those who made this undertaking possible and who have allowed us to mobilize the international, but especially Italian, phenomenologists toward the common effort of clarifying a new aspect of Husserlian thought. In my role as program chairman, I owe, first of all, thanks to the organization committee, that is, to Paul Ricoeur, who has acted as the President of the Conference, Mario Sancipriano, the Vice-President, and Angela Ales Bello, the Secretary General, who have devoted themselves to making this event a success.

Furthermore, our thanks go to: The Honorary Committee, under the auspices of the 'Regione Toscana' /President Avv. L. Lagorio/, presided by Prof. M. Barni, President of the University of Siena, and composed by the following members: Prof. J. Chisholm /Dept. of Philosophy, Brown University, Providence, R.I./, Prof. B. D'Amore /University S. Tomaso d'Aquino, Rome/, Prof. A. Ducci Sindaco of Arezzo/, Dr. E. Eng /Veterans Administration Hospital, Lexington, Ky., U.S.A./ Prof. H. G. Gadamer /University of Heidelberg/, Prof. J. Kockelmans /Dept. of Philosophy, Penn. State University/, Prof. E. Levinas /University of Paris/, Prof. W. Mays /University of Manchester/, Prof. Werner Marx /Freiburg/, Prof. J. N. Mohanty /The New School of Social Research, New York/, Prof. S. Nicolosi /University of Siena/, Prof. H. Nitta /Psychiatric Institute, University of Tokyo, Japan/, Prof. J. C. Piguet /Institute of Philosophy, Univeristy of Lausanne/, Prof. G. Previtali /University of Siena/, Prof. P. Prini /University of Rome/, Prof. M. Rossi /University of Siena/, Prof. M. Salmi /President of 'Accademia Petrarca' of Arezzo/, S. E. Card. K. Wojtyła /Cracovia, Poland/, Prof. D. Wyss /Institute of Psychiatry, University of Würzburg, Germany/, Prof. S. Strasser /University of Nijmegen, Holland/, Prof. H. Tatematsu /Nansen University, Nagoya, Japan/, Prof. L. Tassinari /Assessore all'Istruzione, Regione Toscana/, Prof. S. Vanni Rovighi /Universita Cattolica, Milan/, Prof. D. Verzili /President of Monte dei Paschi, Siena.

Tymieniecka (ed.), Analecta Husserliana, Vol. IX, xv–xvi.

We give thanks also to the following institutions which offered their patronage:

Husserl's Archives of Louvain
Husserl's Archives of New York /Prof. J. N. Mohanty/
Instituto di Filosofia della Religione, University of Rome /Prof. E. Castelli/
Husserl Circle, Waterloo, Canada
Centro Internazionale di Studi e Relazioni Culturali, Rome /On. le A. Salizzoni/
Centre d'Etudes Phénoménologiques et Hermeneutiques de Paris
Husserl's Archives, Freiburg in Br.

THE EDITOR

INAUGURAL LECTURE

ANNA-TERESA TYMIENIECKA

MAN THE CREATOR AND HIS TRIPLE *TELOS*

PART I
THE QUESTION OF THE *TELOS* IN CLASSICAL PHENOMENOLOGY AND OF ITS OVERTURN

At first sight, what could be more unexpected than to raise the question of a *telos* of the human being within the framework of Husserlian phenomenological research! Is it not the first aspiration of this type of philosophical approach to provide for the exfoliation of the constitutive system in which man and his universe emerge, by direct intuition – both certain and indubitable – which would permit "things" to speak for themselves: in offering them, as it were, a tongue by which to speak? Remaining on a par with nature (structure, essence, *eidos* ...) intrinsic to things, this intuition presupposes their exhaustive self-interpretation. On the basis of this double premise and in the context of man's self-interpretation, would there be a need to seek principles and external causes parallel to classical philosophy?

Starting from this double premise, therefore, what reason would there be to raise the question of a principle of finality in the transcendental system of man within the *life-world* assumed as self-explanatory?

Of course, in giving a key place to the progress and development of this philosophical reconstitution in the discourse which is identical, in the thought of Husserl, with the philosophical discovery of the universe of man (but understood at the same time as presenting the constitution of man by himself within his *life-world*), it is natural that this flux of the transcendental constitution itself refers to a principle of orientation, to a purpose, a *telos*. It is not surprising, then, in the conception of the transcendental genesis of the man-in-his-life-world, to discover the idea of a meaning which it tends to accomplish; first of all the direct *telos* of the singular constitutive act, which is the intended object, then of a *telos* of Western man or even of human history as such. Moreover, as we know, phenomenology itself is seen by Husserl as this *telos* underlying the genesis of the Occidental culture. Thus the transcendental genesis *is* conceived as a teleological system.

However, the question arises: "Would the intuition of this teleology, insofar as it emanates from the genetic reconstitution, be elucidated in an adequate manner by this reconstitution itself?"

Husserl himself uses the term *telos* in several ways. He uses it to denote the

Tymieniecka (ed.), Analecta Husserliana, Vol. IX, 3–29.

aim of a particular constitutive process: notably, the final outcome of the constitution of an object, which is the basic segment of the genetic process, by limiting it in itself. The term *telos* appears in the current of Husserlian reflection again in the sense of an ultimate finality of all transcendental genesis as such. It is in this regard that the paradoxical question arises: either the double methodological premise does not suffice in order that the things reveal their intrinsic nature; or this *eidos*, which at the same time presides at every genesis and remains intrinsic to the constituted object, does not contain in itself its ultimate *telos*. And yet this genesis requires it.

Leibniz proposed that in order to exhaustively define a thing, it is necessary to have recourse not only to the knowledge of its intrinsic nature but equally to the mode of its origin.[1] And yet it seems that classical phenomenology in its two lines of development, namely, the eidetic ontology and the transcendental genesis of Husserl, having pushed this demand to an extreme, reveals this demand as still insufficient. It remains also unfulfilled by the thought of both thinkers. But if pursued further, it appears that the phenomenological reflection taken as such does not stop at any point of its genetic development but refers further to postulating the *ultimate telos* of things being constituted. Whether it would be an individual object, the *Man-in-the-World*, or humanity, the disclosure of their respective *telos* seems indispensible for founding a definitively phenomenological edifice of philosophical reconstruction. Attempting to grasp in its fullest extent the object, whatever it might be, the real autonomous individual, the *Man-in-the-World*, we will necessarily encounter the question after its "final cause," its *telos*.

Within the phenomenological perspective the question should then be asked whether objects do not possess an intrinsic *telos* in themselves (for example, a *telos* proper to the real individual or to a purely intentional object, to the *Man-in-the-World*; a *telos* proper to the constitutive cycle of the individual or social development). According to the differentiation of the universe of the phenomenological discourse, there would be a corresponding differentiation of the telic principle. Husserl and Ingarden differ radically on this point.[2]

For Husserl it is apparently from the genetic nature (that of a process in progress) of the self-interpretative method of transcendental phenomenology that his need of a teleological principle arises. Meanwhile, in the eidetic ontology of Ingarden, which consists in a chain of ideal objects, this need remains ignored. Ingarden, who is opposed to Husserl, categorically refuses any meaning to be attributed to the notion of *telos.*[3]

Ingarden's formal ontological research, which provides us with access to beings in their rational structure *par excellence*, aims at fulfilling a two-fold role by means of differentiating their composition. On the one hand, the intrinsic structure of things and beings accounts for their possible *actio et patio* and, on the other hand, it is meant to serve as a foundational principle for the universal order of the human universe. The question arises, first, whether this research has provided us with structural differentiations sufficient in themselves to fulfill these roles. In particular, do the differentiated structures contain the indispensable elements for fusing the particular links of this hierarchical scale of beings, which Ingarden's investigation differentiates in accordance with the continuity of the objective world on the one hand, and, what is more, to the continuity of the human experience in its fullest extent, on the other?[4]

Second, does this formal ontological research offer us at least a basis of unity for these two types of beings: man and the real world? (As we are well aware, the existential claim of the autonomy of the real world was the principal object of Ingarden's preoccupation in his controversy with Husserl on the subject of *Idealism/Realism*.)

As we have attempted to establish previously, a basis for such a continuity either of order or of experience is lacking in Ingarden's vision.[5] Indeed, although Ingarden succeeded in moving beyond the reductionisms of Husserl's transcendental monism by his ontological pluralism, he has been unable to arrive at or discover the continuity of existential links and failed to establish a unified ontological territory on which the empirical reality and the transcendental consciousness would meet.

It is necessary, then, to ask if the ontological structures, the ultimate foundations of being, which on an Ingardenian analysis remain definitively separated, are not therefore perfectly incapable of founding the universal order of things, as we know or understand, in the continuity of human experience. Perhaps the eidetic analysis alone is incapable of revealing the intrinsic unifying factors that would account for continuity.

After having critically examined the Ingardenian analysis, we shall see that, in fact, the ideal structures are not the ultimate factor of the universe of things and beings. They offer merely a system of reference for its rational articulation. It is rather within the *context of actual existence*, existentially prior to it, that the actual individualization of beings together with the arteries of life is established.[6] By stating it, we have moved from the ontological to the cosmological approach; from the eidetic approach we have proceeded to the conjectural one.[7] These two are inseparable in our inquiry insofar as, on the one hand, the ideal structures serve as the indispensable orientation of the conjecture while, on the other hand, only within the *context of actual existence* may the ideal structures prove their validity as well as show their flaws.

Shall we not, following the progressive individualization of the real individual within the world process, see that this latter postulates not only an ontological mechanism of motion and self-sustenance in existence, but first of all, an intrinsic nucleus accounting for the individual's self-governing agency? What else could account both for the intimate transformation of dynamism to structure, and for the intrinsic linkage between the *Elemental Nature* and the system of the hence emerging world of things and beings, if not an *entelechial principle*?

Putting emphasis upon the real individual being we bring together the subjective (Husserlian) and ideally objective (Ingardenian) biases. By the same stroke, distinguishing the individual's entelechial principle, we propose to solve the problem of existential continuity of beingness.

And yet one further question has to be dealt with: Are we going to find the real individual being self-explanatory in his own progress, and of progress of the real world of which he constitutes the foundation, or must we search further in order to account for it?

From this brief outline of the phenomenological aporetic, which appears once phenomenology is confronted with the issue of the *telos*, clearly emerge the great controversies into which the phenomenological project is entangled. Phenomenological investigation also appears ripe to confront them and to revise its classic position.

First of all, when I put into the focal position of phenomenological self-reflection, the question of its intrinsic teleological orientation entails questions concerning "reasons," "causes," and "principles" which the encompassing, hitherto unquestioned network of analysis has hidden but can now no longer disguise. Second, it lets us break out of the closed circle of *man-within-his-life-world* and open the gate for the integral phenomenology of *Man-and-the-Human-Condition*.

In what follows I propose to approach the issue of the teleologies within the phenomenological program directly, that is, to outline the *three stages* of the phenomenological inquiry at which, as is shown in my previous work, the question of the *telos* emerges and is implicitly treated in its own way.

In the first place, searching below the correlation *intentional act – eidetic structure* – representing the narrow sphere of the objectivity of the *life-world* – the real individual being will be approached in its actual existential origins as it individualizes itself within the network of the *elementary forces of Nature* manifesting their own *regulative telos*.

Secondly, breaking the closed circuit formed by the transcendental, specifically human, *life-world* with its complementary zone of the preconstitutive, we will seek to retrieve, through phenomenological discourse, the hitherto lost *subliminal* resources of the human being. The *creative telos* will emerge as our indispensable guide. Will it prove itself self-explanatory? Far from it.

The creative approach to the subliminal ground will lay bare the thread of the specifically human *telos*: the *transnatural destiny of the soul-monad.* Only in it can the other two telic principles of the real individual find their outcome.

The Regulative Telos of the Real Autonomous Individual: Telos and Entelechy

The eidetic intuition instrumental in both the ontological and the transcendental phenomenological analysis gains its clarity and apodictic evidence by focusing upon the object in its intentional, that is, strictly rational, ideal skeleton in which it "appears" with the constitutive consciousness. Although we may retrace the steps of its transcendental constitution, that is, the progress of its origin with the intentional consciousness, yet these origins – or "appearances" – themselves are repetitive. This "repetitiveness" – as opposed to the uniqueness of the actual origination in existence of the real individual being – is the very condition upon which the phenomenological retracing of the intentional constructive progress of the object rests. Indeed, it seems, first, that we may intentionally follow only the genesis of an object preconceived as having been already accomplished in its constructive process. Second, if in doing so we find an intersubjective consensus it is because what we do follow is beyond the *hic* and *nunc* of the concrete existential genesis: what the intentional analysis pursues consists merely of the rational itinerary of its articulations insofar as it retrieves segment by segment the phases of its structures as *reproduced*. In this perspective it becomes even clearer what I mean by claiming that the phenomenological analysis of the transcendental constitution is but a *formal "reconstruction"* of the universal structures of the constructive mechanism, insofar as the *construction* has reached already the clearly objectifying level.[8] The *actual existential genesis* of the concrete individual in its *material process of construction* does not reach to the surface of the formation which "appears" as, or is embodied in, the intentional objectivity. It is then the repetitiveness of the object, in its ontological "universality," preconceived as already completed that, on the one hand, allows for the eidetico/intentional analysis and imparts to it its methodological privileges. On the other hand, however, while taking into consideration merely this phase of the actual genesis in the existence of the individual being, at which its genetic progress has already taken the shape of the structure "appearing" objectified with consciousness, it pays the price of missing the principles of the actual existential origin of the being.

These origins are not to be found in the final stage identifiable by a repetitive, universal model representative of the role of human consciousness in assimilating the givenness, categorizing and establishing universal interconnections of an intersubjective *life-world.* Even if, as seems undeniable, the individual real being progresses by ultimately falling into a universal category of objective structures, yet his origination in existence by *individualizing himself from the otherwise anonymous Elemental Nature* is a unique and unrepeatable instance.

Indeed, if instead of severely narrowing the field of intuition to the sphere in which the already categorized forms appear, we open it, we need not plunge into a precritical attitude. The indubitable evidence of consciousness which was instrumental in their final shaping belongs only to a stage at which they are already individualized from the whole of the elementary processes of Nature, and their appearance with the intentional reflection completes the process of severing them from the same elementary processes. Whereas by opening the expanse of intuition and losing its absolute evidence of consciousness we may make a gain on the part of being. In fact, in the attempt to follow the real individual being in its emergence from the primitive forces as progressively standing out as an autonomous factor of the whole which then becomes the "world," we gain the evidence of his *beingness.*[9] With his beingness in focus, we discover that the real individual makes his route as an *autonomous agent.* That is, he differentiates himself as a singular being from the rest of Nature by virtue of his own regulative devices.[10] He enters the self-constitutive progress endowed with a nucleus of his own material endowment; together with it goes a virtual operational equipment ready to be triggered into action should the external conditions allow it. In this way he may direct from within his constructive differentiation unfolding his potential material as well as using to this effect the matters external to his very own field.

This *inside/outside direction* of his inner workings not only establishes him progressively as a self-directing agent but makes him assume the role of a constitutive factor with respect to the entire realm of Nature. It makes him exercise an impact upon the circumambiant forces of Nature and to become a center in which these forces of the hitherto anonymous Nature converge. While unfolding in his outer/inner directed constructive processes, the agent of the real individual develops his operational mechanisms, and dealing with the outer elements strengthens his very own, unique position. In short, he projects his own itinerary, establishing himself in existence, countering or evading some of the cross-currents he encounters, and assimilating others for his own benefit.

Accomplishing in this way step by step the constructive individualization of his being from within his own inwardly unfolding rule and on the basis of his own initial resources, he establishes himself within *his context of actual existence*, which reaches as far as the radius of his operations.[11]

Our perspective upon the constitution of the real individual being, which we have labeled elsewhere 'cosmological,' retains as its system of reference both the eidetic and the transcendental systems of interrelated rational networks, but it rises above them on its own wings. As Eugen Fink has observed, the philosophical significance of the phenomenological analysis postulates indications from its findings how to proceed further, leaving the analysis itself behind. Verified by these systems of references, the *context of the actual existence* of the real individual yields, indeed, evidences of its own. They corroborate partly those of the classic phenomenological inquiry; partly they enlarge upon them and bring some essential corrections. First of all, the real individual within the context of his actual existence bridges the gap, otherwise insurmountable in Husserl's thought, between the empirical functioning of the human being and intentional consciousness. As we know, Husserl struggled in vain with this problem until the end of his quest and seems to have been ready to give up the initial project of the total hegemony of intentionality. Without it, however, that is, without bringing the empirical zone of functioning within the intentional system, no unity between the two realms can be maintained.

Secondly, starting with the focus upon the real individual in his hazy, changeable, blurred contours from which we may nonetheless "conjecture" the outline of his actual existential process, we avoid with this one initial stroke the Ingardenian predicament.

In fact, Ingarden had, in his ontological analysis of the fundamental types of things and beings, given an eidetic skeleton of the real individual. He has attempted to supply the Husserlian transcendental conception of the real individual being with an ideal intrinsic structure. In the Husserlian conception we see the real individual constituted in his character as a transcendent real object by a series of noematic glimpses (*Abschattungen*) with which the perceptual process of consciousness fills the appropriately left empty spaces within a preposed model. As close as this investigation may come to the ways of coalescing and accounting for the features in which the object "appears" with the transcendental constitutive consciousness, yet it does not reveal its essential intrinsic structure, showing what he is in his own right. On the Ingardenian analysis complementing that of Husserl, it is then the complex ideal structure governed by a nucleus of an essence which contains a constitutive nature that should account for the real individual's existential distinctiveness from other types of beings and his autonomy. To the degree to which the intrinsic essence of the being would indicate the necessary coexistence of structural mechanisms which allow for exchange of externally conditioned properties, internal processes with external outlets are founded in the structure itself. Thus motion, action, and interaction are "ideally possible".

The differentiated structural mechanisms explain the existential status of the real being. That is, they allow him to be a center of action and the foundation for the causal network of the domain of the world. In other words, the temporal spread of his existence is accounted for. However, there is an important difficulty which Ingarden's analysis entails without solving it. The individual cannot remain in a stationary stage; in order to maintain his identity in the temporal spread of existence, the individual could not enter indiscriminately into action undergoing haphazard transformations. To exist as one and the same throughout a succession of temporal occurrences, to be an "individual," means to organize the *actio* and *patio* into a developmental process. The question then emerges whether the Ingardenian ontological structure of the real autonomous individual provides us with the *principle of his intrinsic development*.

This question has to be answered in the negative. Neither the question of the progress of the real being's development nor of his itinerary is raised within the framework of Ingarden's ontology. Hence it is easy to see why the idea of the *telos* seems to Ingarden altogether irrelevant. But can the conception of the real individual do without it?

Furthermore, I consider the incapacity of pure eidetic analysis to find the internal principle of the development of the real individual being as the reason that ultimately prevents Ingarden from establishing existential bonds among the various types of objects that he categorizes. Disconnected as it were, they cannot function as a base of the universal order. The existing world asserts itself as perpetually evolving, and its processlike progress in its continuity has to be accounted for step by step. Whereas without a principle not only of order but also of the dynamic progress intrinsic to the individual – the foundation of the real world – its systematic progress cannot be grasped;

progress and systematic development fade away; and the origin and point of extinction of the individual being fail to be explained.

Could Husserl have provided a plausible formulation of the problem of teleology after having assumed that, while the *eidos* does not hold an intrinsic principle of development, yet it serves as a conduit of the individual object's transcendental constitution? Would the framework of research that he has set for himself allow an adequate formulation of the issues involved? Although he has with time expanded considerably the methodological framework of his research, he has never abandoned its eidetico-intentional assumptions which determine the analytic possibilities of the investigation.

In contradistinction to these two approaches of phenomenology, Husserlian and Ingardenian, I have proposed elsewhere that it is the element of *entelechy* which asserts itself as a decisive factor in the cosmological perspective, within the individual structure of the living individual's autonomous being. The *entelechial factor* of the living individual being appears as the *material nucleus* of his existential mechanism, as absolutely his own, and as carrying out the progress of the being's particular identification.[12] While containing a whole project of its cyclic development, the entelechial factor presides over the overall orchestration of vital processes that allows an individual being to unfold on the one hand, and appropriate ontological structures and the corresponding biologico-psychic "programs," on the other. It is only upon the assumption of an *entelechy* inherent to the development of the real individual that we can give an account of the emergence of the individual real being into existence, as well as of his cyclic progress which leads not merely toward an unavoidable extinction, but primordially aims at a constructive attainment. In fact, while both the eidetic and the transcendental perspectives emphasize either the universal type as such or a universal progress of the constitutive function, it is this latter that implies an orientation. Without this orientation, would there be a life-unfolding which advances from one constructive step to the next? Without progress would there be an order of events? And without order, would life and real existence be at all possible?

At this point, however, a new question emerges: "In order to achieve its fundamental existential role with respect to the real individual, should not the *entelechy* comprise a foreshadowed principle of its final attainment, the *telos*, the development of which the entelechy is planning and to which it would aspire?" With this question, which pinpoints the *conjectural* postulate of individual becoming, referring it to its intrinsic prerequisites, we project the *common network of the existential agent of real being as extending between his entelechy and its telos.*

Concerning the nature and the significance of this established *telos*, we must ask: "Does it comprise the final explication, the sufficient reason for the being of the real individual object?" When I approached this subject in the Leibnizian fashion elsewhere, that is to say, through the bias of the question, "Why is there something rather than nothing?"[12] I have noted that conceived in the cosmological perspective, the real autonomous individual does not contain in the framework of his being his sufficing reason for being; his individual *telos* is custom-made to deal within temporal, contingent, and natural-existential conditions, therefore it is subjected to various circumstances and is partly predetermined by the total life-process. The reasons ordering his origin, those accounting for the emergence of specific means to deal with his existential conditions and those of his having been selected as such a being from other possible beings, remain foreign to him. Asked to account for them, his *telos* refers to the concrete *universal order of things*. This, in turn, refers to the *architectonical plan of the objective constitution* of the universe of things and beings.[13] Once there, the thread of the cosmological analysis, that is to say, of *conjecture* established on the basis of the eidetic and intentional data, breaks down before leading us to a terminus.

These questions remain unanswerable despite our having begun with the consideration of the primitive initial dynamism; we must accordingly approach them from another angle.

PART II
MAN THE CREATOR AND HIS SPECIFIC *TELOS*

The consideration of the development of the real individual oriented toward a *telos* of his natural accomplishment encounters a particular problem in regard to the progress by the specifically human being. But we may wonder whether there is something "specifically human"? However, in contrast to the empirical-reductionist tendencies, any philosophy of phenomenological inspiration seeks to claim and establish a specifically human aspect of this real and autonomous individual being that is man.

In this connection it suffices to mention Dilthey, who had already defined Man by his capacity of self-experience; Jaspers, who saw the specifically human element in man's ability to transcend himself; Husserl, who perceived the specifically human factor in the full expansion of transcendental consciousness; and Paul Ricoeur and Ingarden (in his last years), who seem to point to moral consciousness.

Aside from Ricoeur, Husserl as well as Ingarden conceive of man's freedom as an integral factor of moral consciousness, which is on their analysis confined to the exercise of human faculties within the intentional configuration. This position made us wonder whether, restricted to the basic constitutive system, man would be capable of a complete moral life. Would he be altogether capable of the socio-cultural development recognized in the phenomenological conception of the *life-world*? Furthermore, does not the progress of this development presume a renewal of structures? If man remained closeted in the categorial jurisdiction, all the possible variables of forms would be quickly dried up; as progress becomes stifled, all life would be snuffed out. The successive stages of the constructive progress, according to the variability of the pre-posed set of categories, arrive at a limit at which a new reshaping is necessary. In the postulate of a new reshaping *is included* the request for the renewal of the *types of objects*. As to moral freedom itself, if limited to the exercise within the framework prescribed by an unchangeable set of categories, it would be governed by corresponding pre-established ideal values. What meaning would it have to the ethical man to move solely in the closed circle of preestablished ideal values? The central question here is: "Could man-in-his-world unfold completely by the productive and reproductive function of the intentional constitution alone?"

And yet, throughout the entire span of phenomenological reflection, this assumption of the very possibility of a *life-world-constitution*, that is, the assumption of the *total renewal of the forms* underlying its conception has been ignored. Indeed, this conception appears to have neglected, literally without exception, the *genuine originality* and *exemplary uniqueness* of the masterpieces of artistic creativity and revolutionary scientific and philosophical theories, which have manifested themselves in the present phase of the constituted *life-world* and slowly established themselves and gained rights of citizenship, although at first met with distrust or lack of understanding: they often had to wait a long time until man reshaped new crucibles in which they could be assimilated.

Although in his aesthetic investigations Ingarden established varied forms of "transcendence" of the structural forms of objects with respect to each other – in contrast to the all-embracing homogeneity of the constitutive process in Husserl – he did not acknowledge the unique status of the "exemplary" Work of Art in the world. On the contrary, following Husserl, he accepted explicitly the total domination of the categorial system with "*die oberste Gattung*," and insisted that no new and original object could be introduced into the world. Husserl, in contrast, reserved the privilege to bring forth and introduce into the constitutive genesis a new and original proto-element to the *Urimpression* alone. Once introduced, the constitutive genesis molds and reproduces it into endless variations in its preestablished and universal forms.

Thus, neither of these at least partly correlative conceptions of *man-in-the-world*, takes into consideration this need for the complete reshaping of types of objects on which the very existence in the *life-world* (*Lebenswelt*) *in progress* seems to depend. On the contrary, both of them seem to have erected obstacles to its recognition.

In approaching the world through these two mentioned channels, we are bound to ignore the issue which appears to us concerning the main condition of man's existence in the world. It is obvious that we must then seek to enter directly the path leading to it, that is to say, the analysis of the *creative function of the human being.*

We cannot forget that Bergson was conscious of the necessity of the renewal of the forms; yet, having identified too quickly this need of the *constituted world in progress* with the one of Nature's innermost, he perhaps overlooked some conditions found at both levels of phenomenological consideration. In the final analysis, the two should undoubtedly rejoin each other, although in appearance they rest on two opposed poles: the one of the elementary conditions (blind Nature), and the other of human freedom to go beyond the limits of objectivity. We will now attempt to investigate these poles separately and bring them together precisely by means of an analysis of the human creative process.

Our first step involves the discovery that the *creative function*, guided by its own *telos*, generates *Imaginatio Creatrix* in man, as means, *par excellence*, of specific human freedom; that is, the freedom to go beyond the framework of the *life-world*, freedom of man to surpass himself.

In the second place, by following on the opposite side the networks and elements of its synchronization, we enter into *Elementary Nature*. While recognizing the creative function as a basis of the fundamental human condition, we shall discover, then, that *to be human means to be creative*.

The creative act, which we have analyzed elsewhere, emerges within the *human-being-in-conflict*, in man rebelling against the interpretation that the present phase of the constituted world gives us of the Real. Then, in orienting himself during his first set of operations to the direction to be taken in giving a new shape to his personal and intimate vision of the Real, man finds out that perception of reality which through habit became the surrogate of experience, is disintegrating. Whether it be in poetry, painting, visual art, etc., or in the pursuit of an idea, a thought, a color which are sought to respond to our aspiration for a metamorphosis, the constitutive perception of the object of our attention disintegrates in all its aspects. It falls into pieces, leaving, however, a half-open door through which this searching impulse, unshaped but rallying all our being's faculties, allows us a glimpse of a passionate, emotional depth, fusing together within a *spontaneity* which puts to work all that is within us. This *spontaneity* offers the vision of a new universe.[14] Indeed, starting by disintegrating constitutive consciousness, our efforts do not fall into inertia as in the case of the pathological disintegration of consciousness. On the contrary, all our energies are mobilized in the discovery of means, ways and materials, and of channeling them into a constructive apparatus capable of concretizing this vision. A line of progress is thereby delineated, a path of constructive advance. This path must emerge from a gathering of our faculties of the universal and anonymous constitution that we have unveiled as such. And yet, meanwhile, we pursue a unique and personal objective, within a new *orchestration* of our faculties which become organized under the pressure of the *spontaneity* directed to its own *telos* as foreshadowed by the vision. Is it possible that this *spontaneity*, comprising the blind and dynamic passion of our being, may be itself blind – as Husserl has argued concerning the "passive synthesis", that is, the unfolding of the intentional constitution about which man does not *decide* but with which he flows *nolens volens* onward, or as Kant sees the *Einbildungskraft*? In such a case, could it be capable of giving direction to this progress? In what way then could the *telos* of the vision toward which the creative process aspires be attained?

As we have shown elsewhere,[15] on the one hand, it is in the work of art, of scientific invention, philosophical doctrine, etc. – that is, in an intersubjective object – that the vision will crystallize. On the other hand, this new *orchestration* could not be able to fulfill its role without acquiring its own constructive rules and directives, that is, rules and directives which establish it in its own right, outside the categorial system that controls the eidetic and intentional structurization. Thus it must *create its proper forms.*[16]

In this connection it is appropriate to consider the nature of the vision, emphasizing those features that allow it to fulfill its function of aiming at a specifically creative *telos* on the one hand, and to have intimate bonds with the *particular type of spontaneity* which it may prompt from the specifically human condition, on the other.

In fact, the vision of a "new universe", in contrast to the shape granted to the Real through the present phase of the constituted world, glimmers upon the horizon of every creative impulse. We identify in it again our nostalgias disabused by the actual *life-world*, nostalgias after the "authentic" in contrast to the "fictitious" of this world and our desire to surpass it while undergoing a corresponding transfomation of ourselves. This vision, nurtured and carried by the very passion, the emotive forces and the inner pulsations and impulses of the interior of our being, crystallizes our volitions into a definite objective. It aims at finding or inventing a new form to grasp the Real more adequately and to embody it into the form of an object that can enter into the present phase of the constituted world – provisionally suspended – in order to transform it. In this way the vision foreshadowes the *telos* of the creative process, a process which it guides while prompting it into action.

Prompting the process, the vision appears to stand for that *elementary spontaneity* carried silently by the elementary forces we have mentioned. A doubt still arises: While rising from those blind depths, could not this spontaneity generate itself in a totally "neutral," that is, aimless, outpouring, which may be mute and blind, such as *Einbildungskraft* or the *vital impulse*?[17] If it were the case, how could we explain the nature of the emerging vision which not only seeks an outlet for a constructive fulfillment but also proposes its goal?

Indeed, the deep-seated remaking of our operative faculties in the *creative orchestration*, which the vision initiates and brings about, and the nature of the progress of the creative process into which its dynamism flows and which it leads, bear witness to the contrary. Already at the point of its taking off from the *initial spontaneity*, the vision already carries in it the seeds of this creative tendency and outlines the scope of the creative process and its significance for man: that man through the quest after a form to embody the vision *is seeking to acquire his own form for his self-explication in existence.*

Thus, man's *elementary condition* – the same one toward which Husserl and Ingarden have attempted in vain to break through, by stretching the expanse of the intentional bonds as well as by recurring to the prereduced scientific data – appears to be one of blind nature's elements, and yet *at the same time*, this element manifests itself as possessing virtualities for individualization at the vital level and, what is more, for a *specifically human* individualization. These latter virtualities we could label the '*subliminal spontaneity*.'[18]

However, does the specific human *telos* embedded in the mechanism of the creative vision rest in itself? Does not its own genesis indicate other conditions to be fulfilled in order that man's creative self-interpretation may acquire its specific significance? As to the creative freedom which appears as the source of *specifically human* freedom, is it not necessarily bound to another form of freedom, equally inherent to the human being? The question arises: Is it enough for man to go beyond the boundaries of the *life-world* (*Lebenswelt*) in its present preestablished phase, surpassing oneself by *self-explication within a work unique and original* in relation to this world, in order to satisfy his primitive nostalgia? Does the creative self-explication bestow upon human existence its final significance? Is the transcending creative impulse carried by the vision fulfilled in its concrete completion in a work of invention: art, theory, social life, world-development?

We may conclude this discussion of the *telos* which presides over the enactment of the creative function by stating its crucial role in three major points:

(1) It allows the human agent to break with his imposed survival-oriented patterns and advance toward the orchestration of *Imaginatio Creatrix*. Thus it opens up the exit from the closed horizon of Nature (and of the transcendental circle, for that matter) for specifically human freedom.

(2) Leaving behind the preestablished regulative principles of the intentional system (subservient to Nature) the creative *telos* guides the origin of new forms as fruits of human invention. Thus it leads man from *natural determination* to *creative possibility*.

(3) Finally, the creative vision offers us the much sought system of reference for thematizing the preintentional dimension of the human functioning as its "subliminal" resources: they appear as a *specific endowment of the real, human individual basic to his human condition.*

PART III
THE SOUL'S TRANSEMPIRICAL DESTINY

The Telos and the Sacred

By asking whether the creative *telos* provides the final direction for man's individualization from within Nature and the circumambient *life-world*, we have in fact sought the key to the *ultimate significance* of specifically human existence. The various stages of the individualization of the real individual in his specifically human form may be considered, as we have proposed elsewhere, as *man's self-interpretation in existence.*[19] Endowed with a unique meaningfulness, the creative pursuit may be seen as its crucial stage.

With creative interpretation, the possibility of choice being opened, the question emerges: "What are the ultimate reasons for the principles of choice?" That is, what are the final reasons of man's progress in existence?

It is the specific privilege of men, that is, in being human and in being able to deliberate, select, and invent, to pause for a moment and to contemplate the enigmatic position which man has himself outlined, in the process of weaving the texture of his existence within the otherwise alien forces of the natural universe and the social world. *He will search for the end toward which the ever refining meaningful pattern of his progress tends.* This pattern in itself appears to be advancing within its fragments of intricate plots, itself always fragmentary, without revealing the secret of "what it is all about." He will wonder, "what is the point that the flowing concatenated meanings intend to make? What is the final aim that glimmers enigmatically foreshadowed and which our spontaneities prompt and tend toward accomplishing? Shall man pursue it by himself alone or recur to the Other?"

Indeed, when we pause for a moment to consider it, in a way detached from the flux we flow with, we see that our dynamism will be spent; the works of our invention might outlast it but will last at the utmost only as long as humanity. And yet, caught within his genetic progress itself, man *must seek for its final significance.*

To have failed to find it in the creative *telos* does not mean – as is sometimes concluded – that there is no answer to these questions; it means merely that the thread of continuity of the meaningfulness of the human existence does not stop there but continues at a further stage: *that of the sacred.* The investigation of the primordial role of the creative function in man's existential unfolding has prepared the ground for approaching it.

Indeed, once the absolutism of the constitutive function is disclaimed and contrasted with the creative function, the constitutive genesis of the human being appears as just one basic functional system of the real individual promoting the stereotypic channels of life; that is, when we can clearly see the subliminal ground of the *initial spontaneity*, the territory is ready for the quest after the *transempirical destiny of the soul.*

Where the existential life-oriented processes in which the real individual unfolds come to the end of their course and meaning acquiring, it is into a quest *after the significance of this course itself* that man engages. Having scanned the possible scope of the two previous lines of unfolding and stated that within their *telos* no key to man's self-interpretation in existence is given, we can now search for an ultimate aim of this differentiation of the human being from all beingness.

The question arises: Is it accessible to phenomenological inquiry? Our answer is already given: we may approach it with reference to the creative function.

The process in question shares with the creative function first of all the initial situation that both emerge from a reaction against the constituted world. Both of them surge in the form of a quest. Both of them, working through and with the given elements of the *life-world*, carry beyond the *life-world* as it is experienced in its present phase. However, in contrast to the creative process, the search for the key to the enigma of human existence does not mean a rebellion solely against the present phase of this world. It emerges only when the validity of the human world as such is disclaimed; when we arrive to question the purposefulness of its changing, precarious, never completed course; at last, when its lack of sufficient reason is recognized. In short, it presupposes the loss of the natural belief in the validity of the *life-world* as such as well as of *any* objectivity. The frenetic quest into which the human being enters, once this discovery of the contingency of the natural life within the world is accomplished, does not contain any positive scope. In opposition to the creative impulse, it does not seek a new form in order to incarnate the Real.

Since he has already abolished the outer world, it is toward his inward self that the inquiring subject will turn. Yet the line of questioning he follows, in which he scrutinizes also all objectively constituted aspects of his experience, relies in this negative fashion on the intentional constitution and its product the *life-world*. But while leaning on them, this questioning runs a different and original course. Likewise, it uses the creative function; and yet, while making use of all the operative elements of its orchestration, this quest avoids the crystallization of its aim in the pursuit of a creative work. Thus, lacking the reference to a definitive objective, it proceeds tentatively, in the "darkness of the senses"; yet, we are capable of tracing its progress from the landmarks it establishes on its way.

To be sure, phenomenology cannot ignore these landmarks of experience, once it has retrieved the *subliminal* depth of the human condition.[20] Indeed, they appear *in their self-givenness* authentically *absolute*; which means, with evidence that *is not relativized to the system of objectivity*.

This quest, which we recognize as the road to the experience of the sacred, does not progress without *any* aim. It reveals an inner element of direction, its very own *telos*. If, following Husserlian tendency, we call 'soul' everything in the human monad which represents the mute element of passion, affective pulsations, internal urge, nostalgia, etc. – that is, the material ontological element which flows from Nature's depths, and that represents the empirical resources of human existence – then we can, in parallel as much as in contrast, call this *telos* of the expression of the sacred THE SOUL'S TRANSEMPIRICAL DESTINY. As can be foreseen from our earlier discussion, it is within it that the two *telos* previously discussed find not only their completion but also a necessary condition.[21] I have indicated briefly that man's two basic functions, the constitutive and the creative function, are at the same time utilized and surpassed in molding it. Now it becomes necessary to ask: What would be the purpose of this quest? That is, *of what does human destiny consist*? In which way could it be the fulfillment of the *specifically human* genesis in all its arteries? Lastly, how would it offer an answer to the great problems of phenomenology with which we have been concerned from the beginning of our inquiry?

The Pursuit of the Destiny in a Creative Self-Interpretation Between the Self and the Other

The initial question here is: Do we pursue the quest of our destiny within a soul alone, in isolation from other beings? Or, on the contrary, can we find a meaningfulness in our lives other than its struggle for survival on the short waves transmitted between the self and the Other, by entering with him in the rapport of what may be called "inward creative reciprocity"? The study of the various forms of communication in which human self-interpretation in existence proceeds shows that, apart from the organic and vital interactions as well as strictly utilitarian types of involvements with other human beings, it is in particular in the human relations which have been classically termed as *Eros, Agape, Philia, Storge,* that the specifically personal quest after the meaning of life proceeds. It is carried out by a dialogue between two persons which is essential to the relations just mentioned. There is, however, more to this dialogue than Jaspers has so well described. It remains still to be shown what in the human functioning and condition allows it to take place, that is, in virtue of which the metaphysical horizon of human existence is founded and opened. We propose, in line with our argument, to show that the interpersonal dialogue is founded in an ontological state of the human person: "The inward creative reciprocity." This state itself constitutes but a further stage of man's ontological individualization or self-interpretation in existence.[22] Yet, unlike in other phases of self-interpretation, it does not take the Other merely as a means.

Furthermore, and this merits emphasis, it is a creative process. We will consider from its innumerable manifestations only the innermost profile of the search after the *transempirical destiny* – a search shared by two persons.

In the common pursuit of destiny, self-interpretation has two reciprocal poles. The "reciprocity of love," for example, consists in a simultaneous search after the meaningfulness of events for *our own* existence which results in "ciphering" their significance into the weaving of the thread of our own existence, as well as in trying to find the meanings which *the beloved* gives to them with reference to *his own* thread, that is, in "deciphering" its meaning with reference to our own. Such an interrogation, intimately unique to each, is shared by both members of this relationship. Already man's self-interpretation in the creative work has been directed (in its search for new forms, rules, and principles) toward the other self; the spectator, reader, listener, etc.[23] The message which creative work is conceived to convey is also formulated as the creator's very own meaning of human existence and in respect to another self. In fact, the creative process remains the fundamental vehicle of both.

Indeed, the profile of the self-explication of destiny may be clearly recognized as the thread running through the above-mentioned dialogical relations only in the perspective of the creative function of man, because it proceeds using all the ways and means of creativity. Nevertheless, they differ on several essential points.

First, in opposition to the interiorization and total self-absorption of the artist in the creative process leading to a work of art, in the pursuit of destiny we seek for clues to the meaning to be given to our interior existence, as radically turned toward the Other approached as caught within *his own* identical quest. Second, although evidence of our inventive and constructive ingenuity is present in this search, we do not seek to construct, discover, or project any *object*. Neither do we aim at prompting an event or occurrence of any sort in the *life-world*. In fact, the self-interpretation in destiny is not oriented toward *ciphering* a rational, intersubjective message.[24] In that inner urge to forge a significant thread of our existence in response to the ultimate question which at the point of our discovering the contingency of the *life-world-existence* cannot be postponed any longer – a question which asks how to salvage from the fleetingness of life something of lasting value – we direct our scrutinizing interrogatory to the Other. We seek to discover whether, and how, *he* proceeds. What significance does *he* attach to *his* own existence? We provoke him to reveal *his* secret thread by revealing, without any reserve, our own interior existence. The Other receives the revelations of our quest in his own inwardness, in a spontaneous reaction in which he responds; while receiving it he makes clear to himself what *he himself* is striving for. Inasmuch as he receives our manifestation as the revelation of our virtual self – the one that we tentatively project but have not yet crystallized – he is co-present with us and yet remains absent, because he remains within the network of his own quest after the final meaningfulness of things. From our self-revelation he grasps this virtual self of ours, fluid and fleeting inasmuch as he may compare, confront, identify, or assimilate some of its aspects with *his own* tentative and undefined tender substance.

In fact, what else are we after in this mutual examination of each other, in which we tend to penetrate into the most secretive tendencies and intentions of the other self and the way in which he appreciates their significance confronting it with our own, if not to dig deeper into our virtualities? We seek to scrutinize our resources to feel, to comprehend, and to dig deeply enough in the never ceasing current of experience for some clues by means of which a thread of a perduring validity may be sustained on its course. This thread would sustain the perduring validity of our very self; nevertheless, it is to be wrung from the fleeting segments of existence and spun by their means. Its progress requires that we advance to the discovery of the virtualities of the very undercurrent of our being. In order to transcend oneself – that is, our natural, empirical, everyday self of the present stage of our existence in which we are constituted and seemingly stabilized – within a scheme of a new meaning of the existence's current, we have to reorganize even our vital functioning at its elementary state.

The spontaneity of our quest, which is continuous and carries us relentlessly onward while undergoing innumerable metamorphoses, crystallizes in this new weave by trial and error in the course of mutual self-revelation. My own new self becomes "other" without it ever stopping in its tracks and without identifying itself with a form or mold. Indeed no definite form may grasp it. It flees from and evades its own eye as much as this of the other self who would try to maintain it in vain in any fixed shape, be it for an instant. We would be tempted to consider it almost a "nonbeing," because it may not be subordinated or paired with a definite mold of experience into which our constitutive apparatus would like to force it. And yet, it "exists" because it manifests itself as a *thread* running through our innermost concern, and it is incessantly "molded" and "forged" in its own way in relation to it.

As we have already pointed out, the processes of creative reciprocity emerge from a situation comparable to that from which the creative process *sensu stricto* surges. Although, as we have shown already, their orientations not only diverge on many essential points, but first of all, the quest after destiny decisively takes on the stand of the *final* concern, and of one no more to be postponed on account of pursuing present-day concerns or tasks. Unlike the creative process stopping each time at a new proposal for the answer to the enigma of the Real, the quest after the *ultimate significance of human existence* never stops. And yet it is essentially creative in the ways of its progress.

We address this interrogatory quest simultaneously in a two-fold direction: toward our innermost self and toward the Other, while we attempt to scrutinize the most intimately personal experiences, convictions, and attitudes in their foundations and reasons. Hence we draw clues for the meanings which we should give to the state of things present, and also past. But in neither of these essentially conjugated steps can we expect to find a *status quo*, an already present solution to our perplexities and uncertainties.

Also, it lies in the very nature of the quest, in which it is our own being in existence that is in question, that we do not expect to find within the Other a "sense of life" already established – and one which we could accept. It is from within *our own* resources and upon *our own* evidences that it has to come. But none of them has a ready material to offer, waiting to be discovered. We would reject any such ready-made finding as not being our "very own." We would not stop at any assimilation of anything we did not ourselves deliberately choose; it would mean a merely haphazard, circumstantial occasion, the reasons of which would remain hidden to us. And, first of all, no already available form, or answer, will ever fit the purpose. It is precisely in the awareness of all its reasons that we have to acquire this final answer to our interpretation of existence as giving the key to its complete script; *to know "all the reasons" for our choice, we have to invent the key ourselves.*

This interrogatory is thus not a passive flow; but in the process of scrutinizing and questioning how one's own and the Other's attitudes toward one's own existence are (attitudes which remain hazy, undefined, in the twilight of consciousness of both), we have to probe them by *inventing* ever new ones according to the particular bias which we also invent – as fitting our own, and the Other's presumed, circumstances – and according to the clues that we have already half-established, half-projected. In our tentative approach to his quest, we assimilate, by our interpretation, his own answers provisionally with ours, checking the validity of his presumed stand over against our own.

We address ourselves to the Other also as to a *witness* and a *judge*, seeking his approval or consent for our deepest concerns and dilemmas. We try to introduce him into the very heart of our perplexing attempts to interpret the givenness, by scanning one by one the possible meanings of the given from which we can approach it. We show him the reasons we envisage, weigh, or reject outright, the evaluation of the meanings which our imagination proposes, probing thereby deeper and deeper into their intrinsic virtualities which are suggested to us. We then introduce the Other into the very heart of our creative investigation as a *second self.* We face him simultaneously as an "other self" insofar as we expect to receive from him, into the common current of interrogation, the opening of his own case of being-in-a-quest. We face him as a being-in-a-quest, other than ourselves, over against whom we may measure our own; an "other self" sharing our own self insofar as he receives, observes, and retains for himself the essentials of our own quest, and also develops *his* quest by checking its progress reciprocally in the same way with ours. Each of us, in order to reveal to the Other the meanings which he might propose to facts, events, feelings, decisions, acts, nostalgias, and deepest urges – and to seek with him to understand them and check their relevance in this understanding – scrutinizes one by one all the available and possible systems of interpretation. Trying them each in turn and rejecting each. None of them appears, in fact, capable of transmitting, or holding this unique significance we seek to establish.

This reciprocal inward quest develops and maintains itself through innumerable instances, fluctuating, instantaneous, and yet sustained by the thread they forge. This thread seems to sustain itself on its own. It does not have any definitive foothold in empirical reality nor a pre-posed set of regulative guidelines to follow: it proceeds by trial and error – relying essentially upon human invention it would be accessible merely in disparate instances of spiritual experience, if it could not be understood and accounted for as the process of man's ultimate phase of *creative interpretation. This quest after the transempirical destiny carried on by the self and the Other appears indeed as the quest after the final telos of man's self-explication in existence only when we discover that it proceeds by means of a creative self-interpretation.*

To realize this let us summarize the positive points of our description.

In the first place, as the dynamic thread of our communication with the Other self, it is – like the creative process – constructive. Although it does not construct an intersubjective message, it carries on a unique and unprecedented transformation of the state of our own being. Second, like the creative process, it advances by breaking, one by one, all the joints of the functional intentional network oriented toward the constitution of objectivity. Furthermore, comparable to the creative process, the self-interpretation in the destiny employs to its own aim the means of the human functioning and all the points of reference which the constituted *life-world* and ourselves may furnish to it as useful; and yet, it is by no means subject to its organizing regulations and projects. On the contrary, it is worked out through a functional system devised for its own unique "purpose" by the interrogating process itself. No principle of its interpretative articulation or of its meanings is preposed. All are essentially tributary to our invention with reference to our deepest self in his communication with the Other.

Like the creative process, the search for our destiny in the personal interiority with the Other carries its course through all the functional dimensions and by making use of all the operational mechanisms of the human being, considered as a real individual as much as a person. The questions which the self addresses to the Other, and the answers which he gives in turn appear ambiguous, like the *creative interrogation* – because although they bring to light the most profound and secret longings of our innermost being, yet they are merely *tentatively* proposed; their meaning becomes more precise and nuanced coalescing in their encounter with the response of the other self and of our own probing appreciation of it. In this bringing forth, reaching out, and turning inward movement, the reciprocal quest suspended between its two poles establishes, like the creative process, a specific *orchestration of all human functions*. It draws directly upon the *subliminal* resources of the human being which become liberated, within the *creative context*, from their intentional constitutive ties.

In the search for its own script, the soul lends to this creative dialogue all her means, down to those most rarified which otherwise cannot be partaken of. She addresses the other self with *transparent sincerity*, a sincerity that she is not capable of mastering even toward herself alone. She addresses him beyond the reach of any objectivity, leaving life-pursuits, concerns, and values aside; that is, beyond the vicissitudes of time. Since it does not aim at any particular object, its objective being to conjure the ultimate significance of everything to come to its own, it dwells in the supratemporal, continuing forever, eternal.

And yet, the communication in which its being is crystallized seems to be in danger of breaking down at any moment. The ambiguities of this reciprocal interpretative creation are so extensive that we may even ask: Do the encounters ever come to pass? If so, they are the moment's games. We would have no objective evidence – intuition could not bring it about, since there is no objective content to be sighted – *had not each encounter meant just a new clue that we discovered or invented, weaving a further mesh for our transnatural destiny*. Our quest seems to abandon all of life, and yet it salvages its lasting significance.

But what about the other self upon whom – as it seems – we have suspended our whole being and existence? Do we really ever meet him in *his truth*? Each self progresses, in fact, in separation. Each of them, in the quest after the new interpretative system that would be final, has scrutinized all available cues and rejected them. Nevertheless, it has accomplished the essential task. In the quest of a destiny uniquely his own, each of the selves discarding the interpretative schemes one by one, like the leaves of an artichoke, has in the first place despoiled himself from explicative schemas. Then, it advanced by inventing new ones to be discarded in their turn. We believe in the process that to "encounter" the other self, or to communicate to him this ultimate concern of ours, would occur if we could only find together a meaningful system in which both his search and ours would be explicated. Unreserved radical commitment of both offers the unique possibility to the self and the Other to transgress the closed framework of ourself toward that which he is not in this impervious *élan* which becomes more and more intense as our creative interrogation becomes more demanding and refined. The progress of its "positive" accomplishment, that is, of the "ciphering" of our very own exfoliation of existence, is, however, constantly disrupted; we lose the loops of the thread – the meaningful forms proposed; we lose the meshes which were expected to bring our experience together with that of the other self into the same net. While we believe we have thrown a hook for him to grasp, we see that it has no *meaning* for him.

If for the accomplishment of our endeavor we would expect a genetic "ciphering" of a lasting meaningful system, and a "text" – a task to which our repeated efforts seem to be applied – a text in which the other would recognize himself, this is never attained. At a certain point of its seemingly constructive advance, when we are convinced that our self and the other self is firmly established within the same meaningful text, meaning by this that we have constructed a common "universe" which we both share, a universe that is meaningful with reference equally to us and to him, and when we believe we have grasped the other self in his identity – at that point the entire edifice collapses.

To answer some of the questions that force themselves upon us concerning the significance of this attempt or of the nature of the ultimate self-interpretation in existence, we must state, first, that although no definitive meaningful system could be applied to interpret the progress and attainments of the quest, yet it does not mean that the scanning, discriminating, and temporary adjusting of them were fruitless. On the contrary, it has served several purposes.

First, although we fail to meet the Other in any rationally definitive, objectively "lasting" way, yet the instants at which we communicate together in our ultimate concern with existence are extremely rich in "substance". They expand their significant ways into all hidden dimensions of our being and reveal thereby that our virtualities may serve as cornerstones of its progress. Second, although our quest does not lead to the definitive establishment of a meaningful system for our route, it fulfills the crucial role of transpiercing the objectifying strata of the *life-world* structures which keep us closed upon ourselves and in contingency.

In fact, it is in this progress of liberation from the ties with the empirical, life-oriented structures and processes which run through our being that we do actually uncover its virtualities. In the common search with the Other, the clues are found for their working into a *thread of destiny*.

And yet with the broken communication, this thread itself breaks. Every discourse, every interpretative process, every communication fails. The virtual sense of their genesis for the sake of which we have employed all our resources, escapes a finalization, a completion of its meaning. Does it mean that we face a void upon reaching the breaking point of its pursuit? This would undoubtedly be the case in an isolated, self-centered search for the meaning of life or of human existence – as presented by Kafka, Camus, and other contemporary writers. In an isolated search, missing the *creative reciprocity* of the Other as engine and support, the initial impulse does not crystallize into a constructive process; it does not unfold the *inventive imagination* which would nourish and stimulate its advance; without it, the search becomes sterile and dies.

The *élan* of the reciprocal inward quest to decipher the ultimate significance of the human life-course, on the contrary, is prompted and intensified as it proceeds. When it breaks down in its weaving, the soul, on the one hand, is left free from empirical ties; and on the other hand, the Other having detached himself and vanished from her horizon, she finds herself lost "nowhere" and with "no one" to turn to. Cut off from the world, all her spontaneities flow into the urging impetus of her quest. On the other side of the abyss that opens, the soul discovers the Radical Other staying with her face to face.

In retrospect we come to see that in this quest, working through, but leaving aside, the empirical dimensions, we were uncovering and elaborating the *inward ground of the sacred in man.*

CONCLUSION

Thus, we see how the *telos* of the *soul's transempirical destiny is* the key to the significance of man's specifically human individualization (or self-interpretation in existence) considered within the whole scope of experience. It is ultimate; but constructively achieved, this *telos* is not. The transcending impetus of man's quest stops at the boundary of human experience. All significance that the progress of destiny might have acquired dies in it. However, the impetus goes beyond it and does not fade away. On the contrary, it reaches its apex. Its encounter of the Absolute Other, ineffable, does not need any further explanation or interpretation. Its significance is final and as such appears as the key to man's self-interpretation; so much so, that without it, the specifically human element would vanish. Here, the philosophical discourse ends. If we try to follow it, it would be necessary to perform a total "conversion" of our principles of self-interpretation (or in the words of Kierkegaard, a "radical jump").

We then propose that man's triple *telos* contains an intrinsic interconnected network of the bonds of which man is *the creative agent*. These bonds, at the one extreme, have their roots in *Elementary Nature*, and at the other extreme, tend toward *Transcendence, the Radical Other*, as their ultimate reference point. The emphasis upon his *telos*, then, places the human being within the total expanse of his condition.

Master Husserl's first basic clue for the phenomenological program, that of the "principle of all principles," meets then the ancient formula that can be paraphrased thus:

"To phenomenological research nothing human can remain alien."

The World Institute for Advanced Phenomenological Research and Learning, Belmont, Massachusetts

NOTES

[1] *Cf.* A.-T. Tymieniecka, *Leibniz' Cosmological Synthesis*, Royal Van Gorcum, Assen, 1964.

[2] *Cf.* A.-T. Tymieniecka, 'Beyond Ingarden's Idealism/Realism Controversy with Husserl,' *Analecta Husserliana*, vol. IV.

[3] Roman Ingarden, *Der Streit um die Existenz der Welt*, vol. III, Max Niemeyer Verlag, Tübingen, 1966.

[4] *Cf.* A.-T. Tymieniecka, 'Beyond Ingarden's Idealism/Realism Controversy with Husserl,' pp. 322–341, *loc. cit.* This is my interpretation of Ingarden's analytic work.

[5] *Ibid.*

[6] *Cf.* A.-T. Tymieniecka, *Why is there Something rather than Nothing*? Royal Van Gorcum, Assen, 1966, pp. 13–25.

[7] *Ibid.*, pp. 77–90.

[8] I have proposed the distinction between 'construction' and 'reconstruction' in 'Self and the Other in man's Self-Interpretation', pp. 156–161, *Analecta Husserliana*, vol. VI.
[9] *Cf.* A.-T. Tymieniecka, *Why is there Something rather than Nothing*? pp. 23–71 *loc. cit.* also p. 95.
[10] *Ibid.*, pp. 33–34.
[11] *Ibid.*, pp. 29–39.
[12] *Ibid.*, p. 33.
[13] *Ibid.*, pp. 77–158.
[14] *Cf.* A.-T. Tymieniecka, 'Initial Spontaneity,' *Analecta Husserliana*, vol. V, pp. 17–21.
[15] *Cf.* A.-T. Tymieniecka, 'Imaginatio Creatrix,' *Analecta Husserliana*, vol. III, pp. 3–41.
[16] *Cf.* A.-T. Tymieniecka, 'The Initial Spontaneity,' *loc. cit.*, pp. 21–26.
[17] *Cf.* the previously cited 'Imaginatio Creatrix,' pp. 25–35.
[18] *Cf.* the previously cited 'The Initial Spontaneity,' pp. 20–25.
[19] *Cf.* A.-T. Tymieniecka, 'The Self and the Other in Man's Creative Self-interpretation,' *loc. cit.*, pp. 156–168.
[20] The present writer has developed these ideas in her unpublished book, *The Three Movements of the Soul*, written between 1968–1970.
[21] *Cf.* A.-T. Tymieniecka, 'Hope and the Present Instant,' in *God in Contemporary Thought*, Nauwelaerts, 1977, pp. 1082–1087.
[22] *Cf.* A.-T. Tymieniecka, 'The Self and the Other,' *loc. cit.*
[23] *Ibid.*, pp. 175–186.
[24] *Cf.* A.-T. Tymieniecka, 'Hope and the Present Instant,' *loc. cit.*

PART I

PROBLEMS OF TELEOLOGY IN THE SCIENCES OF NATURE AND IN THE HUMAN SCIENCES

ANGELO CAPECCI

FINAL CAUSALITY AND TELEOLOGICAL SYSTEM IN ARISTOTLE

INTRODUCTION

Although the more recent Aristotelian literature would not seem to dedicate particular attention to the teleological problem,[1] Aristotelian finalism has, however, been subject to interpretations which tend not only to place it in an original context but also to "upset" a reading which, for centuries, seemed to propose as characteristic of the physical and "metaphysical" system of Aristotle, a vision of the world articulated to the end or reason towards which every being tends.

I am referring, in particular, to Wieland's interpretation, which, attacking any identification of Aristotelian finalism with a "universal teleology," interprets the Aristotelian final cause as a purely heuristic point of view and final causality as a veritable *Denkform* or *Reflexionsbegriff.*[2] Such an interpretation has been the target of varied and heated criticism which, to my mind, has definitively limited its weight.[3] The discussion, however, and above all the suggestion provoked by the interpretation has served to highlight the different conclusions now possible in an interpretation of Aristotelian finalism. On the one hand, in fact, it is possible to underline how the final cause fulfils an explanatory function, the power of which is derived solely from the fact that it is best adapted to throwing light upon certain structures of the physical and, in particular, of the biological[4] world. On the other hand, it is possible to insist on the capacity of the teleological explanation to give sense to the entire universe.[5]

Consequently, it would appear to me that an understanding of Aristotelian teleology requires a clarification, first, of the reciprocal implications and of the internal connections which, in the Aristotelian argument, subsist between the discussion of the value and possibility of the final cause, the eventual ontological structure which it involves, and the overall vision of the real which the Aristotelian metaphysics proposes without exhaustively and deductively unfolding.[6]

The scope of the present work is to contribute, within its limits, to the clarification of such a problem, putting forward certain considerations on Aristotelian teleology as it appears respectively in the second book of the

Tymieniecka (ed.), Analecta Husserliana, Vol. IX, 33–62.

Physics (which constitutes the fullest and clearest theorizing on the final cause) and the twelfth book of the *Metaphysics* (in which final causality is discussed in relation to the principle upon which "the heavens and nature depend").

THE GENESIS OF THE TELEOLOGICAL PROBLEM ACCORDING TO ARISTOTLE

In the historico-critical *excursus* of Bk. I of the *Metaphysics* Aristotle, as we know, outlines the progressive discovery of the four causes. The discovery of the various causes is a necessity which emerges from the consideration of reality itself, of the *pragma* which will not allow itself to be led back to one or to certain causes,[7] so that already the first thinkers, having once identified the efficient cause, or "that from which there is movement," "forced by truth itself . . . went to work in search of a further principle" (*τὴν ἐχομένην . . . ἀρχήν*).[8] According to Aristotle, such research was never, in reality, carried to satisfactory conclusions, insofar as it resolved itself as for example with Anaxagoras in the attribution of intrinsically noetic characteristics to a principle whose nature still remained that of an efficient cause.

We are not interested here in the documentary value of such testimony. From it, however, there emerge with clarity some interesting elements for an understanding of the theoretical instances which, according to Aristotle, enforce the necessity of identifying the final cause as a cause specifically different from the others.

In the first place, the necessity for the search for a "further" cause, in addition to the formal, material, and efficient causes, arises from the "fact" that one must "explain why some beings are good and beautiful while others become so," and thereby explain "the order and harmonious distribution of all things."[9] In other words, the motive for which one feels the necessity to look for an explanation which does not avail itself simply of material and final causes, consists in the recognition that in order to explain, to "give the reason for" something, it is not sufficient to refer to the *aitiai* of some of its particular qualifications, or to the fact that the thing is, but rather that it is also necessary to identify the motive whereby such a thing is "beautiful and good," that is to say, the motive for its positivity, for its unity, for its formation as an element in a meaningful whole, in an "order."

Furthermore, the final cause does not present itself as a cause which is against or over and above the other causes, but rather as a form of specific and complementary causation with respect to other possible forms of causation.

In fact, its scope would be to give meaning to the same *pragma* which, in other ways, is explained by the other causes. Consequently, it offers itself as a distinct and indispensable form of explanation, different from that which emphasises the material constitution of an object (*hyle*) or its intelligible decision-making capacity statically considered (*eidos*) or the agency which simply actuates its potential (*othen kinesis*).

This area of thought may certainly have been suggested to Aristotle by particularly obvious necessities arising from the consideration of the realities of biology and already present in the schools of medicine which had, for some time, posed the problem of health as the problem of the equilibrium and proportion of the humours of the body, that is, of the harmonious distribution of the unity of the organism in all its multiplicity of elements.[10] But evidently the problem also arises with direct reference to the preceding "physical" speculation, and, in particular, from dissatisfaction with an interpretation of phenomena as mere compositions of bodies, or as the result of the intrinsic nature of matter, or as a mode of manifestation of the "mother" *physis*.

Such an attitude is manifest, in particular, in the characterization of nature as it emerges in the first books of the *Physics*.

Aristotle, here, refusing a concept of nature as matrix or as the first constitutional element of all things (*τὸ πρῶτον ἐνυπάρχον ἑκάστῳ*) in itself without proportion (*ἀρρύθμιστον*),[11] denies that nature can be the origin and source of movement, either as "subject" of the movement itself[12] or as an "independent entity" which, in itself, can generate movement.[13]

The *physis* is therefore characterized, by analogy with art and change, as a field of phenomena whose specificity lies in their having "in themselves the principles of movement",[14] that is to say, it is outlined as a "transcendental structure," a principle of unity rather than a principle of being.[15]

In addition, nature is defined "in one sense" as "the raw material serving as substratum for each thing which has in itself the principle of movement and of change, in another sense, the configuration and the form corresponding to the definition".[16] This structure which is the *physis*, therefore, takes the form of a structure determined by the twin constitutional principles of things, matter and form, and by the immanent tendency towards movement.

To explain the meaningfulness of natural becoming means therefore to give meaning to a coordinated process – be it of matter and form, but not attributable only to these principles (insofar as matter is only potentiality, and form, in itself, is the principle of intelligibility and thus of thesis rather than of movement), or be it of the dynamic principle which can be referred to the same things that underlie becoming. The problem for Aristotle, therefore, is to explain a reality both unitary and in the state of becoming, an "order" which is intimately connected to the "nature" of the ordered things and consequently not referable in itself to a *deus ex machina* which harmonizes new data by means of a process albeit noetic. Nor, however, is this reality to be attributed to simple "composition" or to mechanical repetition of the same things, in that for the Stagyrite, as we shall later have occasion to see, mechanical composition identifies itself with material necessity, which is always "conditioned" or rather, put into action, by something which is not itself material.

EIDETIC IDENTITY

If nature, as the transcendental structure of things, does not constitute an answer to the problem of order and harmony or to that of things being "good or beautiful," the answer to the problem is however contained in the structure itself.

A natural entity is in fact constituted by the plexus matter-form,[17] but, notes Aristotle, the form is more its nature than the matter – for example, the nature of a bed is not the wood of which it is made, since "the bed, could it reproduce, would reproduce not a bed but the wood," which is to say, something which could take definite form either as "plant" or as bed or any other object of wood.[18] Or rather, a natural entity is such when it takes a definite form as a realised entity, in action, when it has assumed that *eidos* which belongs to it.

In addition, form is a primary condition for the intelligibility of a natural entity, not only insofar as it represents the intelligible principle of the realized entity, that is, the entity actuated in becoming, but also in that it constitutes a veritable "variable constant" in the very process of becoming.

In order to clarify these assumptions it may be useful to refer to a concise passage of the *Metaphysics*, IX, where it is expressly affirmed that "everything which is becoming proceeds towards a principle (ἀρχὴν) or towards an end (τέλος) (in fact the end is the principle and, in the light of this, becoming takes place)".[19] In other words, according to Aristotle, whatever is becoming is, insofar as it is actuating, a potential; such an act is the *telos* of the process and, being thus the condition of becoming and of potential, is the *archè*.

It follows that what is last from the point of view of becoming (γενέσει), is first from the point of view of form and substance (ὀυσίᾳ),[20] that is to say, the direction and meaning of the process is given by that form which actuates itself as *telos* of the process.

The *telos*, in fact, is not only the "result," the form which a thing adopts in its becoming, but the result which is connected to the beginning, that issue which confers meaning upon the entire process. It is, in fact, the culminating point or final state of a real change; "what is born" – declares Aristotle – "insofar as it is born, goes from something to something," but "what is born" is clearly not "what gives birth"[22] so that the *eidos* which actuates itself in the (nature) process is an authentic "constant".[23]

In other words natural becoming is, so to say, to be reconstructed from that determining power which acts last, since it is marked by a form which is the same as that entity from which becoming originated. And the form itself, insofar as it remains as the one constant principle against the changes of becoming, is that which permits reference to a single principle, and therefore the leading back towards unity of the multiplicity of conditions and of the different stages of the process; and then it is what gives meaning to the process itself.

In the final analysis, therefore, what characterizes nature is that, in it, form operates as *telos*. The regularity or order which marks the processes *κατὰ φύσιν* or *φύσει* is given by the normative nature of the end; since the beginning of each process is informed to an end, not because this end is known and therefore wanted,[24] but because, in it, form functions as end, insofar as it is a process which is accomplished and culminates necessarily in the realization of a form already there from the beginning.

As Wagner points out, Aristotle does not thereby attribute a capacity for real determination to the *eidos*,[25] as Plato does, but intends only to define the natural process as a process characterized by an "eidetic identity" which enables him to confer meaning upon the means and stages of the process by virtue of an end which, according to its kind, cannot not be identical to what is at its beginning: "man generates man" and in general "nature produces what always was and is", in a perennial *Wiederkehr*, constituted by the perennial nature of the elements and of the species.[26]

By virtue of such an "eidetic constant", finality becomes the "key" to an understanding of nature and the condition whereby natural becoming constitutes an authentically intelligible phenomenon: the *physis* appears as a field of phenomena directed towards the "formation" or rather the full realization of substances, by means of a process whose sense becomes clear from its very outcome (*ἔσχατον*) which insofar as it is the full realization of the substance (*οὔσης ἔστι τι τέλος*) is an authentic end (*τέλος*) of a "continuous" process.[27]

That does not mean, however, that the *telos*, as the form of a "realized" substance, exhausts the intelligibility of natural reality, or expresses all the conditions by virtue of which a natural process is. In fact – Aristotle stipulates in the very biological studies in which finality assumes a role of particular importance – it is not "possible to say that since this is, this other also is",[28] that is to say, it is impossible in natural production to identify a causal chain whereby one can necessarily infer what follows from what precedes:" it is not at all necessary that, if your father is born, you must therefore be born".[29]

Such a conclusion demonstrates the impossibility of an analytical "demonstration" of natural processes or of the deduction of a "final form" from the process itself, since it is impossible to establish a reversible relation of cause and effect except when we are dealing with entities which are eternal[30] and which are achieved by abstraction.[31] It follows that the recognition of the eidetic identity which distinguishes natural processes involves the placing in evidence of a *telos* which gives a sense to the process. At the same time, it involves the negation of the capacity of such a *telos* to explain exhaustively and also the possibility that such a *telos* might be deducible from something else.

NORMATIVITY AND FINALITY OF NATURE

Having once identified the specific character of natural processes in their structure of processes in "formation," characterized by an "eidetic constant" immanent in the individual processes, and having identified in the *telos* the emergent expression of such a structure, insofar as it represents the ultimate term (*ἔσχατον*) in which such a "constant" form is manifest, Aristotle immediately develops the "epistemological" consequences of this research by drawing a direct comparison with the position previously maintained – a position, to his way of thinking, characterized by a perspective at once ontological and epistemological, materialist and mechanistic.

"Since the scope of research is to know," writes Aristotle, "and since we think we do not know anything before we understand *τὸ διὰ τί* of each thing (and this means knowing the cause first), it is clear that we must do likewise with respect to generation, corruption and any physical change so that, once we know their principles, we may attempt to lead back to three principles each of the objects of research".[32]

But the causes are diverse. In fact, we say cause (*αἴτιον λεγεται*) and *τὸ ἐξ οὗ γίγνεταί τι ἐνυπάρχοντος* and *τὸ εἶδος καὶ τὸ παράδ ειγμα* and *ἡ ἀρχὴ τῆς μεταβολῆς ἢ τῆς ἠρεμήσεως* and again we say cause with the sense of end (*τὸ τέλος* or rather "that in view of which" (*τὸ οὗ ἕνεκα*); for example, good health is a cause of walking. In fact: why take a walk? We say: "to keep fit," and with this reply we think we explain the cause. "In addition, we may classify as cause all those things which act as intermediaries with respect to an end, even if something else has produced the movement; for example, with respect to one's health, slimming is a cause, as well as the taking of an enema or medicine and the use of instruments for exercise in gymnastic; all these operate, in fact, with a view to an end and differ among themselves insofar as some are actions whilst others are instruments".[33]

This is probably the first complete and systematic formulation of the doctrine of the four causes.[34] We cannot enter here into its significance either in relation to its historic precedents[35] or to the theoretical foundations of its elaborations,[36] but this is what we must take for granted: all knowledge is, for Aristotle, knowledge of the causes which make possible the object of that knowledge; such causes as exist in the real and necessary conditions of being of the object under investigation are to be found if we begin with the question "why," that is to say, if we ask ourselves a question to which we can reply in various but well defined and identifiable ways.

Aristotle does not present a real discussion or judgement on such causes, that is, of the possible senses with which we can reply to the question, "why?", probably because of the "originary" character of his research on causation; if all knowledge is knowledge of the cause, a scientific investigation of causation would necessarily finish by presupposing a cause, while it is the research on causation in itself which authenticates and individuates the cause itself. However, Aristotle devotes particular attention to the final cause, the only one he tries to demonstrate as being necessary.

I must immediately say that, to my mind, such particular attention to the final cause is not due to theoretical motives, as if it were the object of a particular justification or assumed a particular meaning within the doctrine of causality. The final cause is not only the only one claimed by Aristotle as his own, original "discovery," but it is, to his mind, *de facto* contradicted by the undertaking of a "physical" investigation which, ignoring the final cause, resorts to a theoretical structure of a mechanistic type which would seem to render it superfluous if not insignificant.

The problem of justifying the final cause arises, in fact, because one might maintain that nature "acts neither towards an end nor for the better, but in the same way that Zeus creates rain not to make the grain grow but from necessity (in fact, what has evaporated must cool, once it has become water it must fall and thus it results that the grain grows when this happens, by accident)".[37]

The position of the problem of the final cause is thus in direct relation to its negation, which consists in the formulation of a type of explanation here defined "by necessity" (ἐξ ἀνάγκης). In fact, for Aristotle, those who resort to necessity as a cause do nothing but affirm that the motive for any process is "the heat, or cold or such similar things",[38] that is, they attribute becoming to elements which can neither be nor act differently, since the elements characterizing the natural becoming are such "by nature".[39]

But in this way, observes the Stagyrite, we merely pose the problem, we do not resolve it, since to maintain that "something is always thus or it reproduces itself," or rather to be aware of the subsistence of constant processes and necessary behavior patterns of things, "is not a sufficient principle" insofar as such an assumption does not proceed to the revelation of a reason but exhausts itself in the affirmation of the *hoti*, and thus does not push research forward but rather gives the illusion of the final, scientific explanation, confusing the data of the problem with its solution (οὐκ ὀρθῶς ἔχει ὑπολαβεῖν).[40]

But the explanation "by necessity," which leads back to the attempt to render absolute the material cause or to the failure to identify or utilize other causes[41] than that which expresses the composition of the object investigated, is destined to compromise the very intelligibility of the natural phenomenon. In fact – and here is the kernel of the Aristotelian argument – an explanation which makes the effect dependent upon a necessity constituted by the material conditions existing by virtue of the preceding existence of material conditions (we think of the cycle evaporation-germination of the grain), when it appears to affirm a succession of natural becoming regulated by an iron law, in reality conceives of each process as a series of coincidences.[42]

Such an explanation, in fact, since it cannot give rationality to becoming in the complexity of its conditions and therefore leads back arbitrarily to one of these conditions, ends up by making the various processes of nature depend upon "fortune" or "change",[43] since, as it limits itself to the consideration of the series of reactions set in motion "mechanically" by the material "nature" of individual phenomena, it cannot explain their regularity, that is, the fact that normally these reactions occur in certain complex situations but not in others.

It follows that the explanation "by necessity," that is, the mechanistic explanation, makes the processes in question depend on events which "happen by accident" (συμβαίνει). That is, they have no other meaning than their "being co-existent," their simply happening at the same time as each other, which is possible but not rationally justifiable. Or it ends up by attributing processes to *tuche* which, by virtue of its indeterminate character, is considered a principle not only ineffable[44] but also intrinsically irrational, as a supreme power which acts for arbitrary motives.[45]

Thus, the explanation (ἐξ ἀνάγκης) is brought back to an explanation in terms of change and, by underlining the irrational character of this explanation, the final cause emerges as a condition of the very intelligibility of the physical world.

This conclusion, it must be stressed, is arrived at not by denying change but by denying that conception of change whereby it is placed, implicitly or explicitly, at the origin of natural becoming. By confuting such a notion and by not presupposing his own doctrine of change, which – as we know – is compatible with final causality,[46] Aristotle can both lay the foundation for, and justify, final causality.

But the confutation is possible on the basis of experience itself. Whatever is generated φύσει is characterized by a fundamental regularity (ἤ αἰει γιγνεται ἤ ὡς ἐπὶ τὸ πολύ), that is to say, it reveals its own normativity and constancy.[47] But "if, therefore, it seems that such phenomena occur either by coincidence or with some end in view, if it is neither possible that they occur by coincidence nor by change, then they must occur with some end in view".[48]

This last affirmation would seem to repeat the alternative character of "change" and finality (this indicates that here Aristotle still has in mind the 'mechanistic philosophers' concept of change), and would seem to conclude simply in opting for one of the alternative members. I believe, instead, that the real meaning of the affirmation is much richer and expresses in an essential form the motives for the Aristotelian refusal of those forms of explanation which refuse the recognition of final causality operating in nature.

The argumentation, in the form of a disjunctive syllogism (things are generated only by change or for an end – but since they cannot be generated by change – they must be generated for an end) demonstrates the intrinsic impossibility of sustaining a theory which places change as the authentic cause, or "in itself," that is to say, a theory which makes change absolute before and against nature. In fact, it seems to me that the sense of the Aristotelian argument can be thus classified: that certain phenomena are identified as "change" phenomena – the designation "change" derives from the fact that they occur in a context in which we should expect a different outcome – in that every phenomenon must be referred back to the totality of the context so that each acquires a meaning from the end to which it appears to be directed. One might give an example: rain is due to the cycle evaporation-condensation,[49] but the frequent rainfalls of winter do not appear to be casual, insofar as they serve, for instance, to make the grain grow, that is to say, the rainfall fits into an order being finalized by another process, whereas it is casual during the hot days. If this is so, it means that we recognize and define a fact as "change" because it does not appear to have a purpose and constitutes an exception to the meaningful finality attributed to every event.

This means that, were there no scope, we could not even say that something was changed. And thus, insofar as we affirm the existence of change, we recognize that nature is regulated by laws of a meaningful finalistic type, and therefore change, as such, is not the authentic principle.

In synthesis, then, we can affirm that the final causality of nature is recognized on the basis of the regularity of the events which characterize nature itself and prevent us from recognizing change as the principle of natural events, and, in fact, final causality appears necessary in order not to compromise or deny the very intelligibility of such events.

Such a denial, in fact, involves the reduction of natural becoming to a succession of moments whose only sense is that of a space-time explanation of the necessity inherent in matter. But such a conclusion, as we have seen, is destined to destroy itself for two reasons: (1) because it presupposes an irrational principle at the beginning of the succession itself (change or fortune) and thus resorts to an explanation which does not explain; (2) because it does not succeed in explaining the regularity of natural phenomena which, if referring to "mechanical" series, are certainly referred each to its predecessors, but not to the complexity of their conditions.

Thus, it is the complexity and regularity of phenomena, the fact that they can be understood only when they are conceived of not as causal or due to some "arbitrary" principle, but as characterized by an intrinsic form, which provokes the necessity of an investigation concerning also the final cause and, at the same time, makes legitimate recourse to such a cause, since any explanation which leaves it out not only fails to explain such regularity and complexity, but actually destroys it.

THE STRUCTURE OF THE FINAL CAUSES

The final cause, finding its configuration on the basis of the recognition of "eidetic form" which characterizes natural processes, and subsequently recognized as a condition for exploring natural phenomena in their complexity, defines itself immediately as the expression of an ontological structure (the *physis*), which is rational and ordered insofar as it is constituted by a definite causal link and by rationally knowable ontological conditions.

In fact, the intelligible structure of natural processes is given, in the first place, by the fact that they normally tend to the "formation" or determination of an entity in itself intelligible over a succession of moments and of "stages" of development, at the beginning and end of which there is an identical form. In the second place, it is given by the fact that the conditions of development and subsistence of each thing (that is, its causes) are not unrelated to each other but, on the contrary, intimately if not formally and analytically connected.

As we know, in fact, each thing is by virtue of causes not only numerically but specifically different, so that a concrete knowledge of the same thing must bring us back to all its causes (*περὶ πασῶν τοῦ φυσικοῦ εἰδέναι*).[50] But "three causes" – affirms Aristotle – "often resolve themselves in only (*ἔρχεται . . . εἰς ἕν*) since "what something is" and its end are one, and that from which the first movement is, is identical to them in species; man in fact generates man and, in general, it is so for all those things which, moved, move".[51]

This last observation, which is not further developed by the Stagyrite, is based on the recognition of "natural" and "factual" relationships, and not on formal implications among the causes.

In natural becoming, the efficient cause which always consists of a real substance[52] resides, as we have seen, in a substance which is of the same kind, or rather has the same form as that which becomes and so is itself the end of natural becoming.[53] The recognition of such implications is therefore sufficient to "lead back to one" three of the causes of natural generation. On the other hand, the matter is conditioned by the "nature" of the end,[54] although it remains absolutely distinct and not reducible to other causes, since, in itself, it is the substratum undefined and existing as pure potentiality.

This observation, while characterizing natural movement as movement which brings to completion the nature of a thing,[55] identifies the epistemological and factual valency of causal implications: the resolution of three causes in one is not explicitly indicated as the resolution of the formal and efficient causes, as such, in the final cause,[56] and hence does not mean that "the end is to be placed above all other causes" but only that they are coordinated among themselves;[57] in addition, such a resolution, if it is valid only for all those things which "having been moved, move," expresses the recognition of a state of things relative to natural generation and has absolutely no ontological significance.

Such causal connection as characterizes natural processes confirms, at any rate, the great value of the end in understanding them. In physical research –

writes Aristotle – we have to reconstruct events according to sequences which permits us to identify "what is generated after something else" (*τί μετὰ τί γίγνεται*), what first moved and what first was moved and so on, forever, in succession (*αἰεὶ τὸ ἐφεξῆς*).[58] But purely chronological succession reduces itself to a connection of Empedoclean or Anaxagorean type, which Aristotle rejects as incapable of explaining becoming[59] which instead needs to be explained by a final cause: only if the end becomes the aim does succession become meaningful and we can explain the process.[60]

Once again, however, the final cause shows itself as the condition for the "normativity" of nature: since the fundamental relationships between natural phenomena are not exhausted in mathematical relationships[61] and since logical relationships which constitute the natural order[62] are not attributable to matter[63] but to that plexus of causes which is re-assumed, from a metaphysical point of view, in the final cause, finality qualifies itself as an expression of the structure and of the intelligibility of the *physis*. "The fixity of the links uniting [the peculiar causes of phenomena] is the solid base on which one may construct the science of nature and, at the same time, found on the real order the instance of what institutes the ordinary course of nature."[64]

THE MEANING OF THE TELOS

The considerations which we have entertained up to now have tried to show how nature is, for Aristotle, regulated towards ends not only insofar as the single phenomena are directed and ordered by an *eidos* which expresses the natural completion of the "realized" substance, but also insofar as each phenomenon seems possessed by an internal "normativity," indicated by the regularity with which the phenomenon takes place, this normativity connects it in a significant way with other phenomena in that it appears intrinsically directed towards an end, which is no other than the realization of a natural, 'perfect', substance, that is to say, a substance which is possessed of that form which most properly belongs to it.

In this way, the regularity which distinguishes the processes *κατὰ φύσιν* or *φύσει* is given by the "normativity" of the end; each natural process is a process informed, from the beginning, towards an end, not because this is known and therefore willed, but since it is constituted by a system of relationships whose successive determining acts on the thing itself subject to generation are always to be referred to its "realization," which is brought about at the end of the process in such a way that its actions, that is, the individual moments and aspects of the process, acquire meaning by virtue of the final outcome.

Thus, the actions of animals as well as the production of the vegetable world show that each natural process is a process determined by an end, not in the sense of being oriented to some preordained scheme[65] but insofar as it is to be conceived beginning from a *telos*, from the "completed" realization of itself (on the basis of its *physis-eidos*), which determines the means and the stages of the process. In other words, the swallow builds its nest and the plant pushes its roots downward because both swallow and plant tend by nature to "realize" themselves and in consequence to reproduce and feed themselves, and such actions constitute a "regelmässige Zweckmassigkeit, die ist nur durch den Ansatz einer Zwecktätigkeit begreifbar".[66]

We may certainly affirm, then, that natural beings are characterized by a "project," that is, subject to a process whose stages and conditions are coordinated with a view to an end, which is the same end as that of the process. But a "project" is not a design or a scope in any way separated from the natural being; it is rather its structure, the originary unity which gives sense to the properties and to the natural functioning of the entity, not reducible to the result of the actions of external forces (nature is the intrinsic principle of movement), though not extraneous nor independent of them.[67]

From this point of view Aristotelian finalism is nearer to the concept of "teleonomy," upon which contemporary biology[68] would seem to insist, than to those forms of "teleology" which identify an end and a general design as the end of a natural process and thus conceive of the end as an expression of the immanent necessity of things, be it the Platonic or Thomistic teleology or the 'consequence' of the ordering activity of an intelligence[69] or the Stoic type of teleology, as an individuation of a central cosmic element by reason of which everything else is.[70] But Aristotelian finalism never assumes the form of laws which express general connections between heterogeneous phenomena. Even in the works on biology, where the use of the "category" end is predominant, "the finalization of natural processes means that they preserve in every case the specific realities in their essential determinations"[71] and not that such determinations are deducible from their *telos*[72] so that knowledge of the end is not knowledge of the *ousia* and of its laws.

Aristotelian natural teleology, in the final analysis, does no more than express the necessity of guaranteeing the true consistency of natural structures, the possibility of understanding and investigating each thing as it *already is*. According to Aristotle, it serves no purpose to conduct arguments as those who try only to save their own hypothesis; they accept any consequence whatever in the conviction that their principles are true, as if some of these principles should rather be judged by that which we derive from them and, above all, from their final outcome (*telos*). But the final outcome of the science which has the scope of doing is the work itself; in natural science it lies in how much from time to time is testified to in incontestable manner by the senses.[73]

What "from time to time is testified to by the senses" is the *ousia*, but only insofar as this is the authentic *telos* is an investigation of its nature possible.

On the one hand, in fact, every process is such when it gives place to a completed substance. On the other hand, only if the substance is the *telos* (and not mere *eschaton*) does an analysis of the causes become possible; the formal cause, or essence, appears clearly only when it is identified as the "end," the material cause is coordinated to it (by "conditional necessity"), and the operating cause is the "principle of change" which, in the natural substance, is similar to the formal cause and still a function, therefore, of the *ousia-telos*.[74]

The *telos* (being that which corresponds to the question *οὗ ἕνεκα*) has, therefore, an essential role to play in the science of nature, not insofar as it can be identified with one of the "terms" of the demonstration,[75] but insofar as it constitutes the structure of relationships, whereby the conditions of becoming are brought together in one meaningful picture and the parts of the living being are brought into their original essential unity.

The sense of Aristotelian finalism having thus been characterized, I think it is possible to share the desire to condemn a certain type of mystification, or, at least, of misunderstanding, which it has suffered in the course of the history of Western thinking with regard to the immediate use not so much of Aristotelian teleology as of the very category of "end" to create a providential conception of the universe and of its history.[76] However, the attempt to present the Aristotelian concept of end as a straightforward *als-ob* does not seem to me to have sufficient foundation, and consequently the reduction of the teleological perspective of the Stagyrite to the use of a category affirmed not for its ontological sense but merely for its "explanatory potential" does not seem to be admissible.[77] To say in fact – as does Wieland – that teleology for Aristotle is not an object of research in itself, but rather a category, a "*Reflexionsbegriff*," is insufficient to prove the merely "regulatory" value of the principle. In fact, the Stagyrite certainly does not propose an absolute and generally valid end for all beings in the same way, but this does not mean that there does not exist a real *telos* for each of them (and, indeed, this is of the same substance and is the condition of any reality). In other words, it would seem legitimate to affirm for the "end" what Aristotle expressly clarifies for the three analogical principles (substratum, form, and privation):[78] while we must exclude the possibility that the ends of all things are individually identical, in the "universal and by analogy," the end is identical for all; that is to say, each thing has an end, which is precisely itself or its form as a completed substantial and determined realization, but there does not exist an end which is the end of all things.

Wagner has conducted a critique, and in some respects a decisive one, against the interpretations suggested above, interpretations, let us say, in the key of Kant. Above all, he shows the character of a real alternative which, according to Aristotle, subsists between change and finality. In fact, the inconceivable nature of a "regular" process attributable to change indicates clearly that the normativity and the constancy of nature are due to a process which actually tends to an end and not to the causal achievment of any end. Moreover, what Aristotle sees as the essential difference between art and nature – both characterized by an identity of causal structures: the fact that nature, unlike art, "produces from itself its own product," gives evidence as to how the natural process is determined and directed by the form which constitutes its inward conclusion, its *telos*, and thus how this is not only a concept or a particular point of view, but a determining power possessed of "full reality and full belonging to nature".[79]

But, above all, in determining what we may define as the ontological *status* of the end, the identification explicitly declared by Aristotle between *τέλος* on the one hand and *ἔργον* and *ἐνέργεια* on the other is, in my opinion, decisive: '*τὸ γάρ ἔργον τέγος, ἡ δὲ ἐνέργεια το ἔργον*' and thus the very name of *ἐνέργεια* is said to be in relation to *ἔργον* and tends towards the meaning of *ἐντελέχειαν*'.[80] In fact, "all that which is becoming proceeds towards a principle (*αρχήν*) and an end (*τέγος*), since the *τὸ οὗ ἕνεκα* is the principle and generation is in view of the end (*τνῦ τέλους δὲ ἕνεκα*)"; but "the end (*τέλος*) is *ἐνέργεια* and becoming takes place by virtue of the end".[81]

To sum up, the end of each process is in act[82] either insofar as it tends to form, or rather to a substantial, complete determination, or insofar as it consists in the *exercise* of a faculty and capacity (*ἔργον*), since, for example, the end of teaching is the exercise in action of knowledge (and not the possession of the capacity to learn and know), and the end of sight is seeing and not vice versa.[83]

The end then appears unequivocally to be identified with the perfection of the substance, either as its state (*ἐντελέχεια*) or as its activity (*ἐνέργεια*), and the exercise of faculty and capacity (ἔργον).[84] And the identification, of end with form, which Aristotle stresses more than once, is no more than a transcription between end and act.

Such a recognition confirms the conviction that Aristotelian finalism is not an explanation of a universal law, of a natural "norm of production," but rather an affirmation of the structural character of natural becoming which, insofar as it can be seen to culminate in a completed substance, in the same measure can be referred to a whole series of conditions closely connected with the being (as "formed" matter) of such a substance, conditions which are yet authentically intelligible and susceptible to investigation by means of a causal series determined numerically and quantitatively.

Aristotelian finalism is therefore the affirmation that nature must be investigated in all its four cases, not only because they are "so many," but because they refer to the same object of inquiry and constitute the same theoretical framework, there being between them a reciprocally conditional communication. Such teleology constitutes, in the first place, a justification of the intelligibility of the real, beginning with the problematic nucleus of all knowing: substance, which requires an explanation in all possible senses and confers a unity which is not metaphysical but structural to the indefinite succession of becoming. This unity, in fact, is not given by the revelation of a supreme unifying cause, but is constituted by meaningful internal relationships which connect the moments of succession as "stages" and means of a process which depends on conditions which are different but not indefinitely unrelated.

It seems clear to me, therefore, that, for the Stagyrite, the final cause is a factual and real condition and not merely a subjective explanatory principle. However, between this position and the assertion that the entire universe is dominated by a general teleological law, given by the tendency of each thing toward a single principle or a single set of principles, there is the same difference as that which exists between the affirmation of the existence of a cause of movement and the affirmation that the general structure of the universe is of a mechanical type, the result of the same movement which comes from the composition of successive relationships between mover and moved.

SUBSTANCE AND END

At this point we might attempt a brief synthesis of the argumentation so far conducted. The theoretical question which gives rise to the problem of the end in Aristotle has been identified, in the wake of the Aristotelian reconstruction of the historical genesis of the teleological question, in the problem of the possibility of a synthetic explanatory form. According to the Stagyrite, in fact, the problem arises from the fact that material and operating causes do not sufficiently explain what makes things be or become "good" and "beautiful," that is, the motives not of the particular qualifications of things, but of their positivity, their unity and order.

The problem is resolved by identifying the specific function fulfilled in natural becoming by the *eidos*. This, in fact, is recognized not only as the intelligible element of a substance, statically considered, but as the "end" toward which the process tends as an expression both of the perfection of the substance in which the process "normally" culminates, and of that synthetic, originary unity to which the functions and "natural" activities of the organic being refer.

In this way the "end" assumes the character of "structure" in the sense of a nucleus around which a series of relationships converge, but not in the sense of a "model" of organically connected relationships. In such a structure, therefore, the subject does not find its resolution in the series of relationships, but rather the relationships are given by the connection between functions, instruments and organs, and the subject. The end, therefore, remains for Aristotle essentially the substance in completed and "perfect" action.

The identification of end with substance, in which the theory of final cause culminates, defines the Aristotelian teleology not as the recognition of a "cosmic" principle which finalistically pervades the entire universe, but as an affirmation of the intrinsic structural unity of natural substance.

The identification of end with the perfection of substance does not however exhaust the Aristotelian problem of the end. On the one hand, in fact, it characterizes the end as *telos*, direction and goal of a process, giving it meaning in its moments of development, structuring its causal co-ordinates, and representing, as *eidos-entelecheia*, an authentic answer to the question *οὖ ἕνεκα*, insofar as completion of the structure of the concrete thing or of the species – unfolded realization of its potential. But on the other hand, it tends to fix the concept of substance with an absolute character which does not belong to it.

Substance, in general may be a self-sufficient principle of intelligibility, but it is not a self-sufficient principle of being.

If all things which move – declares Aristotle – are moved naturally or in manner contrary to nature and by violence, and against nature they are moved by something, and by something other than themselves, and again, if the things moved by nature are moved by something, be they moved by something or by themselves or not by themselves, like, for example, heavy and light objects (in fact, these things are moved by whoever produced them and made them heavy or light, or by whoever removed all the obstacles and impediments) all things moved are moved by something.[85]

Here, we certainly cannot insist on these questions, but we can take for granted that for Aristotle natural substance is not self-sufficient for at least two reasons: (1) because, being "moved and in movement," it owes its movement (and therefore its being) to an unmoving principle; (2) because the very movement due to the "internal principle" needs an external condition which "removes impediments" or constitutes an object towards which it can tend.[86]

Such a situation is also reflected at the level of the final cause. In fact, if the end of the natural substance is its own perfection as realization of its possibilities and exercise of its functions, and if the perfection of the natural substance is its being in action as "self-moving" substance, it becomes possible to admit the existence of an end external to the natural entity which tends toward it.

In order to illustrate this aspect of Aristotelian teleology, we shall here refer to the short treatise *De Motu* which is characterized by the fact that it offers an unusual aspect of synthesis of the Stagyrite's thought, concerning the theme of the "force" holding the universe united in an organic way.

In fact, having recognized the living being as a condition of movement of those natural "entities" characterized by simple, "unidirectional," movement, in that the living being is the cause of the obstacles, or of their removal, which permit or prevent the attainment of the end of each natural being,[87] Aristotle declares that in general the condition of movement is "the unmoving.[88] This conclusion is arrived at without consideration of the requirement of a final cause in the series of moving causes – as happens in the *Physica*, VIII, and *Metaphysica*, XII – but by remaining simply in the framework of final causality.

"Animals gifted with movement all move by attaching themselves to external things," declares Aristotle, in that "all animals move, and move with a view to an end, so that this is the term of their movement: the 'that in view of which' ".[89] In fact, the factors in the natural movement of animals are *νοῦς* and *ὄρεξις*[90] adopted as the most elevated expressions of the diverse animal faculties, and to some degree the most capable of reassuming them, not set against each other, but reciprocally conditioning each other.[91] By virtue of such faculties each object provokes a reaction (of desire or of repulsion)[92] so that the movement of animals is to be related to the *horexis*, as an adequate and possible[93] but not mechanical reaction to an external stimulus elaborated and filtered by all the faculties which characterize each living being.

In this way the object of the *horexis* moves as a factor which is unmoved (not being subject to movement under the aspect for which it is desired), that is, as a term desired for itself, insofar as it is "*property*" which must not be realized but only possessed (or which must not be lost). It is therefore a final cause, but a final cause not in that it is "that for whose advantage" (*ἕνεκά τινι*) but rather "that in view of which" (*ἕνεκά τινος*).[94]

The two senses of end do not appear unrelated here: living beings follow their own ends *τινί* (that is, they realize themselves), in that they tend to an "external" end (*τινός*) (that is to say, they act with a view to something already realized): the living being is perfect in that it is living and fulfills that activity belonging to it as such; but such activity is possible not only in the definitive possession of the "*property*" but in the tension toward it.

The *horexis*, in fact, does not simply depend on an unstable equilibrium which always stimulates the animal to seek or avoid something which momentarily escapes or threatens it, but on the ontological structure which places it in a relationship of dependence or tension towards the "*property*".[95]

On the one hand, then, Aristotelian finalism must lead back, as to its central nucleus, to the doctrine of substance, as the fully revealed and unavoidable mode of being of any intelligible determination. But rather than rendering substance absolute by proposing it as the ultimate and self-sufficient scope of any becoming, this finalism expresses also the relationship of correlation between substances constituting the order of the universe; a relationship which does not violate the "specific" autonomy of each substance and which does not consist in a hierarchical placing of species and substances one in function of the other, but which is given by the intentional specific "term" of their tendency and of their operating.

Such a term, whose unity would seem here denied (being the end adequate to the *horexis* relative to the species and thus different for each species), is at any rate placed in an order of value[96] which Aristotle will never explicitly clarify. And for this motive also the problem of the presumed teleological order of the universe as universal tension with a single principle remains an open question – one of the thorniest for Aristotelian exegesis.

ON THE TELEOLOGICAL SYSTEM OF ARISTOTLE

The problem of the universal teleological connection, or rather the problem of the eventual general ordering of the entire universe in relation to general final causation, has recently provoked a most meaningful reply, both for the radical nature of its interpretation and for the extent of its documentary throughness, in a volume by H. Happ.

He maintains that Aristotelian teleology is both based on and forms the foundation for "the continuity of becoming and of process" in a "general order" of the universe characterized by a hierarchy of ends. This hierarchy, in addition, would be constituted by virtue of the fact that all things tend toward the best, that is, toward being, but in such a way that the fullness of being possessed by the various things is to be classified according to its distance from the supreme *archè*.

From this derives the tendency for each level of being to aspire to a superior stage in a relationship of imitation which gives rise to a sort of universal *ordo amoris* leading from the low to the high. Thus the sublunary world tends, imitating its movement, to the world of the stars, which is granted a greater degree of completion and of being, and the world of the stars tends toward the unmoved move.[97]

Happ's thesis, which clearly belongs to the "deductive" and "academizing" readings promoted by the so-called "Tübingen School," appears to be documented by a series of general references rather than by an articulated discussion of Aristotelian texts. It would seem, however, that it is primarily based on a passage of the *De generatione et corruption*, II, 10, and therefore we will discuss his thesis on the basis of this text.

In the passage under examination Aristotle affirms:

> in all things is projected towards the best, and it is better to be than not to be (it has been explained elsewhere in how many meanings we use the term 'to be'), and since it is impossible that being is present in everything because some things are too far removed from the principle, the god has completed the universe in the only manner possible, that is, rendering uninterrupted the course of generation; in this way, in fact, reality can attain the maximum possible cohesion since the perennial becoming of generation approximates more than anything else to substance.[98]

The teleological foundation is clear: nature tends toward the best – that is, toward being.

It must be said immediately that the being toward which it tends is not the principle considered as *esse ipsum*, but rather that being which is "said in many senses" (strangely Happ ignores the clarification of Aristotle who evidently is here recalling his own conception of being), and thus things which are by nature (or nature which is in all things) tends, as we already know, not to participation *in being* but towards the realization of their substantial being. On closer study, this interpretation is confirmed precisely by that part of the passage which would apparently state an ontology of participation ("it is impossible that being is present in everything, since some things are too far removed from the principle"). In fact, the "being" in question cannot be considered as being in itself or being as substance (in the neoPlatonic sense, unless we want to attribute to Aristotle, without any real necessity, precisely that conception of being as a transcendent kind or as a "substantial universal" against which the Stagyrite constantly argued).[99] To my mind, it should be considered as the first of the senses of being, the substance which is often defined by Aristotle as "simply being" or par excellence as "that which has greater reason for being".[100]

The passage in question, therefore, does not justify an interpretation of Aristotelian teleology in a "deductivist" context, as an affirmation of the tendency of all things toward a "unity," which is "the being" and "the good," but means rather that things tend to their own proper substances). At the same time it underlines that the things themselves cannot be without a principle which is always active. Their own movement as sublunary substances, therefore, could not take place without a mover not subject to generation, and such generation must be "continual" and "perennial" (since it depends on a continuous movement which, in turn, depends on an unmoved first mover always acting).[101]

THE FINAL CAUSE AND THE FIRST UNMOVED MOVER

If the final cause, far from exhausting the substantial reality in the absolute character of each substance, implies an ontological connection among substances; and if this connection does not give place to a constituted system on the basis of a metaphysics of the "deductivist" or neoPlatonic type, from the relationship of "imitation" which every "grade of being" has with the grade above, the problem remains, however, of the possible unifying function that the unmoved mover can fulfill in the order of final causes.

A propos of this, we may note immediately that the first unmoved mover, though being the final cause, is not the final cause directly – neither as the one end of a multiplicity of natural "entities" nor as the "first" and supreme principle in a "consecutive" series of final causes. In fact, not only is there no argumentation in the Aristotelian text which can echo the Thomist "fifth way" (which would seem to presuppose an intelligent and consequently providential principle), but neither is the first unmoved mover demonstrated by a causal chain of the final type, that is, by investigating the *διότι* of things in the sense of *οὗ ἕνεκα*.

Such a substance does not imply, in any case, that the final cause is rejected or "put to one side" at a certain point in the evolution of Aristotelian thought; but it does seem coherent with the theory of final cause here outlined and with the conception of being which forms its basis. In fact, when investigation assumes a "transcendental" character, that is, comprehending all experience, or when it becomes "metaphysical," there arise the following difficulties with regard to the utilizability of the final cause: (1) we cannot identify a *οὗ ἕνεκα*, which consists in the *telos* of a substance realizing itself in it, that is to say, arriving at its proper perfection: the final cause of being cannot be "the being itself," which is not a substance or a kind, but *πολλακως λέγεται*; (2) becoming, though still conceived of as the realization of potentiality oriented teleologically, that is, towards a "term" which is the *telos* of the becoming substance, does not have sufficient motive for being: but, considered in its totality, it has as a condition a non-becoming principle, always in act, which is not the "form" of the becoming substance; (3) among substances which are becoming there is no relationship of subordination of kind, that is to say, there is no hierarchical continuity constituted by a degree of differentiation within the *κοινόν* which can be the common "end" of a substance, even if in different ways participated in;[102] (4) therefore, a series of ends *ἐφεξῆς* such that one requires the other, all of them constituting elements in a hierarchical series, does not exist.

In fact, the first unmoved mover is discussed, in relation to final causality, not as the "ultimate end" required by things, but as "entity" which, though unmoved, is the principle of movement. As a matter of fact, the final causality of the first unmoved mover is introduced after the proof of its necessity, or existence, in a discussion which does not touch on its reality but on the mode of its causation.[103]

A propos of this, Aristotle wants to demonstrate that it moves in the same way as "the objects of desire and of thought do." Such a demonstration proceeds in two stages: first by showing how the objects of desires and intelligence coincide in the "optimum," then by showing that the end belongs even to "non-moving things." The substance of the demonstration consists in indicating "the good" in its highest degree as the characteristic which belongs to an "entity" and as condition of the tendency of each thing toward its own good.

Aristotle proceeds, adopting a typical academic scheme, to lead all "entities" back to a series of opposites and thus to a further unification by virtue of common elements which are proper to the final terms of the two series. For this purpose he affirms, first, that the objects of desire and of thought, radically considered (*τούτων τὰ πρῶτα*) are identical (*τὰ αὐτὰ*), since, while the *ἐπιθυμητὸν* is "that which seems beautiful," the *βουλητὸν* is what is objectively beautiful from the moment that the latter is determined, not by mere sensible appetite, but by thought, which is the principle (ἀρχὴ γὰρ ἡ νόησις).[104] In other words, the object of desire (ὄρεκτον), that is, the "global pulsation" of man,[105] since it cannot be reduced to mere concupiscence but is *also* determined by reason, is not 'beautiful' because we think it so, but is rather desired because it is objectively beautiful.[106] Therefore, "what is supremely desirable (desirable with rational desire) is intelligible".[107]

But, on the other hand, what is supremely intelligible is also supremely desirable. In fact,

> the positive series of opposites is in itself intelligible; and, furthermore, within the framework of substance, most important is that substance which is simple and in action (the one and the simple are not the same thing; unity is a measure, simplicity is rather the way of being of a thing); now, the beautiful and also what is desirable for itself are in the same series and what is first is always the optimum.[108]

Thus, the "simple substance," that is, the first unmoved mover, is a *horekton*, indeed, that which is most desirable and desirable for itself. But the *horekton* in general is just "that in view of which" the living being acts, the object which moves insofar as it is desirable and thereby "motive" of decision and action. Therefore, it is connected to the very structure of the *horekton* being final cause, that is to say, a principle of movement insofar as it is the "term" toward which everything that moves tends, without thereby being implied in the movement.

The continuity between the position of the *De Anima* and that of the *De Motu* is transparent, thus confirming a thesis already expressed here: Aristotelian teleology is not directed toward guaranteeing a cosmological continuity given by the subordination of all things to a single principle, but consists essentially in the rediscovery of an explanatory structure (and thus an ontological one) given by the substance as realization "in act" of functions and activities culminating in the realization of the substance itself. Such functions or activities are fulfilled not by an "unconditioned" development of the substance itself, but insofar as each substance tends to an "adequate" object of desire – adequate to its nature (and therefore different according to different species) so that both the complete realization of the substance and the object of desire constitute the real *οὗ ἕνεκα* for which movement is instituted.

The unmoved mover can be the end (*τὸ οὗ ἕνεκα*) only insofar as it is the "object of desire," being "simple" substance, not "the one" (that is, being a transcendent entity and not a common form), and thus being necessarily removed from any form of becoming.

The clarification is repeated by Aristotle: the "final cause (*τὸ οὗ ἕνεκα*) may belong to unmoving things" only as "that in view of which" and not as "that for whose advantage".[109]

Having thus clarified that it is in the "nature" of final causality that the final cause can consist in an unmoving reality, not insofar as this is the "ideal" causal form to be imitated but insofar as it is a condition of the expression of the potentiality of the moved, which moves assuming the unmoving as an object of desire; and having in addition affirmed that the causality of the first unmoved mover is a causality of this type, it appears clear how the first unmoved mover cannot in any way be understood or assimilated (to use Hegel's formula) to "an ultramundane intellect" which imposes its ends on nature, but must be affirmed as the ultimate "term" in a structural relationship which beings (and in particular living beings) reveal as a necessity of their natures: "the moving principle, living, moved, is the faculty of desire" – declares Aristotle – but the first mover, "that which moves the same faculty of desire, and is not moved by anything, but, unmoving and unmoved, moves, is the object of desire".[110]

Having established in what sense we must understand the final causality of the unmoving first mover, it remains to be seen how we explain this causality. Aristotle explicitly declares that the unmoving first mover is (1) the object of love,[111] (2) the "optimum" as an entity "whose way of living is most excellent",[112] and (3) that without which there would be no good.[113]

Now, the unmoving mover is final cause as "object of love" really only of the first heaven, that is, of the desire and of the intelligence of the soul.[114] One might therefore suppose, reasonably enough, that it is the final cause directly of the first heaven, and indirectly, of all the universe, in that the grades of being inferior to the first have this last as the object of their love, and so on. But the causal activity of the first heaven is essentially an efficient causality,[115] while we have already seen the difficulty of admitting a universal *ordo amoris* in Aristotelian thought.

The sense of the final causality of the first mover, in the order of final causes, may appear clear if we evaluate the two attributes of the first mover (the optimum and the principle of good).

> All things [affirms Aristotle] are ordered together in a certain way, but not in the same way. All things are in fact in relation to one (ἀλλ' οὐχ ὁμοίως), but in the same way as in the house it is not fitting for the free men to behave in an arbitrary way, but all, or the greater part of their actions are ordered, whilst slaves and animals contribute a small portion to the common good and their actions are generally unreasoned.[116]

The "good of the universe lies then in the order of the things which compose it, but above all in the principle of the optimum which is the condition of such order",[117] as unity which is not a requisite possessed by things, common substance, but is a term of tension; and therefore the good and the order of all things is given by the relationship of each thing to the entity which expresses its unity.

Such a relationship is a relationship πρὸς ἕν, that is, a relationship which does not consist in the dependence of a kind and its consequent subordination (ἐφεξῆς), but is constituted by reference to a single principle of realities which are constitutionally different.[118] Whereby the principle is the common term of ever-differing relationships so that it is given neither by an element common to all "goods," nor by an "idea," nor by "the end of human action",[119] but rather by a reality (or substance) in relation to which all the other "goods" exist.

Such an affirmation implies that there is not a single "good,"[120] and that the first unmoving mover is not the end (in the sense of perfect substance in action), because in this case it would be the common good for all (and, incidentally, would no longer be transcendental). The unmoving mover is rather the aim (in the sense of ἕνεκά τινος) as the optimum.

In fact, the existence of each thing, as we know, is tied to its conditions, which are matter, form, and the efficient cause; these conditions, in any case, unfold only when they "converge" in a structural relationship (or rather when they are placed as coordinating operative factors to "actualize" a potential). And so life itself, like the movements of the animals, is not dependent on change, but has a view to an end, which is the "good." This "good" is the one proper to each substance and to each kind but in its turn it can be identified as good insofar as it relates or can relate, in a way which is not univocal, to the supreme good which is the condition of life of the unmoving mover.

In other words the multiplicity tends to a good which is proper to it (its own completed realization), that is, to a particular good and not to the good in itself. But the good in itself has a particular sense insofar as it can be led back (πρὸς ἕν) to the absolute good, to the optimum. It follows that particular ends are authenticated as such (that is, recognized as positive terms of a process or as "objects of desire") insofar as they are in view of what is absolutely positive, the good in itself, the optimum, even if the latter is neither properly the end of things which tend toward their own particular good nor an end from which it is possible to deduce particular ends as if these were implicit or contained in it.

In conclusion, we may, therefore, speak of a universal Aristotelian finalism, but not as a general connection of means and ends, not as the position of an absolute "purpose" from which the specific end of each thing can be deduced, because the supreme good is not a common kind, but only as an affirmation of principle, one in itself but meaningful in different ways for each thing, a necessary condition for the natural tendency of each thing towards its own perfection, that is, towards its own unity of "formed" being, and hence it is a condition of the intelligibility (and existence) of all natural things.

University of Perugia

NOTES

1 In the wide panorama of contemporary Aristotelian literature, studies dedicated specifically to Aristotelian teleology are noticeably rare. Among such we should note: C. Werner, 'La finalité d'après Aristote,' *Rev. de Théol. et de Philos.*, 1931, pp. 5–16; A. Sandoz, 'Le rôle de la cause finale dans l'expérience chez Aristotle,' *Rev. de Philos.*, 1934, pp. 46–75, 271–82, and 1937, pp. 104–22, 210–31; some articles by G. Bortolaso in *La Civilità Cattolica*, 1954, v. II, pp. 27–39, 249–60, 630–41; 1960, v. IV, pp. 461–70; 1961, v. I, pp. 571–80, and v. II. pp. 465–75; and above all M. P. Lerner, *La notion de finalité chez Aristote*, Paris, 1969. Nonetheless, naturally, the teleological problem has been widely referred to in numerous studies dedicated to other aspects of Aristotelian thought or in general to Aristotelian philosophy.
2 W. Wieland, *Die aristotelische Physik*, Göttingen, 1962 (reprinted 1970).
3 Cfr. for all H. Wagner, *Einleitung*, in Aristoteles, *Physikvorlesung*, Berlin, 1967, pp. 337–60.
4 A propos, see M. Vegetti, 'I fondamenti teorici della biologia aristotelica in *De Partibus animalium*', in Aristoteles, *Opere biologiche*, Torino, 1971, pp. 489–553.
5 Lerner, *op. cit.*; H. Happ, *Hyle*, Berlin, 1971.
6 With regard to the "problematic" and non-systematic configuration of Aristotelian Metaphysics as it emerges from recent Aristotelian literature, cfr. E. Berti, 'La nouvelle métaphysique d'Aristote', in *Akten des XIV Internationales Kongresses für Philosophie*, Wien, 1970, V, pp. 447–56 (now in E. Berti, *Studi aristotelici*, L'Aquila, 1975, pp. 135–42).
7 Cfr. *Metaph.*, I, 3, 984 b 8–11.
8 *Ivi*, 984 b 8–11.
9 *Ivi*, 984 b 11–19.
10 See for example M. Pohlenz, 'Hippocrates de prisca medicina', *Hermes*, 1918, pp. 396 ss.; P. Kurcharski, 'La méthode d'Hippocrate dans la Phaedre,' *Rev. des Etudes grecques*, 1938, pp. 301 ss.
11 *Phys.*, II, 1, 193 a 9 ss.
12 "There is no movement outside" single and determinate "things" (*Phys.*, III, 1, 200 b 32–201 a 1).
13 Cfr. Wieland, *op. cit.*, p. 239.
14 *Phys.*, II, 1, 192 b 13 s.
15 For a justification of this assertion see my *Struttura e fine. La logica della teleologia aristotelica*, to be published by Japadre, L'Aquila.

[16] *Phys.*, II, 1, 193 a 28–31. The meaning of such a definition seems to me to have been classified perfectly well by St. Thomas who shows how nature is here defined as matter and form not so much because the substance of matter and form is one but because in order to be, or to become, that particular thing which it is, a thing must acquire a form which determines it in action, while matter without form is simply an indeterminate impossibility (Thomae Aquinatis, *In octo librum physicorum expositio*, ed. Maggiolo, Taurini-Romae, 1954 lib. II, lect. II, 1534 *passim*).

[17] From the sphere of "nature", according to Aristotle, are excluded both things which are "by chance" or "by design" (*Phys.*, II, 1, 192 b 16–18) and mathematical realities (*Phys.*, II, 9, 200 a 15 s.) and "the non-physical principle ... the unmoved mover" (*Phys.*, II, 7, 198 a 31–b 5).

[18] Cfr. Wieland, *op. cit., passim.*

[19] *Metaph.*, IX, 8, 1050 a 8–9.

[20] *Ivi*, 1050 a 4 s.

[21] *Phys.*, II, 1, 193 b 16–18.

[22] "Man is generated from man, but a bed is not generated from a bed" (*Ivi*, 193 b 14).

[23] A propos, cfr. K Oehler, 'Ein Mensch zeugt einen Mensch', in *Einsichten*, Frankfurt a.M., 1963; and Wieland, *op. cit.*, p. 237 and *passim.*

[24] The attribution to Aristotle of an anthropomorphic vision of nature now appears practically indefensible. For a precise confutation cfr., for all, A. Mansion, *Introduction à la physique aristotélicienne*, Louvain-Paris, 1946(2), pp. 106 ss., 253 s., 261 ss.

[25] Wagner, *op. cit.*, p. 481.

[26] Mansion, *op. cit.*, p. 279.

[27] *Phys.*, II, 2, 194 a 28 ss.

[28] *De part. anim.*, I, 1, 640 a 4–6.

[29] *De gen. et corr.*, II, 11, 338 b 8 s.

[30] *De gen. anim.*, II, 6, 742 b 25 ss. *passim.*

[31] *Metaph.*, III, 2, 997 a 28 *passim.*

[32] *Phys.*, II, 3, 194 b 17–23.

[33] *Phys.*, II, 3, 194 b 23–195 a 3.

[34] It coincides almost entirely with that provided in *Metaph.*, V, 2 and is taken up again in various other salient works in nearly every period of the development of Aristotelian thought.

[35] The relationship existing between the Aristotelian doctrine of causality and Platonic-Academic teaching was noticed even by the very first commentators and highlighted by the greater part of contemporary scholars.

[36] For some, the Aristotelian doctrine of four causes finds justification in an adhesion to an entire tradition of thought (F. Solmsen, *Aristotle's System of Physical World*, Ithaca, 1961, pp. 116 s.); for others, it is founded on immediate evidence (W. D. Ross, 'Introduction', in Aristotle's *Physics*, Oxford, 1936, p. 37); for others, it consists in the transposition into the field of "physics" of distinctions relevant to the fields of human activity (J. M. Le Blond, *Logique et méthode chez Aristote*, Paris, 1970(2), p. 399); for yet others it is founded on linguistic analysis (Wieland, *op. cit.*, p. 179).

[37] *Phys.*, II, 8, 198 b 17–21.

[38] *Ivi*, 198 b 12 s.

[39] *Ivi*, 198 b 13.

[40] *Phys.*, VIII, 1, 252 a 32 ss.

41 Almost all commentators think that the explanation "by necessity" is that of the physicists; as we know, they considered the material cause, only and in the first place, as a principle of explanation of the universe. The individuation of the "physicists," though referred to by Aristotle now and again, remains a subject of controversy (Empedocles, Anaxagoras, Democritus).

42 "These things [that is, in winter it rains frequently, or incisive teeth are sharp at their birth] and all those things according to nature – writes Aristotle – always happen, whilst they never happen with those things which are derived from fortune or chance" (*Phys.*, II, 6, 197 b 11 ss.).

43 For distinction of "fortune" and "chance" according Aristotle, see *Phys.*, II, 6.

44 *Phys.*, II, 5, 196 b 23 ss.

45 A propos, cfr. Solmsen, *op. cit.*, pp. 10 s.

46 Chance, as we know, is not denied by Aristotle, who considers it in fact a principle of generation (along with nature and art), but is led back into the sphere of final causality and is ultimately subject to it (cfr. *Phys.*, II, 4).

47 *Phys.*, II, 8, 198 b 34 ss.

48 *Ivi*, 199 a 5.

49 Cfr. *Meteor.*, I, 11, 327 b 12–20; *De part. an.*, II, 7, 653 a 2–8.

50 *Phys.*, II, 7, 198 a 22–24.

51 *Ivi*, 198 a 24–27.

52 *Metaph.*, VII, 1032 a 13–18; *Phys.*, II, 6, 198 a 3 s.

53 For the identification, according to species and formal and final causes, cfr. sp. E. Zeller, *Die Philosophie der Griechen in ihrer geschichtlichen Entwicklung*, t. II, abt. II, Leipzig, 1921 (Hildesheim, 1963), p. 328, which also collects numerous passages of the thesis (note 1); and Oehler, *op. cit.*

54 *Phys.*, II, 9, 200 a 32 s.

55 Cfr. S. Thomae, *op. cit., loc. cit.*, II, XI, 242.

56 In the Aristotelian argumentation in question the final cause occupies the position of syllogistic "medium" (first the form is led back to the end, then the end to the mover), but without performing its function; the implication of the other three causes is not formal but has "general" (ὅλως) value for "everything which moves, having been moved" (*Phys.*, II, 7, 198 a 27).

57 Cfr. Wieland, *op. cit.*, p. 263; for the opposite interpretation (and the traditional one), S. Thomae, *op. cit., loc. cit.*, II, 511.

58 *Phys.*, II, 7, 198 a 33–35.

59 *Ivi*, 198 b 10–199 a 8.

60 "In all things where there is an end (*τέλος*) by virtue of this (*τούτου ἕνεκα*) there is a before and an after (*πράττεται τὸ πρότερον καὶ τὸ εφεξῆς*) (*Ivi*, II, 8, 199 a 8 s.)

61 Cfr. *Phys.*, II, 9, 200 a 15–22.

62 Cfr., for example, *Phys.*, I, 5, 188 a 10–16 in which the order of nature is compared to harmony in that it expresses unity.

63 Cfr. *e.g. Phys.*, II, 9, 200 a 5; and IV, 8, 200 a 15.

64 Le Blond, *op. cit.*

65 Cfr. *e.g.* W. D. Ross, 'Introduction', in Aristotle's *Metaphysics*, Oxford, 1931; J. H. Randall, *Aristotle*, New York-London, 1962, p. 215 *passim.*

66 Wagner, *op. cit.*, p. 480.

67 Wieland, *op. cit.*

[68] Cfr. M. Grene, 'Aristotle and Modern Biology', *Journal of the History of Ideas*, 1972, pp. 395–424.
[69] Cfr. *e.g.* Plato, *Phaedo*, 97; *Tim.*, 46 d; S Thomae, *Summa Theology.*, ed. Caramello, Taurini, 1950, I a., 103 a 1.
[70] Cfr. *e.g.* Cicero, *De natura deorum*, II, 113.
[71] M Vegetti, *op. cit.*, p. 506.
[72] *Ivi*, p. 518.
[73] *De Caelo*, II, 7, 306 a 13–17.
[74] Cfr. also *De part. anim.*, 640–642.
[75] In *An. Post.*, II, 14, Aristotle underlines the difficulty of the use of *telos* as a syllogistic medium (which after all is the only function it can perform in a demonstration by means of syllogism). A propos see D. Ross in Aristotle's *Prior and Posterior Analytics*, Oxford, 1949, pp. 642; and Lerner, *op. cit.*, pp. 48–52.
[76] Cfr. Wieland, *op. cit.*
[77] Such a thesis, advanced by Wieland, is shared, with different emphasis, by Vegetti, *op. cit.*, Grene, *art. cit.*, Randall, *op. cit.*
[78] *Metaph.*, XII, 4, 1070 a 32 s. A propos see W. Leslz, *Logic and Metaphysics in Aristotle*, Padova, 1970, p. 311.
[79] Wagner, *op. cit.*, pp. 356 *passim.*
[80] *Metaph.*, VIII, 8, 1050 a 21–23.
[81] *Ivi*, 1050 a 6–10.
[82] Cfr. G. Reale, *Note* to Aristotle, *La Metafisica*, Napoli, 1968, v. II, p. 85.
[83] *Metaph.*, VIII, 8, 1050 a 17–24.
[84] The distinction here proposed between *entelecheia* and *energeia* is not always to be found expressed in precise terms in the works of Aristotle – and therefore has no general value.
[85] *Phys.*, VIII, 4, 255 b 31–256 a 2.
[86] A justification of the assertion may be found in Aristotle, *La Fisica* (a cura di A. Capecci) Padova, 1972, *Introduzione.*
[87] Cfr. *De Motu*, 5, 700 b 26 ss.; 700 b 8 ss.
[88] *De Motu anim.*, I, 698 a 7–10.
[89] *Ivi*, 6, 700 a 15 ss.
[90] *Ivi*, 6–7. The term *horexis* which is introduced for the first time into "technical" language probably by Aristotle, indicates any process which having as its point of departure a living being tends towards an object. It therefore contains the ἐπιθυμία, the θύμος and the βούλησις.
[91] Cfr. *De Anima*, 413 b 32–414 b 15.
[92] *Metaph.*, V, 11, 1018 b 25; *Eth. Nic.*, VI, 2, 1139 a 31.
[93] *De Motu anim.*, 6, 700 b 23–25; *De Motu*, 6 *passim.*
[94] *Ivi*, 700 b 15 s. With regard to the distinction between the two senses of "end", cfr. above all E. Berti, *La filosofia del primo Aristotele*, Padova, 1962, pp. 335 s.; Lerner, *op. cit.*, p. 122; K. Gaiser, 'Das zweifache Telos bei Aristoteles', in AA. VV., *Naturphilosophie bei Aristoteles und Theophrast*, Heidelberg, 1969, p. 102.
[95] Le Blond (*op. cit.*) rightly points out how the presence of desire derives from the fact of being imperfect.
[96] Cfr. *e.g. De Motu*, 6, 700 b 26 s.
[97] Happ, *op. cit.*, pp. 509–12 and *passim.*

[98] *De Gen. et corr.*, II, 10, 336 b 27–34.
[99] Cfr., for all, E. Berti, 'Il problema della sostanzialità dell'essere e dell'uno nella *Metafisica* di Aristotele', in *Studi aristotelici, op. cit.*, pp. 181–209.
[100] Cfr. *e.g. Metaph.*, VII, 1, 1027 a 25–31.
[101] Cfr. C. Natali, *Cosmo e divinità in Aristotele*, L'Aquila, 1974, pp. 67 ss. and *passim.*
[102] Cfr. *Metaph.*, III, 3, 998 b 22–27; VIII, 6, 1045 b 6; XI, 1, 1059 b 27–33; *An. Post.*, II, 6, 92 b 13.
[103] The problem of final causality of the first mover is discussed by Aristotle – as it is known – in chapter VII of *Metaphysica*, XII, after the proof of its existence had been given above all in chapter VI.
[104] *Metaph.*, XII, 7, 1072 a 25 ss.
[105] Cfr. also *Eth. Nic.*, VI, 2, 1139 a 17–b 4.
[106] *Metaph.*, XII, 7, 1072 b 28 ss.
[107] Here, I follow Reale's interpretation (Reale, *op. cit.*, v. II, p. 284), which does not differ much from more famous interpretations.
[108] *Metaph.*, XII, 7, 1071 a 31–b 1.
[109] *Ivi*, 1072, b 1 s.
[110] *Ivi*, 1072 b 3.
[111] *Ibidem.*
[112] *Ivi*, 1072 b 14.
[113] *Ivi*, 1072 b 10 s.
[114] Cfr., for all, Natali, *op. cit.*, p. 101.
[115] Cfr. E. Berti, 'Aristotele era un pensatore dualista?', in *Studi aristotelici, op. cit.*, p. 242 ss.
[116] *Metaph.*, XII, 10, 1075 a 16. Here the order of the universe is compared to that of an army. "This in part resides," writes Aristotle, "in the order, in part it consist in its commander, in fact in great measure it resides in the latter since the commander does not exist because of the order, but the order because of the commander". (*Ibidem*, 1045 a 14 s.)
[117] Cfr. *Metaph.*, XII, 10, 1075 a 10 s.
[118] For the problem of the relationship *pros hen* (focal meaning) cfr. above all E. Berti, *L'unità del sapere in Aristotele*, Padova, 1965; and Leszl, *op. cit.*
[119] These are precisely the three "opinions on the good" which Aristotle criticizes in *Eth. Eud.*, I, 8, 1217 b 2–1218 b 27.
[120] *Ivi*, 1218 b 30–32.

ERLING ENG

THE CONCEPT OF EVOLUTION AND THE PHENOMENOLOGICAL TELEOLOGY

I

It is a striking fact that, since Husserl's work, phenomenological reflection on science has been oriented toward Galileo and Descartes, and on psychology toward Locke and Hume through Stumpf. Yet it was Darwinian evolution, which Haeckel – not to mention Nietzsche – made a religion, that might have been expected to enter phenomenological meditation. Not only has this interest been lacking but it has been generally ignored, presumably in view of its apparent naturalism. A notable exception was Bergson, but his work unfolded within the French tradition of evolutionary thought. This lack of reference to biology on the part of phenomenology is the more significant because of the pervasive influence of positivist evolution as an ideology. Could it be that this accounts for its exemption from phenomenological consideration? In any event, the present paper proposes to take a step toward remedying that deficiency.

To be sure, when we turn to Husserl's writings with this interest in mind, we notice what might be considered a turn in this direction in his notion of the life-world. But this would not prevent him from being considered a vitalist in his own way, if it were not for other features of his thought. For the later Husserl also shows an increasing interest in the instinctual sphere (*Triebe*), the unconscious (with unmistakable reference to psychoanalysis), and in "egology" as well as in "purpose" (*telos*) in manifold contexts. But another development is even more important by virtue of the fact that it also relates Husserl to the archetypal biological philosophy of Aristotle. Briefly, this is a shift in his understanding of the notion of "hyle," from its first mention in the *Logical Investigations*, through his late working notes, where he also reveals his interest in instinct, the unconscious, and purposiveness. The meaning of the term "hyle" changes from "matter," as contrasted to "form" in the conventional Aristotelian sense, to something like "originary affectivity" in Husserl's later years. It is this shift that accounts for the usefulness of the Husserlian "correlational a priori" in developing the implications of evolutionary thought for a structuralism of consciousness. Thus what I hope to develop in this paper is an "hyletic phenomenology" in a double sense, i.e.,

Tymieniecka (ed.), Analecta Husserliana, Vol. IX, 63–71.

as a phenomenology of "hyle" and as a phenomenological perspective on evolutionary biology.

It would also be possible to consider the topos of evolution in relation to the thought of Husserl by starting from his elucidation of part-whole relationships in the "Third Logical Investigation" or, just as profitably, from his analysis of inner time-consciousness. But an approach from the changes in Husserl's understanding of "hyle" seems more to the point in a consideration of evolution.

As Seebohm has reminded us, the Austrian context of Husserl's thought links him with the British tradition of philosophy to which we can relate his interest in Locke and Hume.[1] It was this British tradition in which science apparently replaced philosophy by becoming natural philosophy pursued in a spirit of wonder that was represented in the golden age of British science from Newton to Darwin. Thus this may help to explain the deeper ground of an affinity between Husserl and Darwin, as well as with other nineteenth-century evolutionists. Until now a phenomenological elaboration of modern evolutionary thought has been most fully realized by psychoanalysis, a possibility that may be traced back to Darwin's contemporary, Herbert Spencer. But the naturalistic allegiance of psychoanalysis with its traditional suspicion of philosophy seems to have brought its progress to a halt, save for refinements of therapeutic technique and reformulations to fit changing intellectual fashions. Thus, a phenomenological meditation on themes that form the matrix of evolutionary thought may also be expected to renew the fruitfulness of our musings on the riddle of the Sphinx, without which, after all, we cannot imagine Oedipus. That riddle, a riddle of time, is, I think we would all agree, preeminently "biological," even though that term was invented only at the opening of the nineteenth century.

A phenomenological biology begins of course *in medias res*. Characteristic of the living creature is direction, direction at once of development into a mature type, and then, within this development, changes of direction or quality of movement toward features of the world or "other," accompanied by changes in direction of process within the living being for itself, Bernard's *milieu interne*. Thus, the notion of goal, and with it the notions of "attainment," "norm," and "perfection," is intrinsic to an accounting of living beings.

But if we view this development and these changes not simply within an "environment" conceived of as "external" (consider the naturalistic duplicity of Bernard's "internal environment"), but rather in the context of Husserl's "correlational a priori," we are compelled to recognize that movement toward a goal is inseparable from its meaning horizon as something else. Such a situation is imperfectly represented if it is thematized in terms of an opposition between purpose and chance, creation and destruction, life and death.

In terms of the distinction of noesis and noema within intentionality itself, this directionality of life may be initially understood as an emphasis on the noematic pole. Biologically, the ingestion of food, the development of adaptation like natural camouflage, and learning in general, represent an emphasis on the noematic, if we start by taking noematic here to mean the world of which the living individual is part. This represents, as it were, a movement to assimilate to the world a telos which accounts for the dwelling, the habitat, or the house inhabited by living beings. This direction or teleology I will refer to as the *oikeiotic*.

But to what would the "noesis" correspond in terms of the "correlational a priori"? This would be represented by just that movement of opposition to the surrounding environment which makes possible animal movement, and which characterizes human consciousness above all. The enhancement of such a movement of opposition, of aseity, with its tendency to assimilate the world to the self, I will call "idiotic" as being characteristic of the uniquely individual.

It would be strange if we did not discover in the major evolutionary thinkers some awareness of this "correlational a priori." I will try to show its presence in both Darwin and Lamarck. Then I will proceed to sketch the way in which Husserl's change of thinking about hyle itself represents an evolution within his own thought.

In Darwin's work an implicit awareness of the phenomenological starting point can be seen in his perplexity about the priority of the male or female choice in sexual selection, a priority which interestingly enough, Darwin settled in favor of the female.[2] The pertinence of the oikeiotic and idiotic moments for understanding his conception of evolution may be seen from his hesitation whether to call his theory "Natural Selection" or "Natural Preservation." In correspondence with friends he wavered between the two terms before finally returning to his initial choice, "Natural Selection".[3] The choice of "Natural Selection," by its emphasis on the negative, omitted any reference to an original nature. "Natural Preservation" on the other hand, emphasizing the survival of adaptive types, had the disadvantage of eliminating any reference to death as the agent of selection.[4]

Darwin's final choice was for the "idiotic" term "Natural Selection" rather than for the more "oikeiotic" one "Natural Preservation," placing the emphasis on contingency rather than "domestication." However, Darwin appears to have eventually grasped the possible reciprocal implication of "selection" (of individuals) and preservation (of species) in accord with Huxley's and Asa Gray's agreement that Darwin had succeeded in harmonizing the claims of mechanism and teleology.[5] Thus, for Darwin "Natural Selection" conveyed the sense of "Natural Preservation," placing the emphasis on the individual as over against the world, even though as species it formed a part of that world. This accounts too for the priority of mechanism in his thinking, an emphasis aggressively asserted by Thomas Henry Huxley to the point of virtually obscuring his own philosophical starting point. This, however, he preserved by his "agnosticism."

Lamarck, whom Darwin initially adopted as his imaginary scientific adversary, but whose views became increasingly necessary for him, forms an interesting counterpart to Darwin. The notorious Lamarckian principle of the "inheritance of acquired characters," actually a traditional premiss from antiquity, stresses, like the ancient world-view itself, the oikeiotic moment. Creatures survive by becoming molded to the world. This oikeiotic moment Darwin needed after placing primary emphasis on the individual as idiotic. In Lamarck the idiotic moment differs from that of Darwin; it is marked with death from the beginning, in sharp contrast to the sphere of the oikeiotic. Taking limestone and coal as his model, all of the minerals of the earth, in Lamarck's view, derive from the remains of once living creatures. But from this detritus the germs of life spring once more spontaneously into oikeiotic existence. The idiotic movement is present in the Lamarckian "life-world" as the immanent death of each living creature.[6]

Through this contrast between Darwin, with his emphasis on the moment of separation and opposition to a world order, and Lamarck, with his emphasis on man as part of a cosmic order, we may better understand both the nineteenth century's rejection of Lamarck as well as Darwin's recognition that Lamarck's cosmic portion of evolution is implied within his own more idiotic perspective.

II

Let us now turn to Husserl in order to review a comparable set of transformations of the meaning of hyle as these develop from his meditations on meaning as conscious.

Husserl's interest in "hyletic data," "hyletic phenomenology," and eventually "hyle" as such, may be traced from his references to *Stoff* and *Materie* in the *Logical Investigations* through the *Ideas* and *Phenomenological Psy-*

chology to his posthumously published *Analyses on the Passive Synthesis*. This development may be characterized as a gradual shift from a focus on the actively intentional constitution of identities within polar organized experience to an increasing awareness of a passively constituted identity as an expression of an *Urtranszendenz* of anthropocosmic scope. (Husserl has provided something like a sketch of this in his "somatology" in Addendum VIII to *Phenomenological Psychology*.) This development also involves a shift from a more simply intellectual analysis of consciousness to a more actively "listening" reception of what Husserl terms *Affektion*, a shift from an emphasis on perception and its objects to remembering and affectivity. As Husserl makes clear in his asides on *das Unbewusste*, these changes of interest occur without loss of continuity, so that earlier stages of analysis are not contravened by subsequent stages of discovery which are once more subject to analysis.

Let us turn now to a series of typical texts from different periods of Husserl's work to illustrate his elaboration and enrichment of meaning of the hyletic. In the *Logical Investigations* we find the following passages:

> Matter [*materie*] stands for that in the act which first of all endows it with relationship to something objective, and indeed this relationship is of such perfect definiteness, that through the matter not only the objective, which the act intends, is definitely determined, but also the manner in which it is meant.[7]

> Thus matter is, however, according to our law, matter of an objectivating act and is only able by means of such an objectivating act to become the matter of a new act quality grounded in such an act.[8]

> Matter for us means that moment of the objectivating act which results in the act presenting this object and in just this way, i.e., in just these articulations and forms, with special relationship to just these determinations and relations.[9]

Here it is the objectivating act which is taken as primary in potency as well as in time. Matter rather serves as a passive determination, becoming determinative only in and through the objectivating act. In the context of the *Logical Investigations* the objectivating act is inseparable from consciousness, if it is not consciousness as act. This distinguishes Husserl's position, among other ways, from that of Aristotle. We may observe in passing that this early position of Husserl would not be inconsistent with the idealistic aspect of Lamarck's doctrine of evolution in which matter appears subordinated to a protean vital metamorphosis.

In *Ideas* I we may notice the following:

> We have already indicated [when we characterized the stream of experience as a unity of

consciousness], that intentionality, quite apart from its enigmatic forms and levels, also resembles a universal medium that finally bears all experiences, even those considered as nonintentional, within itself.

On the level of observation to which we are provisionally committed, which does not allow descent into the dark depths of ultimate consciousness constitutive of all experiential temporality, but rather takes experiences as they present themselves as integral temporal events in immanent reflection, we must distinguish in principle:

1. All experiences which were termed "primary contents" in the *Logical Investigations*.

2. The experiences or experiential moments which bear what is specific to intentionality within themselves.

To the first belong certain experiences which are unitarily "sensual" in terms of their highest genus, "sensory contents" like color data, touch data, acoustic data, etc., which we no longer confuse with the phenomenal moments of things, coloredness, roughness, etc., but which rather by their means manifest themselves experientially. Similarly, the sensual sensations of pleasure, pain, and tickling, etc., and certain sensual moments of the sphere of "instincts" [*Triebe*] as well.[10]

It is now that Husserl introduces the term "hyletic data" to refer to the dimension common to sensual and sensory spheres of experience alike.

Sensuousness [*Sinnlichkeit*] in a more limited sense characterizes the phenomenological residue of what is mediated by the "senses" in normal outer perception. This discloses after the reduction, an essential kinship of the respective "sensory" data of the outer viewings, and corresponding to it a distinct genus or basic concept of phenomenology. But in a still further and essentially unitary sense, sensibility [*Sinnlichkeit*] also comprises the sensual feelings and urges, which have their own generic unity, as well as an essential kinship of a general character with those sensuousnesses in the narrower sense – all of that quite apart from the commonalty, expressed moreover by the *functional* concept of hyle. Together both compelled the earlier displacement of the originally narrower usage of sensibility onto the sphere of temperament and will, namely, onto the intentional experiences, in which sensory data of the spheres referred to function as "stuffs." Consequently, we need a new term, which expresses the unity of function of the entire group and have chosen for this the expression *hyletic* or material [*stofflich*] data, or simply stuffs [*Stoffe*].[11]

In this rich and complex meditation on the theme of "sense" and its variations, sensibility, sensuousness, and sensuality, we notice that the widening circle of the meaning of sense comes to include the non-intentional sphere as one of self-manifestation, however much hidden in the penumbra of the intentional sphere as such. And to this widened circle of sense now bounded by sensuality as an apeiron, Husserl applies the notion of "hyle" as functional, perceptually relevant as "hyletic data." Compared with his emphasis in the *Logical Investigations* there has been a distinct shift in the direction of elaborating the implicit potency.

If we were to view the evolutionary conception of Darwin in this context, we could say that just as in Husserl there is no necessary connection between the sphere of intentionality and that of nonintentionality, despite the commonalty of the hyletic, neither is there any necessary connection between the adaptive behavior of individuals as such and the hyletic dispositions represented by the mechanisms of heredity. Only within the world as actual, within the life-world, can we speak of accord or discord between "sense" as it appears in the "sensory" and "sensual" spheres respectively.

It is in *Ideas* II that Husserl reveals the extent to which the affective life is opened up by being considered in terms of intentionality. In return, the depth of this affectivity repays conscious intentionality with an enriched awareness of its potencies, along with a deepened sense of indebtedness to the life-world. In terms of the problematics of evolution the analogous terms to Husserl's "ego" and "affectivity" would be that of "individual" and "species," or using the terms introduced earlier in this paper, the "idiotic" and "oikeiotic."

From *Ideas* II the following passages may be noticed:

Let us now discuss sensibility [*Sinnlichkeit*] : we understand it as the "psychic" underlay of the mind, and of the mind at all conceivable levels, or the underlay of mental acts at all conceivable levels, from the lowest, which we picture to ourselves as the simple act of radiation of the mental glance toward something affecting the subject, which ends in it, grasping, contemplating it, or of a mere turning toward it in active approval, and so forth, up to the highest rational acts of theoretical thought, artistic creation, or social-ethical action.

Now what is this psychic underlay? Primal sensibility [*Ursinnlichkeit*] consists of the sensile data, color data in their visual sensory field, which is a unity prior to all "apperception" and which can also be discovered subsequently as a moment of an apperception, namely, as a moment of the visual aspect. Likewise, the sense feelings are based on these sensible data, as well as the sensual data of the urges [*Triebe*], as urges hypothesized as entities transcending consciousness, but as primal experiences, always part of the psychic underlay. That is a primally given sensibility [*Sinnlichkeit*]. . . This however does not arise out of association, primordial sensuality, sensation, etc.; it does not develop *out of immanent grounds*, from psychic tendencies; it is simply there, makes its entry. The ego presupposes sensibility as affect, as excitant, first of all as primal sensibility and then as secondary sensibility. The ego always has something which is had. Primal sensibility is what it originally has. . . ."[12]

What is striking about this long passage which I have quoted in its entirety because of its importance for the present argument is the way in which the different primacy of the hyletic *Sinnlichkeit* is brought out, without disestablishing the primacy of the ego as noetic. This instinctuality represents a kind of bodily counterpart to the world as opened in consciousness as intentionality. It is literally an *Untergrund*, the word Husserl uses to refer to the underlay of the mental acts at all conceivable levels. This is related to evolutionary thought in the way it makes clear the necessary primary reference to the world as the setting of all experience. This is more direct in the thought of Lamarck, with his emphasis on the inheritance of acquired characters, while more indirect and fraught with accident in that of Darwin. For the world rules even more powerfully the vision of Darwin than of Lamarck, in which Neoplatonically there is still the presence of the self-transforming power of eros. To be sure there is an eros in the vision of Darwin, but in the sublime form of a world transformed by it into the arbiter of life and death. This vision of the life-and-death world of Darwin is most vividly depicted in the effect on him of the teeming jungles of Brazil over against the bleak landscape of Tierra del Fuego in his diary of the *Voyage of the Beagle.*[13]

To focus on the ties between evolutionary thought and the phenomenology of Husserl is also to rediscover the ancient affinity between Hermes and Eros.[14] To include Hermes in phenomenology is to include movement and change of meaning, hence requiring a phenomenology of every hyletic phenomenology. It is just this which generates a counterteleology within every specific, particular teleology. In this paper I have tried to show how Husserl's hyletic teleology represents an indirect reference to the encompassing transcendence of the life- (and death-) world, over against any egoic teleology seeking fulfillment of its intentionality. That transcendent subjectivity of all subjectivities is, like the world in Darwin's youthful vision, the arbiter of existence itself, existence which includes as subordinate instances the empirical ego and the own body. It is the same world which rules the vision of Lamarck, save that for Lamarck the power of transformation is represented by the vitality of the world and body rather than by the world as a throne of judgment.

The continuity of Husserl's thought with this implicit sense of Darwin is mediated in part by the British philosophical tradition, more deeply stained with the tinctures of the Christian Platonism of the Cambridge Platonists than is generally realized. From Newton through Darwin it is this subterranean tradition that is carried on by Husserl, represented by his treatment of parts and wholes, inner time-consciousness, and above all, by his Christological understanding of hyle, from the *Logical Investigations* to the *Crisis*. Recalling the Aristotelian character of terms like *telos* and "hyle," it is tempting to consider the possibility that in Husserl's phenomenology, as in Darwin's thought, we recover through Aristotle the meanings of Plato in a world no longer divided into two realms but become historical in the modern sense.[15]

NOTES

[1] Cf. Thomas M. Seebohm, 'Reflexion and totality in the philosophy of E. Husserl,' *Journal of the British Society for Phenomenology*, 4 (1), 1973, pp. 20–21.

[2] Darwin's emphasis is oikeiotic. *Cf.* the present writer's paper 'Darwin's Struggle Between Reason and Imagination,' *Diogenes*, summer, 1976.

[3] Francis Darwin, ed., *Life and Letters of Charles Darwin*, New York: Appleton, 1888, I, 508; II, 111, 138–39, 160. Cf. *ibid., More Letters of Charles Darwin*, New York: Appleton, 1903, I, 161.

[4] "That which takes the place of the breeder and selector in nature is Death." Cf. Thomas Henry Huxley, 'Time and Life: Mr. Darwin's 'Origin of the Species,'' *Macmillan's Magazine*, 1, 1859, p. 147.

[5] Francis Darwin, ed., *Life and Letters of Charles Darwin*, New York: Appleton, 1888, I, 554; II, 367.

[6] Thereby it is not inflicted by the world. Cf. J. B. Lamarck, *Zoological Philosophy*, Hugh Elliot, transl., London: Macmillan, 1914, pp. 194, 264.

[7] Edmund Husserl, *Logical Investigations*, II/1, 415.

[8] *Ibid*., p. 494.

[9] *Ibid*., II/2, p. 86.

[10] Edmund Husserl, *Ideen* I, *Hua* III, 207–8.

[11] *Ibid*., pp. 209–10.

[12] *Ibid., Hua* IV, pp. 334–35.

[13] Charles Darwin, *The Voyage of the Beagle*, Bantam Books, 1958, pp. 23ff, 176ff, 436.

[14] Kerenyi, 'Eros is a divinity bound up with Hermes through an intimate kinship,' in *Humanistische Seelenforschung*, ed. K. Kerenyi, München: Langen Müller, 1966, p. 101.

[15] Thomas M. Seebohm, 'Zur Genese des Historismus,' in *Bewusst sein: Gerhard Funke zu eigen*, ed. Alexius J. Bucher, H. Drüe, and T. Seebohm, Bonn: Bouvier Verlag, 1975, pp. 111–25.

AURELIO RIZZACASA

THE EPISTEMOLOGY OF THE SCIENCES OF NATURE IN RELATION TO THE TELEOLOGY OF RESEARCH IN THE THOUGHT OF THE LATER HUSSERL

It is significant that Husserl, in his lecture he gave at the *Wiener Kulturbund* on 7 May 1935 with the title "Philosophy in the Crisis of European Humanity," referred to philosophy as the foundation and conclusion of all scientific argument.[1] Thus, from the teleological point of view, one can anticipate the hypothesis that for the German philosopher there exists a continuity through the philosophy, science, culture, and self-consciousness of mankind as regards its historical function.

In this light, therefore, the crisis to which Husserl referred involved humanity in its theoretical consciousness, representing a teleological commitment to the research, which should develop along with the technological advancement of civilization. The matter then at issue is the role and responsibility of both philosophy and science, since they produced the crisis – and could eventually start a process of conversion capable of overcoming the crisis itself. My present concern, however, is only to point out the relation between the epistemology of the sciences of nature and the teleology of the research, as it appears in Husserl's *Crisis*.

At the outset it should be noted that in twentieth-century Western culture, philosophy, while striving to become more and more exact, has adopted a fundamental analytic dimension and has often come near to science. This approach to science, however, has only been methodological, since scientific research, in its turn, has become more and more technical, and has adopted an ever more specialist language. By doing so, it has made its criteria of control even more complex. Thus an overwhelming breach between philosophical argument and scientific research has been created. Between the two comes epistemology, seen as a philosophical undertaking in the critical analysis of scientific methodologies.

However, the methodological change which modifies the relation between philosophy and science, initiating the independent development of epistemology, is a consequence of the crisis of the monist systems of philosophy, both of idealist and positivist traditions. The overcoming of the former, in fact, diminishes the controversy of philosophy in the face of science, while the overcoming of the latter leads, in the philosophy of science, to a shift of interest from content to method.

Tymieniecka (ed.), Analecta Husserliana, Vol. IX, 73–84.

In this way, then, epistemology, as much in the neopositivist tendency as in the analytical one, concentrates its attention on the elements which make scientific argument exact. The examination of criteria for recording evidence, and also that of the reflections on the empirical, logical, and linguistic methods of control, are examples of this.

For the sciences of nature in particular, the fall of the nineteenth-century positivist conception has caused, for contemporary epistemology, the discarding of the objectivity of scientific laws, and, at the same time, has allowed for the inventive meaning of the hypotheses of explanation. In order to understand this, it is enough to consider the change which has taken place in classical physics brought about by the theory of relativity. The importance of the subjective intervention on the part of the scientist in the building up of explanatory hypotheses has become an inseparable element of all methodological reflections on the scientific thought of the twentieth century, mainly because of the effects of "indeterminism," "conventionalism," "operationalism," "verificationism," and "falsificationism." After all, in epistemology, the idea that has gained ground is that the theory is nothing but an ever more exact approximation for the purpose of solving problems of explanation or prevision inherent to experience.

In addition, this methodological change has been helped along by the perfecting of logical and mathematical instruments which, beyond any Platonic type of intuitionism, and by representing nothing but the inventive interventions of human thought, which inclines toward the effort of building up symbolic systems ever more consistent with its hypotheses of explanation and control.

The epistemological development as outlined here finds, in spite of the diversity of the philosophical premises, some significant correspondences in the thought of the later Husserl. He is, in fact, perfectly convinced of the nonontologic function of the mathematical systems which are part of logic. Besides, the discarding of positivism leads the phenomenologist to attribute to the sciences of nature an objective function which is independent of ontology. As we shall see, for Husserl the sciences of nature are only a stage in the theoretical working out of human thought which, in its intentional teleology, is open to unlimited research.

This is exactly the philosophical and scientific meaning of the phenomenology undertaking which, throughout successive periods, moves away from the subjectivity of human conscience and, going toward objects, makes for their essence, and then returns (by means of a more profound awareness) to the pure transcendental intentionality of the consciousness.

From what has already been said, it should be clear enough that Husserl puts forward a philosophy that can be regarded as a universal science operating with exact methods, in which the basic theoretical view joins the philosophical stage with the scientific one in a sole and infinite task of research.

It is also clear that Husserl's thought, which concentrates the interest of the researcher alternately on human subjectivity and on the objectivity of the world – thereby achieving the continuity of the cognitive process – investigates the methodological *a priori* both of the research taken as a whole and of the separate investigations constituting the significant stages of the former. Along these lines, both the phenomenologist and the epistemologist, even if in different ways, come to bring to light the mental aspects of all theoretical constructions; and, while the latter finds an explanation of the scientific research, the former, in a sense returning to Kant, shows the teleology of human knowledge both in its different aspects and in the general framework of the thought of man seen as the author of civilization.

Starting from the above premises we shall consider the teleology of the sciences of nature in the work of the later Husserl, bearing in mind, however, that the eludication of these sciences only constitutes one of the stages in the whole process of phenomenological philosophy and perhaps not even one of the most important. Our argument will be outlined in seven, closely interconnected points.

1. In the first place, we can say that, from the phenomenological point of view, the sciences of nature represent a self-enclosed systematization, so that one can easily examine their logical structures, methodological characteristics, and historical genesis.

After all, for Husserl, the scientific quality of the sciences of nature is not disputed nor is it disputable, since this quality is shown in the exactitude of the results produced. From this point of view, the sciences of nature have a particular theoretical function and a particular technical function of their own. The former, which is provisional, places science in the series of stages of phenomenological reduction; the latter, on the other hand, explains the origin and development of scientific processes in our Western civilization. Thus the two functions mentioned fuse together and become an indistinct part of the sciences of nature themselves; in fact, for Husserl, the synthesis of theory and practice takes place in the teleological perspective, hence science directs humanity to action.

Returning to the historical aspect, we realize that the later Husserl places scientific work in the continual process of its forming and perfecting; in this way modern science embraces all the theoretical and practical ideas of the science of the past. Thus the cultural foundation of the theories formulated by the scientist is inherent to the traditions of both Greek and European worlds, where it has its roots.

This explains the historical interest of Husserl, who, seeing the teleology of scientific research in general, examines some specific models of technical and theoretical systematization. For technical systematization, the philosopher's attention centers on Greek civilization; while, for the theoretical one, it concentrates on Galilean physics. Moreover, this attempt to find the ideatic meaning of the historical models is also used by the philosopher to trace the genesis of the crisis which today separates the sciences of nature from the sciences of man. In particular, it focuses upon the dualism which derives from Cartesian philosophy as its possible source. The *Crisis* denounces the inadequacy of scientific knowledge to respond to the intellectual needs of humanity, because sciences are unable to go beyond their objective systematizations of the rules of nature. This is what happened, in fact, to psychology, which, taking physics as its model, ended up distorting the teleological continuity which was supposed to inspire its own researches. For Husserl, in fact, the crisis of the sciences has reached its apex in scientific psychology. The reductionistic distortions in which it conceives of man deny the twentieth-century scientist all possibility of establishing a bridge with philosophical psychology.

In order to understand the crisis of European sciences – which is the crisis of our Western civilization – at the level of its profound methodological substratum, we should then consider Husserl's conception of the sciences of nature, examining their crisis with reference to their systematization.

2. Furthermore, Husserl realizes that the crisis of scientific research pervades all its most meaningful aspects, such as its contents and methods.[2] The crisis then has an overall meaning; Husserl, however, recognizing the epistemological criteria of validity of cognition, realizes the importance of method for the legitimately scientific character of any research. In point of fact, for him, one of the first requirements of the scientific character of any research lies in the recaptured legitimacy of its operations. Consequently, the central importance attributed to the clarification of the method allows us to appraise the phenomenological reflections concerning the nature of the scientific argument, by means of the logical and semantic criteria related to epistemological issues.

3. From Husserl's historical remarks we can even infer the epistemological insights which preside over his conception of the sciences of nature. This is particularly clear in the change of the epistemological orientation of particular

sciences, which took place during the shift from Platonism to Galilean mathematization.[3] From what has been said, we can easily gather that Husserl forsakes the philosophico-ontological spirit of Greek science, and moves to a completely ideal scientific formulation which has no ontological assumptions. However, as we shall see, this is only a temporary stage in the phenomenological reduction which will be overcome in the teleological perspective when it comes to thematizing the life-world; from there a truly new ontology will emerge.

To find the model by which Husserl interprets the birth and development of the sciences of nature, we must go back in history to the Renaissance when Leonardo and Galileo renewed science by coordinating it with technology. Here, in the mathematical relations and causal bonds which link the different phenomena, the phenomenologist finds the elements that constitute the structures of the sciences of nature. The two structural elements mentioned remain unchanged in the epistemological assumptions of Husserl, and even when he interprets the process of modern mathematization, hence pointing out the logical sense of symbolic relations, he remains at the level of a mathematical logic of the Leibnizian kind with a steady interest in the *mathesis universalis*; this is because the fundamental interest of the writer is more philosophical than scientific.

Continuing along these lines, the causal bonds between different events and the invariability of regular transformations in nature take on the gnoseological sense of *a priori* elements.[4] It is then clear that Husserl believes in the objectivity of scientific research and in the fact that the structures found by knowledge perform a descriptive function of the world. In fact, in this respect, he has doubts neither about causality nor about induction – even if their ultimate validity to account for reality is questionable in the renovating aim of his philosophy. From the epistemological point of view his position to start with links his thinking to the conception of classical physics, although ultimately the latter is without doubt more congenial to the meaning of the science of nature as given to it by Kant than by Husserl's own basic concern. It is clear, however, that for Husserl the epistemological limit just mentioned is a consequence of the fact that scientific research is part of the theoretical kind of teleology of the phenomenological argument, which is open to the ideatic ontology of the different fields of being.

We can, however, understand the role played by the sciences of nature in Husserl's philosophy, if we keep in mind that for him mathematics is a privileged science, capable of performing a philosophical and epistemological function from which it is possible to derive the ideal foundation of the other sciences. Indeed, the importance of mathematics not only plays a fundamental role in modern science; according to Husserl, in classical Greece, science and technology, founded on geometry, were created for technological needs of calculation. In other words, we can say that, for Husserl, the science of nature as formulated by Galileo tends toward the individuation of causal bonds between events through the "quantification" of the sensible qualities of material *plena*; such "quantification" is seen as calculation. This is, after all, the essential feature of the "mathematization" of the world of nature.

However, if "mathematization" constitutes the characterizing stage of scientific systematization, Husserl leaves a way open for the work of the scientist who has the tasks of formulating the hypotheses central to science and of undertaking the infinite methodical process of verification.[5]

We can then say that Husserl's approach to the sciences of nature, with its idea of the objectivity of principles and its insistence on induction and verification, is very close to positivism in spite of the fact that he has rejected its philosophical consequences. And yet, scientific research, as conceived by him, reaching beyond positivism, is meant to undertake a theoretical task of a metaempirical character; in fact, it follows that in this conception the "true nature" is, for science, a governing idea and not an attainable stage in the historical process of hypotheses and verifications.

What has been said so far leads us to recognize a real paradigmatic character of the actual development of science. It stems from Galilean physics. As a matter of fact, with his investigations Galileo discovers but also conceals the sense of the researches into nature.[6] Galileo then represents for Husserl a con-contradictory figure in the history of Western philosophical and scientific thought, since, along with epistemological advantages, he also brings about the possibility of a theoretical impoverishment of the research; thus, in a sense, the reasons for today's crisis in European sciences, as denounced by the founder of epistemology, are already present in Galilean science.

In the perspective of a phenomenological critic of Husserl, the methodological function of science consists in taking some ideal models which go beyond the life-world, in order to initiate ever more reliable previsions. As a matter of fact, in its progressive retreat from the life-world, science unknowingly takes on more and more a technological dimension. But the scientist should be aware, according to Husserl, of the first preliminary postulate of scientific research, that is, of the quest after truth, which in a philosophical and theoretical sense brings the philosopher back to consider the life-world from which he had departed. In addition, the technological function frees science from positivistic ontology while keeping it close to the real object of its explanations and previsions. Science itself, however, has a theoretical meaning, although in the phenomenological perspective of the reduction its results are put "in brackets."

4. Husserl's reflection, dealing with the crisis of European sciences, finds in the life-world the *telos* which has been systematically forgotten by the sciences of nature in the progress of their historical genesis.[7]

With this last point, science takes on in Husserl's interpretation the character of a cultural production building up the spirit of European humanity, of which the life-world constitutes a primary factor.[8] This also confirms the practical function performed by scientific research.

5. In the fifth place, in Husserl's theoretical reflection the teleology of the research outlined here performs a role which is mainly theoretical, and philosophy and science are conceived by him as intentional in nature.[9]

It has already been frequently mentioned that the objective sciences constitute for Husserl only a temporary stage in the process of research. This is because the phenomenologist's interest does not stop at either an epistemological or scientific plane; it probes deeper into the ontological substratum. In point of fact, the phenomenological method consists in an *epochè* from the objective sciences, in view of the "superior science," the *telos* of which is the exfoliation of the life-world.[10] Thus Husserl's project involves a truly continuous process of research which, through a series of phenomenological reductions, tries to get further and further away from the *doxa* and convention of the teleology of research and closer and closer into the *episteme*.

In this perspective Husserl places his various arguments. Thus, in particular, it can be said that the attempt to direct teleologically the phenomenological method toward the thematization of the life-world itself leads to an ontological turn, which deviates from the epistemological direction of the reflection which has developed in today's current debate on the function and meaning of scientific research.

If we try to see Husserl's approach in terms of semantics, we find that life, the object of natural science, never becomes the object of a "thematization" in the sense given to it by phenomenological research.[11] It is therefore obvious that terms such as 'life,' 'world-already-given,' and so on, are used by Husserl in a prescientific sense; but Husserl's ideal would be to thematize them phenomenologically to go beyond the objective simplifications given by scientific systematizations.

Consequently, in the phenomenological approach the exfoliation of the world becomes a goal toward which ontological teleology directs itself. Provisionally the world is seen as the "totality of things placed in the spatial-temporal form."[12] This definition is designed to grasp the real in its essential state in order to give a starting point toward a vision of the world and life; however, this latter is not accomplished by Husserl's later researches.

In this way, Husserl sets the problem of the foundation of a universal science suitable to the understanding of the life-world. Then, by trying to explain the function of the subjectivity present in all cognitive experience of the world, Husserl establishes the need for this new philosophical science; it consists of an investigation of the "universal 'how' of the 'being-already-given' of the world."[13]

However, in this perspective, the inquiry into the meaning of the science of nature – which, according to Husserl, aims at determining the limits of competence – already represents a theoretical task. Thus in outlining the relation, in which the attention of the knower concentrates alternatively on the "very idea of the real as pure form," and on the various kinds of experience which the researcher encounters, Husserl determines the problem of the theoretical role of the sciences of nature.[14] For Husserl, then, critical thinking designed to theorize the sciences of nature does not mean producing an epistemology suitable to improve scientific research; rather it means producing theoretical results capable of advancing phenomenological philosophy as a system of knowledge, representing the real which itself becomes the object of investigation. This seems to coincide with the idea of philosophy as a strict science.

This project is developed in two phases. One consists in the determination of the objectives of the purely mathematical approaches to the sciences of nature; the second consists in the "science of the method of scientific determination on the grounds of the data of experience."[15] It can therefore be seen that the sciences of nature, both in their objectives and in their methods, can be considered within a broader context of research; as a result, such sciences, considered in themselves, present some inadequacies. Husserl is then doubtful about the way in which objective natural sciences guarantee the intersubjective concordance by reducing the qualitative differences to quantitive ones by using calculation. For him, in fact, it would be better to have an ideal ontology of nature, from which the methodological consequences for the foundation of a real natural science could be drawn. It is hardly necessary to recall that the epistemological insights assumed by Husserl for the sciences of nature are still bearing the traces of the classical distinction between primary and secondary qualities; however, we can state without doubt that Husserl's criticism, which points out the inadequacy of the sciences of nature themselves, is framed in a wider argument related to the idea of the ontological foundation of nature.

6. Nevertheless we cannot understand the role of the sciences of nature in the crisis of European sciences without considering Husserl's distinction between the sciences of nature and those of the spirit. In fact, the former represent an objective systematization, while the latter in general, and psychology in particular, determine with their inadequacy the crisis of the sciences. This happens above all since the mathematical and causal model of classical physics is generalized.

Indeed, in phenomenological perspective, on the one hand, nature represents an organic unity in a prescientific sense; Galilean, Newtonian, and Cartesian systematizations, on the other hand, mathematize and mechanicize it, introducing the dualism between nature and thought from which both the material and the psychic realms of the world derive.[16] Thus from the dualism some equivocations in the results follow, since the methodological distinction between the two phases of the research is not clean-cut. Furthermore, mathematization cannot be applied profitably to the sciences of nature and taken as a model to follow, since it brings about specific equivocations identifiable as the loss of the meaning of the scientific research. This, according to Husserl, is also the effect of the dualism; the loss of meaning, resulting from mathematization, is present in the technicalized sciences of nature.[17]

At this point, Husserl, denouncing the equivocations that follow from the technicalization of science and which spring from the use of mathematical formulae, makes way for the reformulation of the notion of scientific research suitable for grasping in depth the meaning of the life-world.

Science is also contested because of its positivist consequences which are responsible for the crisis. Husserl proceeds to develop his arguments, placing at the center the role of human existence; postulating a theoretical foundation for it, he goes back to the realm of subjectivity.[18]

7. If we want to trace Husserl's teleology of the research, we must keep in mind that the sciences of nature constitute only one of the many chapters of the phenomenological investigation of knowledge. However, focusing our interest on the sciences of nature, we may assess the spread of the various phases of knowledge. Beginning with a prescientific, natural gnoseology, Husserl passes on to the objective natural sciences, in order to reach then an ontological philosophy of nature. If we scrutinize the intermediate phase of the sciences of nature, we can detect there with Husserl an epistemological view of the following type: *sensible objects* have primary qualities and secondary ones; *science* places in the foreground the primary ones through an operation of calculation; there intervenes *mathematization* which, through geometrical entities and arithmetical confrontations, establishes an ideal

structure of nature; this is finally integrated by the process of *rationalization* from which the causal relations established among the different events of the spatiotemporal world emerge. We are thus at the Galilean level of science which has become the model for all objective sciences.

Husserl thinks that from the point of view of a knowing subject, a prescientific nature, constituted by the various objects of the world, must be given to the ordinary perception of everyday life. Such nature is then transformed by the interpretative theories given by scientists in their respective objective sciences. To make it clearer, we can say that whenever we refer to the transformations of the life-world, we construct a "natural history"; on the other hand, when we work with the mathematized logic of objective theories, we work out a "scientific biology." However, Husserl does not go so far as to point out all the particular distinctions of the various specialized branches of the science of nature, since for him natural science is, in general, the science of physical things. Besides, on the methodological level, his epistemology of the objective sciences of nature is but an extension of Galilean physics into the whole natural field.

If we ask what meaning the term 'nature' has for Husserl, we must refer to Cartesian considerations. From there derives the conception of a 'rationalized nature' which lies between the demands of science and those of philosophy.[19]

Husserl realizes that the conception of nature, as produced by a certain kind of science – such as Galilean and Cartesian – and as then recaptured by the following centuries, can lead to philosophical naturalism.[20] At this point, without doubt, Husserl's thought abandons the scientific level for his own eminently philosophical perspective.

Husserl separates naturalistic behavior from the task of the objective sciences, whether they pertain to nature or to the spirit.[21] Then, from the theoretical point of view, according to Husserl, there is no difference between the objective sciences of nature and the objective sciences of the spirit; the difference exists only on a methodological level.[22]

Concerning the numerous distinctions made by Husserl, there is a need to elaborate conventionally the system of semantic notions for the science of nature. This need appears in relation to the subjective task of the researcher; with its own semantics we could venture to say that the science of nature finds full recognition among other sciences.[23] With the above conception of natural science Husserl goes beyond the theoretical and ontological field of the phenomenological perspective, and meets with the most advanced conceptions of the linguistic conventionalism of contemporary epistemological reflection.

To terminate our analysis of Husserl's *Crisis*, which was carried out for the purpose of establishing the role and methodological meaning of the sciences of nature, we can conclude by restating the eminently philosophical sense of phenomenological theory.

From the perspective chosen by the present writer, on the one hand, the sciences of nature lose importance if considered as self-enclosed; on the other hand, from the methodological point of view, there emerges the profound significance of the teleology of the research. This teleology appears as the endless commitment to a theoretical task which allows no respite, since each result is temporary and must be kept aside while looking for the successive result. It is then possible to restate synthetically the basic thesis of this study by saying that, for Husserl, the teleology of all scientific knowledge is of a *systematic and philosophical* kind.

In phenomenology, teleology proceeds methodologically, through successive *epoché*, in an indefinite and transcendental development of the intentionality of consciousness which, going from the subjectivity of man, opens on to the world by "going to objects" first and then "going to their essence," to rediscover in man the root of scientific knowledge and cultural reflection. Thus the supreme teleological postulate to thematize the world of life itself, along with the spirit of humanity in its inexhaustible theoretical and practical aspect, is produced.

Viterbo

NOTES

[1] E. Husserl, *Die Krisis der europäischen Wissenschaften und die transzendentale Phänomenologie*, Husserliana Bd. VI, p. XIII.
[2] *Ibid.*, par. 1, p. 1.
[3] *Ibid.*, par. 9, p. 20.
[4] *Ibid.*, par. 9 b, p. 29.
[5] *Ibid.*, par. 9 e, p. 41.
[6] *Ibid.*, par. 9 h, p. 53.
[7] *Ibid.*, par. 9 h, pp. 48–49.
[8] *Ibid.*, par. 33, pp. 123–124.
[9] *Ibid.*, par. 9 d, p. 39.
[10] *Ibid.*, par. 35, pp. 138–139.
[11] *Ibid.*, par. 38, p. 148.
[12] *Ibid.*, par. 37, p. 145.
[13] *Ibid.*, par. 38, p. 149.
[14] *Ibid.*, p. 283.

[15] *Ibid.*, p. 283.
[16] *Ibid.*, par. 10, p. 61.
[17] *Ibid.*, par. 9 g, p. 48.
[18] *Ibid.*, par. 2, pp. 3–4.
[19] *Ibid.*, p. 294.
[20] *Ibid.*, p. 295.
[21] *Ibid.*, p. 305.
[22] *Ibid.*, p. 306.
[23] *Ibid.*, p. 311.

FRANCO BOSIO

THE TELEOLOGY OF "THEORESIS" AND "PRAXIS" IN THE THOUGHT OF HUSSERL

I

The praxis-theory relationship is one of the most controversial problems of contemporary thought. The aim of this study is to offer, on the basis of phenomenological reflection, a contribution to the solution of this problem. In most of its expressions, contemporary thought is characterized by a marked anti-intellectualism and an intolerant refusal of any form of "theoresis" aiming at first principles.

In opposition to classical thought and the Western metaphysical tradition, according to which the "*theoresis*" of foundations is the main aim of philosophical research, various currents of contemporary thought proceed in many ways to devaluate *theoresis* in order to favor *praxis* and man's practical behavior in relation to things. Of these forms negating the supremacy of classical *theoria* the most remarkable are Anglo-American *pragmatism* and, in Europe, *historical materialism*.

According to the former, every theoretical human attitude towards reality is valid only if it works to the solution of contingent practical problems concerning man's adaptation to his environment and enabling him to restore an equilibrium which had been threatened. Therefore theoretical problems are often determined in a contingent way, in situations which are different for every individual and in every circumstance. On this view, there are no philosophical problems endowed with ultimate significance. A philosophical problem has a meaning only if it responds to an emotional exigency of life and is the sign of a break in the balance between man and his environment. But if thought emerges from practical life and has functions merely of a practical character it follows that the ultimate and lasting needs of human existence lose all their meaning; in fact, philosophical thought so conceived cannot long dwell upon problems which have no solution answering to practical needs in determinate circumstances. But the traditional problems of philosophy are exactly of this nature. Therefore, according to pragmatism, the main request of life will be represented (satisfied) by technical-scientific research, that is, by the thought that advances hypotheses and verifies them in a satisfactory way so as to enable man to find solutions to contingent issues as they emerge in everyday life.

Tymieniecka (ed.), Analecta Husserliana, Vol IX, 85–90.

In its turn, historical materialism represents the other aspect of the supremacy of the practical moment over the theoretical one. The fourth gloss made by Marx to Feuerbach says that the question whether man can achieve the objective truth of things or not cannot be resolved in a theoretical way but only in praxis. The world of praxis in Marx's historical materialism constitutes the totality of the relations of production and exchange in which man, or better, every man maintains objective and real relations with other men and with nature. A theory is only the expression, at a conscious level, of the practical-sensitive situation of relations of production; the purpose of theory *is* to reflect the production and exchange relations that are typically of a certain social order, and to try, moreover, to justify and to rationalize them. It is well known that, according to Marx, there is concealed in this rationalization the danger of ideology and "false consciousness." Ideological theory is deceiving and mystifying, since it seeks to justify the existing sociohistorical order in an effort to perpetuate it in an absolute metahistorical present: Marx points out the urgent need for a demystifying "theoresis" which may reveal the conditioned ideological matrices of the relations of production and power in order to create in history a praxis aiming at the abolishment of these relations.

There is a substantial difference between these two most important philosophies based on praxis which have become established that we cannot fail to mention: in fact, the *praxis* of pragmatism is understood as starting from a theoretical pattern governed by a biological matrix, and has an individualistic tendency, whereas the *praxis* of historical materialism has as its foundation a totalizing vision of history. Nevertheless, beyond their differences, the unity of their root is clear: the exhaustion of "theoresis" in "praxis," the finalization of theory for the satisfaction of man's practical needs. The operative, scientific and technological exigency of pragmatism ends, however, in the aporiae of a reason reduced to instrumentality, in which means have an absolute supremacy over purpose. It is known that, on the one hand, for Dewey, purposes are only means for the attainment of further purposes that have not as yet emerged but may in the future. In historical materialism, on the other hand, praxis, which is expected to resolve the contradictions of reality, gets its force and charge from its postponing to a utopian and chiliastic expansion the time in which a "false" and "imperfect" history will come to its conclusion and a new history, true and perfect because free of historical contradictions, will have its beginning. But the price paid for this delay is too high: the unverifiability of the definitive (final) historical and temporal moment, which has, consequently, a dimension of unattainable transcendence. Therefore, it is necessary to return from praxis to "theoresis." In fact, if the former offers the means for action, only the latter can determine the means' conformity with the ends and the value of the ends that we want to pursue. Our purpose in this study is, therefore, to show how the contradictions that we have pointed out can be resolved by phenomenological reflection.

II

Husserl's *Crisis of European Sciences* and *Transcendental Phenomenology* are inspired by a powerful impulse to recognize the need for a philosophical foundation capable of offering man the meaning of his acts and of their ends. About the purposes of life, Husserl gives us a radical warning. He maintains that the purposes of life are generally of two kinds: some meant for a limited time, some for ever; the former are useful for the improvement of ourselves and of our contemporaries, and the latter for the improvement of our posterity up to the furthest generations.[1] It is the definite task of science to find out the purposes which are timelessly valid.

European sciences are, according to Husserl, undergoing a distressing crisis without resolution just because they have lost the sense of these purposes, and so have upset the relationship between facts and the idea. Instead of measuring the value and meaning of their progress by the criterion of the universal idea of science in its radical unity, they have rendered knowledge subservient to success and therefore to all that is merely practical. Husserl emphasizes that

> the sciences of mere facts make men of mere facts ... In the distress of our life, some say, this type of science has nothing to tell us. So conceived science disregards the most burning problems for man who, in these restless times, feels himself to be at the mercy of fate: the problems of the significance of futility of human existence on the whole.

Writes Husserl:

> Do not these problems in their generality and necessity require also general considerations and a well grounded solution for all men? After all, these problems concern man in his conduct, man as endowed with free will, man as free to mold rationally his own self and his environment.[2]

But positive science has nothing to say on this matter and, for this very reason, the sciences, in spite of their uninterrupted series of successes, are now in crisis.[3]

It is well known that Husserl fixes the first signs of that process of concealment of the foundation from which the sciences proceed as early as the beginning of the modern era, in the constitution of Galilean science. The substitution of an abstract physical-mathematical structure for the originary concrete, which is identified with it, is the first sign of the loss of meaning in scientific operations. The subsequent Cartesian theorization, aiming at offering an unconditioned metaphysical foundation and affirming it in the absolute substantiality of the *cogito*, confirms the concealment and heavily conditions philosophy and the sciences. At this point the cause of the crisis of the idea of knowledge seems clearly singled out in the renouncement of the supremacy of "theoresis" and its ideals in favor of the practical moment of success and man's control of reality.

Yet in Husserl's subsequent research and in its developments there occurs a change of perspective which seems to be, but isn't, in contradiction with all that has come out so far. Husserl points out, as we know, in the "life-world" the forgotten ground and the neglected source of sense of scientific operations. More than that, he wants to join the reflective and foundation-recovering route of his return to the *cogito* with the pure *Ego* as principle and foundation of every scientific theorization, and with the return to the anonymous, unexamined depths of every performance and action as indissociable routes. We admit that the "life-world" is the very dimension of the praxis *par excellence*, the ground on which praxis and theoresis co-exist from the beginning, in which the problems of knowledge spring from praxis. We cannot, therefore, tax Husserl with a unilateral intellectualism that sees in knowledge an activity disdainful of any contingent purpose of life and proudly detached from it. Far from it, the third section of the *Crisis* vehemently attacks any form of preconstituted intellectualism that leaves out of its consideration the problems of the life-world. In paragraph 33, Husserl clearly speaks of "praxis" and "practical life"[4] as peculiar to man's way of learning and proper to a pre-scientific attitude towards the world; he writes:

> If we keep in mind all this, we realize that an unequivocal clarification of the objective validity and of the general task of science exacts first of all a re-examination of the world as given. It is already naturally given to us all, to us as persons in the horizon of mankind in every actual connection with the others; it is the 'world common to all.' Therefore . . . the world is a constant ground of validity, an ever disposable source of obviousnesses, and we, both as practical men and as scientists, are always concerned with it.[5]

The "life-world," which for the contemporary scientist is a contingent, subjective, and relative problem, becomes an absolute problem for the philosopher. In it there is no sign of the conceptual contrapositions between physical and psychical world, spirit and nature, because dualizations and oppositions come out of the praxis of an objectifying and categorizing theorization. Husserl recognizes in the *praxis* of the "life-world" the true originative praxis in comparison with which the praxis of sciences, with all its value, is derivative and secondary.

At the same time the praxis of the "life-world" is quite different from the praxis of work and production in Marxism or the praxis of the biological and organic metabolism between individual and environment theorized by pragmatism.

In this study we do not want to examine Husserl's research on the *essential structures* of the "life-world" because we prefer to consider, for our ends, the themes pertaining to the *spirit*. The amazing discovery about which we cannot cease to wonder is that the "life-world" is neither nature, nor the irrational, nor the antithesis of consciousness and reason: it is actually in its own self *spirit*! In the *Ideas* (especially in books II and III), Husserl recognizes the essence of spirit in the originary world, a world originally pre-existing any theorizing objectivation of the naturalistic sciences, as the one of interpersonal and spiritual relations in which not only intellect or pure reason, but also the psychophysical totality of the concrete man, both individual and intersubjective and therefore *personal*, comes into play. In this world the intentionality of consciousness operates in its naive spontaneity. The risk of a theorizing objectivation is to betray this intentionality by concealing it in a net of categories that alter its substance and falsify it. Husserl writes:

> Among the things which are around me, this one, which is here, attracts my attention towards itself, and its peculiar shape 'catches my eye'; I choose the cloth for a suit because of its softness . . . in short: in my emotional, practical and theoretical behavior, in my theoretic acting and thinking, in my taking a definite position towards pleasure or delight . . . I feel conditioned by things. Which evidently does not mean I am conditioned in a psychophysical way.[6]

The daily behavior of man in everyday life, in the life-world, is the very thing that the philosopher perceives when he opens the dimension of spirit as interpersonal spirit and as the totality of interpersonal relations. The world of praxis is therefore *spirit* in the meaning of the supremacy of the being of a person over any other reality, both over the reality of the living natural organism, and over the reality of the *homo faber* as the subject of the relations of production and exchange. Persons expressing ourselves in personalistic behavior are what we "always are when we live together, when we speak to one another, when in greeting we hold out our hands, in love and in repulsion, in meditation and in action".[7]

Hence if we now ask after the purposes and ends we see that the teleology of praxis is exactly like the teleology of theoresis: they both try to find anew the originary meaning of our life and of our acts, a meaning which exists even if we are not aware of it. And that is an end which is also an origin, motivation and foundation.

Theory and praxis, seen in this perspective, do not stay in a relation of dialectical subordination, as in historical materialism, but they constitute themselves as modalities of existence proceeding from the same root. Theoresis is *logos* of the will to illuminate those significant existential aspects which develop inside praxis. Praxis, in its turn, is the whole of the dark anonymous background of life from which the impulses to self-enlightenment in theoresis proceed. Phenomenology, unlike pragmatism, succeeds in acknowledging knowledge also beyond the dimension of a praxis because instead of limiting praxis to a mere sector of man's functions, it envisages it as constituting a vast omnicomprehensive totality.

This teleological movement common to theoresis and praxis is not, as we must remember, peculiar to Husserlian thought. It is the living substance of all phenomenological endeavor; we find it also in Scheler, in his often dramatic conception of the tension between life and spirit.[8] We find it again in the Heideggerian and Gadamerian teleology of the "hermeneutic circle". We see it at work in the Heideggerian meditation that teaches us, in an incomparable way, how going back to the origins means also reopening the dimension of the future in an original way.

NOTES

1 *Philosophie als strenge Wissenschaft*, French tr., Paris, 1955, p. 13.
2 *Die Krisis der europäischen Wissenschaften*, Ital. tr., Milan, 1961, p. 35.
3 *Ibid.*, p. 33.
4 *Ibid.*, p. 150.
5 *Ibid.*, p. 151.
6 *Ideen zu einer reinen Phänomenologie*, vol. II, Ital. tr., Turin, 1965, p. 534.
7 *Ibid.*, p. 579.
8 Cf. by the present author, *L'idea dell'uomo e la filosofia nel pensiero di Max Scheler*, Rome, 1976.

ENRICO GARULLI

THE CRISIS OF SCIENCE AS A CRISIS OF TELEOLOGICAL REASON

A series of Husserl's texts, from the early *Philosophie der Arithmetik* to *Crisis*, may be used to deduce how his reflection on the sciences became characterized in relation to the development that took place in his thought on the phenomenological idea.[1] The link between the two problems, that is, the state of the sciences in relation to the manifestation of teleological reason – seen in the different modalities that consciousness takes on from subjective *telos* to *telos* of the entire history of humanity – constitutes the *leitmotiv* of Husserl's thought in its development, resumed and deepened in the various, though continuous, phases in which phenomenology becomes characterized.

Indeed, in regard to the origins of Husserl's thought, there remains an open controversy whether it was his reflections on science, and in particular on mathematics and geometry, that turned his interest toward philosophy, or whether this should be explained by other causes, among which may be included the happy encounter with such themes as that of "intentionality" (which had been reintroduced by Franz Brentano). The question seems particularly interesting if we take into account the texts of the first period, in which the conflict between psychology and logic seems to be particularly acute, especially in regard to the idea of number, interpreted in the form of *kollektive Verbindung.*[2]

It is well known how, in this first group of investigations by Husserl, problems that concerned the so-called "preempiric" world were more in question than specific solutions on the state of the formal disciplines of the "first level." The specific theme of the procedure of numeration (*zahlen*) as a development of processes of mental contents and not simply as "demonstrative technique" of signs forms a background for this problem. Substantially this idea is already alive starting from the early *Beiträge zur Variationsrechnung*, in *Über den Begriff der Zahl* (later incorporated in the abovementioned *Philosophy of Arithmetic*); but this type of approach to the problems of science will also be continued in the so-called proto-phenomenological texts: for example, in the review of E. Schröder's project of an "algebraic logic" and in the *Darlegungen* on the "logic of logical calculus" of A. Voights. Furthermore, his contemporary polemics with Schröder on the "deduction of calculus" and "logic of content," and later with Frege,[3]

Tymieniecka (ed.), Analecta Husserliana, Vol IX, 91–104.

certainly has opened Husserl's eyes to the importance in mathematics of the question of the "structure" of numeration, and led him to the shift of emphasis from its presumed psychological origin, to the conceptual mode of thinking that specified formal activity.

The manner with which Husserl entered into the discussion of these problems, with the work of Lobachevski, Riemann, Weierstrauss, and Cantor before him – to cite the most important names – indicates the presence of an intuition in the young philosopher different from that which was to be followed by the exponents of mathematical formalism. Contrary to the latter, in fact, Husserl continued to be guided by the union of psychology and phenomenology, and thus according to the formulations that this orientation took in his thought.

That this hardly orthodox procedure in the exact sciences (as revealed by the developments that came about in mathematics with Frege and Russell) had a reason for being, is demonstrated, in part, by the resumption of its motivations within logical positivism, as documented, for example, in Carnap's first project for the *Aufbau,*[4] where in regard to the grades of the *Konstitutionstheorie*, explicit references to the structure of a phenomenological character are not lacking. In contrast to Carnap, as well as to other exponents of the Vienna Circle, Husserl's design asserts itself by means of its own horizon: not only by referring to the already mentioned specificity of acts of consciousness – from which Husserl expected a specific characterization of the function of thought – but also by its confrontation with the essential stages of Occidental thought. The problem from which emerged the reflection on the mathematical aspects of science thus united with that of the history of philosophy, within which the phenomenological ideal became delineated.

The second period of Husserl's research, that begins with *Logical Investigations* and, continuing with *Vorlesungen* (1907) and the essay on *Philosophy as strict science* (1911), goes up to *Ideas I* (1913), also includes many parts in which the problem of reflection on science continues to be the *stimulans* of the first formulation of phenomenology, as elaborated in the typical descriptive procedures of the analysis of conscious acts (*Erlebnissen*), to which will be added, very soon, the analysis of "intentionality," of "epoché," and of "reduction." It would certainly be erroneous to see in these developing conceptions and methodological devices a different formulation of phenomenology; instead it is rather a deepening of the same grounding and clarifying commitment.[5] Precisely by means of the conceptual instruments mentioned, phenomenology pursues a purpose unknown to the philosophies of the period; this purpose reveals itself in the necessity that the process of self-givenness (*Selbstgebung*) by which things manifest themselves to the ego, be placed both in relation to the phenomena of consciousness in their free "effectuations" (*Leistungen*), and in relation also to the so-called "fulfilling" (*Erfüllung*) which takes from these "effectuations" the character of images or pure products of the imagination. Substantially this was the way in which Husserl proposed to initiate a "return to the things themselves." These procedures, which developed gradually in Husserl, are revealed in the programmatic and radical intention of a return to consciousness (*Rückgang auf das Bewusstsein*) whose latitude will acquire a very different meaning from that of the epistemological idealism or of the subjectivism of Cartesian origin.

The *Prolegomena*, that precedes the aforementioned *Logical Investigations*, contains a text in point. In this work Husserl attributes the state of "theoretical imperfection of the particular sciences" to the lack of intrinsic criteria, or better, from a "lack of clarity and rationality."[6] And yet in spite of these lacunae – states Husserl – the sciences have become an irreplaceable instrument of dominion over nature; they, however, "cannot satisfy us from the theoretical point of view."[7] In a succeeding passage Husserl compares the relationship "science-philosophy," correctly distinguishing two levels of considerations: the first regards the already asserted necessity for reflection upon the terms that empirical science habitually uses ("thing," "event," "cause," "effect," "space," "time," etc.); the second regards the need for the specific examination of each of these terms in its essence in such a way that – says Husserl – one can examine "the marvelous affinity that this essence maintains with the essence of thought, in order that it may be thought; of knowledge, in order that it may be known; of the signified, in order that it may be signified."[8] Thus a vast field of investigation is opened to be explored by the *Logical Investigations*; the gnoseological horizon of these problems was then affirmed in Part VI of the *Logical Investigations*[9] opposing their empirical or transcendental conception; in its turn, the field of the meaning opened to philosophy an endless field of analysis. Conjoined with its subjective intentional aspect, it brings forth, finally, the need for a comprehensive and regulative *a priori*.

In specifying the relations between science and philosophy, Husserl writes also that "the *Ars inventiva* of the specialist and the criticism of knowledge of the philosopher are complementary scientific activities that alone permit the construction of the well-defined theoretical vision that underlies all ontological relations."[10]

Therefore a new horizon of problems is opened for consciousness: rooted not only in the new signification proposed for "science," but also in the phenomenological concept of "consciousness," and called to resolve, by means of reflection, the contents and the references of the essences and of all the acts (*Erlebnissen*) – a program that called for a progressive emphasis to be placed upon reflection about subjectivity as the "seat of all meaning." These investigations, with the reflection on science in its center, allow phenomenology to avoid the extreme of idealist formalism as well as to approach "reason" in its constitutive and teleological significance, throughout the stages in which the incessant resumption of philosophy advances in an ever more general and comprehensive way.

It would be useless here for our theme to follow the itinerary of Husserl's reflection as it developed in the texts of the Göttingen period,[11] during which emerged the various trends within the movement inaugurated by Husserl, each with its particular orientation.

In delineating in 1911 his position with regard to the discussions, then rather lively, about the relations between sciences of man (*Geisteswissenschaften*) and natural sciences (*Naturwissenschaften*), Husserl returned to reaffirm the close link between the "idea of science" and "subjectivity." "The idea of science," he writes in *Philosophy as strict science*, "thought in its ideal perfection is nothing other than reason itself, which can recognize no authority beyond and above itself."[12] In Husserl's view this means that philosophy must oppose the natural or ingenuous attitude of science: but it means above all that philosophy should demystify the unquestioned data, the current *Selbstverständlichkeiten*, to which science had often assigned a rigorous and scientific significance, without treating them adequately from the point of view of consciousness. This omission holds, for Husserl, both for the exact sciences and for the natural sciences. In fact, in both the same design is manifest, directed toward making of philosophy a "conception of the world" instead of conceiving of it as a rigorous type of investigations, capable of dealing with its own methods with consciousness itself, or rather – says Husserl – the different ways in which consciousness organizes the objects which it constitutes, according to their intrinsic structures, taking into account their indifferent modes of evidence, presence, directness, or mediation, etc.[13] In this text such an observation is particularly suggestive because of its programmatic character; for Husserl here affirmed the necessity that all the "givenness" (*Gegebenheiten*) should be able to rise to the level of "consciousness." By means of this affirmation he is able to oppose the natural and objectivistic attitude of the modern science and, under its influence, philosophy also. But later, in further investigating the problem of consciousness, Husserl stated clearly that "all types of cognition . . . must be actuated in, so to speak, a 'teleological mode,' under the title of 'knowledge.'"[14] The significance of this text is generally stressed by all scholars, as it represents not only the conclusive point of the first phase of phenomenology, but because it also contains indications which will be developed in the immediately successive texts, published and unpublished, that form a link with the works of the last period.

In particular, the text mentioned above seems to insist upon three things: in the first place, a notion of consciousness that involves all its operations (in *Ideas I* this structure of consciousness will be more explicitly indicated as a complex of noetic and noematic acts); in the second place, it indicates, for consciousness, the necessity of a resumption of a "beginning" in the Cartesian and Kantian sense; finally, it also seems to bring forth, for the purpose of science, that its "organic unfolding," together with its development by "successive addition," should be considered. The specific thematization of these aspects, which seems to draw specifically upon gnoseological assumptions – although conceived in a new way in respect to that of modern philosophy, given Husserl's preoccupation with the problems of subjectivity – will very soon acquire a different and much more radical turn.

The cycle of *Ideas* constitutes in Husserl's program an effort toward providing an image of spatial reality, withdrawn from the dominion of the empirical sciences. It lays down the main features of phenomenology insofar as the stress upon "essences," according to the method of the eidetic insight, gives way, especially in *Ideas II*,[15] to that of the discovery of "sense objects." The "cycle" begins with the theory of the "constitution," and proceeds to the specific analysis of the originary constitution of objects, from which are formed the spheres of objective reality (*Sache*), of the soul (*Seele*), and of the spirit (*Geist*). With this turn at the constitutive or genetic level, as Husserl says, phenomenology acquires the character of "transcendental *organon*,"[16] since it has the right, with the method of clarification, to determine the relationships among the various sciences. It is above all in *Ideas III* that the problem of the relations between phenomenology and the sciences are most carefully investigated; first, of the relation of phenomenology to the exact sciences and, second, to psychology, which remains the target of Husserl's controversy. The exact sciences seem to be constructed on a few evident principles from which the various axiomatic systems develop; phenomenology, in contrast, seems to be conceived as an infinite recrossing of intuitively evident objects. The evidence of the physical sciences is particularly suspect: "The advances of science have not enriched our patrimony of evident knowledge. These advances have not rendered the world more comprehensible;

they have simply rendered it more useful".[17] Husserl's interest here was to arrive at a clearer delimitation of the tasks of science from those of phenomenology, which has, as its supreme objective, that of making itself into strict science. The limitation of the sciences, that seem to lack a "general noetic," was thus stressed by Husserl: "in all sciences the foundation leads, beyond the sphere of thought, to vision and finally to the originarily giving vision, that can be an experience only if other objectivities are valid as objectivity of the experience".[18] Science therefore needs to be preceded by what Husserl calls "*a priori* of the phenomenological constitution." In front of the particularized tasks of science, Husserl stressed the necessity of an analysis of this specific *a priori*, to which *Crisis* later gave structure in the form of the phenomenological "correlation": "it should be recognized" – Husserl observed – "that the whole range of the phenomenological *a priori*, the entire multiplicity of the essences and the essential laws, must be systematically elaborated".[19] With that the specific field of phenomenology is further differentiated from the field of science. If that of science is a field of deductions, then that of phenomenology is instead "an infinite field of essential analyses and descriptions of essence".[20]

Stressing these characteristics in the foreword to the English edition of *Ideas I*, Husserl again claimed for phenomenology the radical character of reflection in contrast to the method of the sciences, which start from the "already given" (*Vorgegebene*), that is, from the "experience of the world," conceived as obviously existing. The judgment of Husserl about this procedure seems equally drastic on another occasion. He claims that from this precritical assumption it follows that all the sciences are harassed by difficulties regarding their foundations and that they get entangled in paradoxes which they seek to remedy by means of a theory of knowledge. This latter, however, instead of preceding them, is devised from the result of scientific researches. Consequently the positive sciences are aphilosophical. Furthermore, since they cannot find their legitimate justification of their procedures in an ultimate cognitive foundation, they are not absolute sciences.[21] It is phenomenology that becomes thus the philosophical foundation of science, since it leads science to a more universal and foundamental horizon of problems. The ideal of phenomenology is characterized as follows:

> phenomenology does not study objects that are considered by the specialist of other disciplines, but the total system of the possible acts of science, of the possible operations, of the meanings that are precisely in relation with these objects.[22]

The problem for Husserl remains the same throughout the whole span of his pursuits, namely, not to lose sight of the point of view of the consciousness with reference to which every givenness is verified, and thereby to assure to phenomenology the title of "first philosophy."

In Husserl's final period, beginning with the works of 1929–1930 (*Formal and Transcendental Logic, Cartesian Meditations*, and parts of the posthumous *Experience and Judgment*), the themes of the relations between the sciences and phenomenology are resumed at a level that Husserl himself defined as in the "strong sense,"[23] that is, when in the meaning of a question its logical dimension is distinguished; or rather, the specificity of the relation that science has maintained with logic, to which Husserl was now preparing to give a significance capable of embracing its two traditional dimensions, the formal and the transcendental. To Plato, and generally to the Greek world, Husserl gave credit for having provided science with an autonomous foundation; this and its close relationship with logic ensured the passage from the mythic world to the world of reflection.

In fact, at the opening of the logical treatise of 1929, Husserl recalled that "science, taken in a new sense, is born from the '*Platonic foundation of logic*".[24] In Husserl's thought logic is the science or area of research designed to answer the exigences of 'real' knowledge and 'real' science. As such it puts norms in evidence, in conformity to which a science may be established. A science that aims at a universal normative legitimacy may explicitly justify its own method and its theories.[25] In fact, Husserl saw in the conflict between "theory" and "method," between possible knowledge and the methods actually practiced by the sciences, the first form of opposition between the ideal advocated by science and that proposed by phenomenology. When the first predominates, science progresses: according to Husserl, this was the case in Greek philosophy because of the impelling new demand for rationality which emerged in antithesis to the skeptical tendencies of the Sophists. This also confirms a view that investigation of principles lays down the conditions for an affirmation of the sciences: again, this is the case with the "Platonic dialectic," which – Husserl argued – "had the mission of making existing science possible as well as of guiding it from the practical point of view".[26]

The situation in modern times is completely different. The sciences have freed themselves increasingly from logic as a normative system, with the consequence that logic has lost the sense of its historic mission. It did not happen at once. The first stage of this transformation, in which the sciences affirmed their independence of philosophy, occurred in the seventeenth century with Galileo and Descartes, both responsible in equal degree for an autonomous affirmation of the sciences. The process that will have its developments in various forms of modern objectivism and empiricism, finishes, according to Husserl, by involving logic itself. In the next period the "sciences make themselves independent, become specialized sciences,"[27] but that rebounds against logic which, as a result, loses its own distinctiveness and its normative character in order to assume the methods and ends of the particular sciences. As is known, Husserl saw in this transformation the sign of the "tragic character of modern scientific culture".[28] Husserl made the explanation of this phenomenon coincide with the loss by science of the links with that more ample ontological and metaphysical horizon already called by Descartes *sapientia universalis*. The autonomous affirmation of the sciences developed in proportion to the loss of the directive or normative role of logic.

For Husserl, the sense of the crisis of science was also proved by the progressive loss of the role of science, particularly felt today. Whereas in other epochs, such as the Enlightenment, science functioned in antithesis to the moral and religious beliefs.

In the second part of the foreword to *Ideas I*, Husserl further investigated the problem of science in relation to that new task of philosophy that was embodied in the idea of phenomenology. The question arises, how to effectuate this radicalism to which Husserl constantly referred? The *Logica* of 1929 seemed to allude to two possible ways, the first of which was that of a resumption of the Cartesian *ego cogito*, which seems to furnish an "absolute foundation of philosophy" and an absolute foundation for the sciences. The other way is to propose an authentic idea of science referring to the norms of logic. In this manner, Husserl remarked, the distinction is made between the authentic meaning of science in general and their essential possibility and only "presumed sciences".[29] This distinction is important since, according to Husserl, we cannot ignore nor neglect the fact that the sciences and the various logical doctrines have already produced in their historical progress, as their sedimentations, given cultural forms. Theres forms must be accounted for; and it is a question, above all, of investigating whether and in what sense sciences have an ultimate, teleological meaning toward which they could tend in all their efforts.[30]

Husserl defined in this manner the task of reinvestigating the nature of logic and, first of all, of formal logic. For he saw in this task the first step of an attempt at an elaboration of a "general doctrine of science," or rather of a theory that related to the conditions of the essential possibility of sciences in general. In a broader sense, it is also a question of an "originary explication of meaning"; that is to say, that the task of phenomenology is to explicate meaning in terms of essential clarity and possibility. Yet the specific project of a reconstruction of logic does not exclude the other task, which to Husserl seemed to take necessarily precedence over all undertakings: that is, to incorporate in its structure the "subjective *a priori*." From this postulate transcendental logic is born.

Phenomenology takes up, then, the treatment of the basic problems as "original explication of sense." On its ground is developed "the authentic meaning of a logic understood as theory of science."

In substance, the reflection on logic and its diverse levels of predicative formations confirms, according to Husserl, the attribution to logic of the role of a *mathesis*, though without its deductive and axiomatic intent; it reveals, moreover, a fidelity to the very spirit of the science; science is then conceived not only as a doctrinal body of the various scientific theories, but as including the study of the unity of its principles – a dual character in which the real "epistemological" attitude (*wissenschafttheoretisch*) is conveyed.

This new level of investigation, however, implies, as we have pointed out, a particular attention to the subjective pole, which constitutes the "guiding thread" of phenomenology. In the *Cartesian Meditations* this pole comes to assume a crucial role in relation to the problem of the constitution of the sciences and of the normative role attributed to logic. The problem of the constitution of the *ego* around a nucleus (*Kern*), from which radiate all the evidences, also suggests a need for a deeper analysis from which the ego would emerge in the entirety of its *habitus*, that is, as monad or person. Phenomenology in this manner proceeds in the regressive and progressive method, showing itself capable of rediscovering the meaning of all the sciences. To understand the real meaning of science – Husserl states – we have to leave out of consideration or put in parentheses the question of the theories (doctrines) of their presumed validity or exactness; it is necessary instead to radicalize the difference between sciences actually given (as belonging, that is, to a cultural phenomenon) and possible: "sciences in the strict sense."

Husserl insisted that from the new point of view of phenomenological reference to the ego sphere, the new scope of scientific investigation was designed "to discover the constitutive elements of the general teleological idea proper to each science."[31] At this point of his reflection Husserl had not yet thematized the historical meaning of this affirmation, but limited himself to stressing the significance of the link between consideration of science and the diverse levels of the constitution of subjectivity. He claims in point of fact the necessity of tracing the meaning of the tendency of science in its development in order to explicate the foundation of its judgments. This applies to the meaning of the act of founding (testing), that is, the act in which the exactness or truth of a judgment, or on the contrary its inexactness or falsity, must demonstrate themselves (*sich ausweisen*).[32]

If the task of science is to surpass commonsense knowledge, it aims in its turn at an order of universal truths. In this way science manifests its deeply seated contact with philosophy. Both – Husserl wrote – imply the existence of two orders of cognition: one anterior with respect to the other and yet interrelated; finally, both imply a *commencement and a progress*; a commencement and progress not fortuitous, but, on the contrary, founded in the nature of things themselves.[33] The idea of a teleological orientation of both comes into the focus of the argument.

Without entering into a discussion of the particular sciences, the reflection on science in general brings up certain fundamental elements of the "teleological idea of science" – reflection that in the end concerns, as stated, the very nature of phenomenology itself and its role in the ambit of history of thought.

The various texts which comprise the last great work of Husserl – the so-called *Crisis* – constitute a continuity of the different criticisms and postulates claimed by phenomenology in relation to science in its historical development. The issue pervading this terminal part of Husserl's thought is, however, that of a teleological reason that finds its exfoliation in history. One could approach the diverse moments of exploration of the same nucleus of problems by showing how Husserl passed in his reflection from a "psychological genesis" to a "phenomenological genesis." The beginning stage of this development is to be found, as is known, in the texts and courses of the years 1915 to 1920, in the passage, that is, from a "static constitution" to a "genetic constitution." The first phase of Husserl's thought can be considered as dominated by the "descriptive method"; in *Crisis* this method itself finds its application with respect to history. In point of fact, for a "worldly" history, Husserl substitutes an "intentional" history, within which the conditions of the particular sciences are delineated, that is, a universal history conceived as the objective horizon of all meanings.

Husserl thus arrived at the ultimate meaning of phenomenology, in which the *telos*, which comes to the fore as an essential factor of the development of the sciences, is in the first place intersubjective (according to the model of the *Wir-subjectivität*). However, ultimately it assumes the significance of an infinite *telos* which Husserl sometimes calls 'God,'[34] even if this term is not elaborated toward a positive and definitive meaning but remains rather to signify a regulatory principle in the Kantian sense, that is, an idea of "pure reason."

Let us return to the principal trend of reflection of the *Crisis*, to consider science from the angle of a "philosophical form of existence." Inasmuch as the sciences depart from this aspect, Husserl maintains, a phenomenon of separation is effected in the goal of the sciences themselves. In *Crisis* the problem of the "immanent sense of history" at large is thus raised, the problem inherent to the emergence and the development of universal ideas – the idea, finally, of a universal science; relevant not, indeed, to the *factual* world, but to the world as retrieved intentionally. In this universal perspective, the structure of science appears as necessarily to be founded in relation to the teleological ideal: "the founded science, which must be founded in a universally apodictic manner, becomes thus the highest function of humanity."[35] This is the conclusive point of view emerging in the famous paragraph 73 of *Crisis*, but it permeates the whole work.

As is demonstrated in the "draft for a continuation" of *Crisis* made by Fink, Husserl intended to resume the general problem of science after the *Crisis* work. For our purpose this draft contains two important indications: (1) "the reassumption of all the sciences in the unity of transcendental philosophy," and (2) "the unity of the sciences as unity of a universal system of correlations,"[36] in which consists the main task of phenomenology.

Recapitulating, we may say that in this way the set of problems unfolding within a continuous sequence is oriented toward the recovery of the *telos* of science and human history. It emerges in its intentional nature in contraposition to the "objectively" given mode of being of positive science. In contrast to the preceding texts in which Husserl posed explicitly the problem of consciousness, in *Crisis*, as in other posthumous texts, the problem of the *telos* is displaced, since for Husserl it is a question of recovering the *telos* in the formations of the consciousness, starting from the precategorial world or the life-world. To this end the two principal parts of the "third part" of the *Crisis* are oriented toward: (1) the access to transcendental phenomenological philosophy through the "reconsideration" of the life-world and (2) "the access to transcendental philosophy starting from psychology."[37] It is a question therefore of a diverse ideal from that affirmed in modern times, in which Husserl sees above all the tendency of science to fix on particular ideals, as for example, that of "prosperity." The new ideal of science, which he proposes, is instead that of a radical reflection (*Besinnung*), in which the entire history of philosophy should be reexamined.[38]

Thus with Husserl's phenomenological approach the problem of the "theoretical-scientific substructures" (*Ideenkleid*) is reborn. It is formulated with reference to a world which is given in perception and, even before, in the life-world. The phenomenological ideal of the crisis of the sciences and culture at large, at this point, is seen in the perspective of the subjective propositions according to the model of "phenomenological correlation."

It is precisely on account of this conclusive section that emerge criticisms and reserves about Husserl's method. In fact, in spite of the polemic with the axiomatic methods, it seems that Husserl, in order to arrive at his apodictic intent, reproposed a methodology very similar to this type.

The paramount difficulty is that of a correlation between scientific theories and a phenomenological-ontological basis. *Crisis* does not for this reason depart from the preceding solutions; it stresses instead the point that the lack of an explicit theorization on the part of particular sciences, as well as of an internal motivation, during the course of the Occidental thought, is in itself the most evident sign of the "crisis" not only of science, but of an aspect of philosophy itself.

The major problem raised by Husserl appears permanent: namely, that of a relation of the framework of every scientific investigation to a final reason or *telos*, so that sight be not lost of the continuing emergence of reason within this framework.

University of Urbino

NOTES

[1] Cf. E. Paci, *Funzione delle scienze e significato dell' uomo*, Il Saggiatore, Milan, 1963.

[2] Cf. M. Farber, *The Foundation of Phenomenology*, Paine-Witman Publishers, New York, 1962, pp. 3ff.

[3] G. Frege, in *Zeitschrift für Philosophie und philosophische Kritik* **103** (1894), 313–332.

[4] R. Carnap, *Der logische Aufbau der Welt*, 1928. Cf. E. Paci, 'Fondazione e costruzione logica del mondo in Carnap', in *Logica e analisi* (AAVV), Archivio di Filosofia, Cedam, Padua, 1966, pp. 97ff.

[5] See in particular L. Landgrebe, *Itinerari della fenomenologia*, Marietti, Turin, 1974, pp. 9ff.

[6] *Richerche logiche*, Il Saggiatore, Milan, 1968, section 4.

[7] *Ibid.*

[8] *Prolegomena*, section 71.

[9] *Elemente einer phänomenologischen Aufklärung der Erkenntnis*, reprinted separately in 1922. Cf. the testimony of M. Heidegger in *Mein Weg in die Phänomenologie, Zur Sache des Denkens*, M. Niemeyer, Tübingen, 1969, pp. 86ff.

[10] *Prolegomena, op. cit.*, section 71.

[11] Whose most important moment is represented by the cited *Vorlesungen* of 1807 (published later with the title *Die Idee der Phänomenologie* (ed. by W. Biemel), M. Nijhoff, The Hague, 1950).

[12] 'Philosophie als strenge Wissenschaft', in *Logos* 1 (1910–1911). Cf. *La filosofia come scienza rigorosa*, Paravia, Turin, 1958, p. 15.
[13] *Ibid.*, p. 23.
[14] *Ibid.*
[15] Exemplary is the analysis of this work by P. Ricoeur, *Analyses et problèmes dans Ideen II de Husserl*, nel vol. *Phénoménologie – Existence* (AAVV), Colin, Paris, 1953; also cf. J. Wahl, *Les aspects qualitatifs du réel*, Centre de Documentation Univ., Paris, 1955.
[16] In a text successive to *Ideen I*, with the title: 'Rapporto fra la fenomenologia e le scienze', published in *Ideen II, IV Beilage*. Cf. the three volumes of *Ideen*, corresponding to vols. III, IV, and V of *Husserliana*, M. Nijhoff, The Hague, 1950–1952. Citations are from the Italian edition (Einaudi, Turin, 1965).
[17] *Idee per una fenomenologia pura e per una filosofia fenomenologica*, p. 871.
[18] *Op. cit.*, p. 805.
[19] *Op. cit.*, p. 847.
[20] *Op. cit.*, p. 838.
[21] 'Postilla alle "*Ideen*" ', in *Idee per una fenomenologia pura, op. cit.*, p. 934.
[22] Cf. note 16. The same idea is found in *Husserliana*, Bd. VII: *Erste Philosophie*, 1923–1924, erster Teil, *Kritische Ideengeschichte*, hrsg. von R. Böhm, M. Nijhoff, The Hague, 1956, pp. 230ff.
[23] In the "introduction" to *Formale und transzendentale Logik* (published in vol. X of Husserl's *Jahrbuch*). The citations refer to the French translation (with double numbering) (ed. S. Bachelard), Presses Universitaries de France, Paris, 1957.
[24] *Op. cit.*, 1/3.
[25] *Ibid.*
[26] *Op. cit.*, 2/4.
[27] *Op. cit.*, 3/5.
[28] *Op. cit.*, 8/13.
[29] *Op. cit.*, 8/13.
[30] *Ibid.*
[31] *Méditations cartésiennes – Introduction à la Phénoménologie* (transl. by G. Pfeiffer and E. Levinas), Colin, Paris, 4 and 5, pp. 7ff. The text was published by S. Strasser in *Husserliana* 1, M. Nijhoff, The Hague, 1950. Citations from French edition, *op. cit.*, p. 8.
[32] *Op. cit.*, p. 8.
[33] *Op. cit.*, p. 10/11.
[34] Cf. L. Landgrebe, *op. cit.*, Citations are from the Italian edition, *La crisi delle scienze europee e la fenomenologia trascendentale*, Il Saggiatore, Milan, 1961.
[35] *Op. cit.*, p. 287.
[36] *Op. cit.*, p. 543.
[37] *Op. cit.*, pp. 113ff, 216ff.
[38] Under this aspect the importance assigned to the position of Galileo (sections 8–10) is important. With Galileo a model for interpreting reality on a purely formal level comes into existence. Mathematics and geometry tend to surpass the limits of the finite Euclidean space, creating a model of rationality that tends toward the constitution of a "nature mathematical in itself." Ancient mathematics knows a "finite Euclidean *a priori*": modern mathematics aims instead at the formation of a "universal *a priori*."

It is in this transformation that the ideal of a "rational and omnicomprehensive science" is born. It ensues that the terms of interpretation of this mathematicization of nature look forward to the possibilities of calculus.

ROMANO ROMANI

"ERLEBNIS" AND "LOGOS" IN HUSSERL'S *CRISIS OF THE EUROPEAN SCIENCES*

I

"Die Welt ist alles, was der Fall ist – Die Welt ist die Gesamtheit der Tatsachen, nicht der Dinge".[1] These statements from Wittgenstein's *Tractatus* do not define the world in its immediate existence, but rather the structure of the totality of cognizable objects in front of the cognizant subject. This statement is presupposed by the aim of the *Tractatus*, which is to establish a criterion of knowledge based on logic. The subjective activity which the world is facing is not that which *occurs in the world*, but that which knows *it*; therefore *the world is seen not as the totality of things, but as the totality of facts*. The world of the events (*Fall*) is not the world of arbitrariness: "Die Tatsachen im logischen Raum sind die Welt – in der Logik ist nichts Zufällig: Wenn das Ding im Sachverhalt vorkommen kann, so muss die Möglich des Sachverhaltes im Ding bereits präjudiziert sein,"[2] but is grounded in the logically structured object. This structure is independent of the cognitive subject, of its *present* will and projects, of its hopes and desires, and is enclosed in the past, impervious to all relations with the subject other than purely cognitive ones. As a structure of meaning, logical space has only one temporal phase: the past. By *Tatsache*, in fact, is meant an event which has taken place, *ein Geschehenes*, as Wahrig explains.[3] The continuation of the past into the present, in the concept of 'state of affairs' (*Sachverhalt*), which explains the verbal form *ist* followed in Wittgenstein by the expression 'totality of facts' (*Gesamtheit der Tatsachen*), cannot but suggest the Aristotelian definition of substance (*ουσία*): *τὸ τί ἦν εἶναι* (*quod quid erat esse*).

That all becomes that which is, however, is the presupposition of every science understood as knowledge which is itself an object. Aristotle demonstrates this in the *Metaphysics*: and the whole history of philosophy shows that it is impossible to escape the necessity of pure theory. Pure knowledge is only a relationship of presence of the object to the subject, which might be defined in different ways: as content of consciousness, possession of the intellect, etc. (The analogy of this relationship with visual perception, however mistaken it may be, and it definitely is, has been the most commonly adopted from the time of Greek philosophy.) This conception of science,

Tymieniecka (ed.), Analecta Husserliana, Vol IX, 105-114.

more or less implicit in the foundations of the great philosophical systems, as well as in the great religious myths and the most widely spread social prejudices, assumes that intellectual activity aims at the contemplation of objects and that science represents the embodiment of this tendency. Pure knowledge by its content influences practical life, since it determines the mores of the individuals; its being is posited also as its own end because its "nature" does not consist in solving problems, but in having already solved them.

Furthermore, pure or absolute knowledge is *always* "science of being" and foreign to becoming. In its perspective, becoming is either a result of an imperfection, of a lack of "being" (Aristotelian materiality), or it is interpreted as pure appearance.

Finally, pure knowledge is adialectical and based on the principle of non-contradiction: logical tolerance has no place in pure knowledge, and dialectic cannot cross the boundaries traced on the territory of pure science.

Kant's distinction between phenomenon and noumenon already prefigures dimly a problem that has never been thoroughly examined. In Kant's view, absolute reality of things in themselves – beyond time and space – the knowledge of which is absolute knowledge (and the knowledge of God which, according to Aristotle, is to some degree granted to men) is unattainable for human beings. Consequently, the certainty of human knowledge does not rest on the form of things in themselves, but on the rules of intellect and the principles of reason.

In both ethics and speculative philosophy the *noumenon* seems to Kant to be the necessary postulate of cognition as laid down by science in general and by philosophy in particular. If it is impossible for man to know intuitively, immediately, the atemporal and aspatial realm of being, to be simultaneously an active potential intellect, the assumption of an active intellect is indispensable in justifying "negatively" a type of cognition that is available and which consists of the presence of the object in the subject in an act which is an end in itself and which determines the *cognitive* (rational) nature of human beings.

And yet, the objectivistic apriorism and Aristotelianism is not completely overcome in Kant's subjectivism. It recurs in the notion of the 'noumenon' seen as a negative limit of sensibility and in that of 'categories' of intellect. Reality, in fact, plays an essential role as the first condition of pure knowledge seen as an autonomous, spiritual, and higher function of man, in two distinct forms in intellectual cognition: (1) as noumenon, and (2) as categories. The "I think" is predetermined by the logico-categorical structure and only thus does it reach a "pure" result: objective knowledge. It is certainly no coincidence that the Kantian categories are identical with the Aristotelian ones; they are the positive counterpart of the noumenal negativeness (they transform the sensitive manifold into cognition) by virtue of which, albeit within the scope of the possible and therefore of the temporal, the sensitive manifold is given as an object to human science.

In Kant's thought time and nontime, space and nonspace, reality and possibility, seem to combine into an equilibrium, open to all problems and all solutions. This equipoise of opposites in which the largest possible number of issues is taken into consideration establishes a conception of knowledge that attributes to the concept of rationality a purely cognitive role.

However, the conception of 'rationality,' as autonomous cognitive activity, allied to sensibility as an indispensable element, cannot stop the query of the mind from seeking further, namely, for things in themselves. Indeed, if the distinctiveness of man consists in knowing, and if this knowledge comprises its own end, why should that which "constitutes" the real object of perfect cognition of the real object escape it? Furthermore, if human thought involves temporality, change, becoming, and possibility, why should human cognition be founded in a "negative" way on something which is immutable and atemporal, rather than in a "positive" fashion on something changable and temporal?

From Fichte to Hegel, German idealism is consistently committed to the founding of a science that, while taking into account temporality and possibility, remains nevertheless concerned with the immediate presence of an object in front of the subject without raising critical problems with respect to an uncognizable noumenal reality. That the meaning of the 'world' is assumed as constituted within human thought indicates how fertile Kantian criticism has been for contemporary philosophy; yet for Kant the position of the thing in itself did not *affirm* autonomous existence of the concrete world beyond thought, but situated the true sense of the 'world' and 'existence' beyond human comprehension. It is not from the starting-point of sensitivity that we can justify Kant's exigency of noumenon but from that of ethics.

Thus we may conclude that possibility is ultimately indispensable in the explanation of reason.

In my opinion, in this conclusion lies the historical significance of the idealistic hypostatization of the transcendental ego. The act of cognition always maintains, in idealism, the Platonic-Aristotelian characteristic of perfect intuition: becoming is appearance, and reality which causes this becoming does not lie outside but inside human thought: it is its essence. Ego/non-ego, being/nonbeing – these dialectical opposites, which found temporality and possibility, are misconstrued in idealism to offer the basis of a definitive reaffirmation once again of spiritual autonomy, and the "divinity" of the cognitive act.

But temporality conflicts with the pure cognitive act. Hegel first saw in time and history and manifestation of dialectic and in dialectic the foundation of temporal and historical becoming. He confronted dialectic and time with the necessity of finalizing history into "science." He wrote in the concluding pages of *Phänomenologie des Geistes*:

In dem Begriffe, der sich als Begriff weiss, treten hiermit die *Momente* früher auf als das *erfüllte Ganze,* dessen Werden die Bewegung jener Momente ist. In dem Bewusstsein dagegen ist das Ganze, aber unbegriffene, früher als die Momente. Die Zeit ist der *Begriff* selbst, der *da ist* und als leere Anschauung sich dem Bewusstsein vorstellt; deswegen erscheint der Geist notwending in der Zeit, und er erscheint so lange in der Zeit, als er nicht seinen reinen *Begriff* erfasst, d.h. nicht die Zeit tilgt. Sie ist das aüssere angeschaute, von Selbst nicht erfasste reine Selbst, der nur angeschaute Begriff; indem dieser sich selbst erfasst, hebt er seine Zeitform auf, begreift das Anschauen und ist begriffenes und begreifendes Anschauen.[4]

The whole which precedes as substance self-knowledge of consciousness is the static cause of a circular time. Time is itself a concept but devoid of a content; in its simple being there is a necessity. Whereas in the fullness of the pure concept time is demoted to the status of appearance.

From the standpoint of *theoresis* as an end of movement, and of movement as an empty schema of awareness of consciousness (this word in German contains in it the image of a circle that encloses itself in the self-consciousness of being: *Bewusst-Sein*), time and becoming are attributed the consistency of the nonbeing – as Plato would say. By conceiving of becoming as existing through self-consciousness, time through eternity, and movement through the rest – when the concept regains the content the substance of which, however, is manifestation – the Hegelian dialectic aims to justify the origin and nature of the given content in its form. In his vision of reality Hegel focuses upon the element of conflict which he sees at the heart of the world of fact as well as in the logos as signifying a significant experience. The conflict manifests itself through contrasts and antinomies. Hegel works with the linguistic structure, its forms and meanings, as they are currently given, and it is in this givenness that he studies dialectic that has as its aim, the reconciling of every type of conflict.

Like time, dialectic lasts until its internal movement between *the opposites* in self-awareness or self-knowledge reaches its utter completion; this completion annuls both time and dialectic. What was movement is stasis, because stasis was movement.

In the capacity in which I *am*, good and evil, *Insichsein* and *Fürsichsein*, the consciousness of the servant and that of the master do not stay any more in reciprocal opposition, but in reciprocal justification. In point of fact, opposition becomes its own justification, and this justification, in turn, seems to be the soul of opposition, the basis of dialectic itself. Thus, we find in Hegel not a foundation for a new "dialectical" science, but rather a science of dialectic. A dialectically founded science is not yet born.

History is represented by Hegel as a dimension of vital development which comprises temporality and dialectic. Both temporality and dialectic have, as may be seen, an existence only in the logos. The conviction that life with all cultural phenomena, in its dialectically and temporally determined development, takes its meaning from the concept, from the logos, is the basis of Hegelianism, and, in general, of idealism. Remaining in the sphere of immanence, unity of science and love of science is actually achieved in a particular synthesis between life and knowledge; *life is seen as if it existed for the sake of knowledge*, and philosophy is the priestess that performs the task of retrieving in the cognitive act the fullness of being which in life is present as becoming, temporality. Thus philosophy descends, like the owl of Minerva, at sunset, and science is always *post factum* knowledge.

In this way Hegel does not abandon the Platonic and Aristotelian assumption of the autonomous spiritual act of knowledge – all becomes that which is. Indeed, following Hegel's assumptions one cannot avoid the Hegelian conclusion that man can have science only of what has been, that his temporally determined being constrains him to a knowledge of what has already passed, a *post factum* science. There is in this no distinction between realistic and idealistic, realistic and positivistic positions, in the Wittgensteinian thesis: "die Welt ist die Gesamtheit der Tatsachen, nicht der Dinge".[5]

II

The position that thought partakes of the exchange between man, the living being, and his environment, and endows with meaning the products of his activity only because it is an element of this exchange, is a radical negation of the autonomous spiritual act of knowledge and reverses Hegelian dialectic. This position erupts in the history of philosophy and science, with the revolutionary thinking of Marx. The Cartesian problem of the existence of the world beyond thought does not appear in Marx's conception of *experience*.

He starts his quest not from the individual situation in which thought is expected to give reasons of the cognizing subject's existence, as well as of the existence of other beings, but from the social situation. In this perspective the intellectual activity of the living individual works for his own life-expansion.

Knowledge in its form of an autonomous spiritual act is branded by Marx as 'ideology.'

Die Frage ob dem Menschlichen Denken gegenständliche Wahrheit zukommt ist keine Frage der Theorie, sondern eine praktische Frage. In der Praxis muss der Mensch die Wahrheit, i.e., Wirklichkeit und Macht, Disseitigkeit seines Denkes beweisen. Der Streit über die Wirklichkeit oder Nichtwirklichkeit des Denkens – das von der Praxis isoliert ist – ist eine rein scholastische Frage.[6]

Marx maintains that the function of thought is to be concerned simply and concretely with planning. The dimension of logos, of language as form and content of thought, remains in the service of planning. Only in relation to human plans does the depth of the past as human history, of the present as experience, and of the future as operative intention, take its meaning as experiencing temporality. Subject and object stay, according to Marx, not in a theoretical but in an "operative" relation. Knowledge functions as an inside production either within the plan as a relationship between its elements or in the execution of the plan. Also, the relationship between subject and object being used may be called knowledge; its meaning, however, is derived from the relation to the use – that means that ultimately it is determined by the nature of consumption itself. Consequently, one may never speak simply of the relationship between the subject and object of knowledge, but instead of the relationship between the subject and object of production or of consumption.

Marx's conception of scientific method does not differ from his views on scientific action, inasmuch as Marxian science is conceived not as a science of dialectic, but a dialectical science. But the question arises: "how is dialectical science to be learned as such without becoming a science of dialectic?" "Could a science of dialectic remain outside of a dialectical science?" This problem is very complex, and was briefly touched on by Marx in short texts (the introduction to *Kritik* [1957], prefaces, unedited economic writings, and letters) which in their development have been so far only inadequately studied.

III

The knowledge of nonknowledge implies, always, rejection of knowledge historically determined. This premise is indispensable for an understanding of the Socratic "Know thyself." Between the extreme of the logical restoration of the myth and that of skepticism, it is a declaration of impotence to face the problems of man as a being historically, socially, and individually determined. Between the extreme of the logical restoration of the myth and that of skepticism there are only few paths; they are difficult, and rarely followed.

We will now direct our attention to the path outlined by Husserl. The problems posed by Husserl are strictly bound to the philosophical tradition and constitute an attempt to rediscover its fundamentals, to recover its ends. But it would be a grave omission – one which has too frequently been made in the interpretation of the philosopher – to think that Husserl's transcendental phenomenology tends toward a form of restoration of the autonomy of theory in opposition to praxis. Although Husserl concentrates upon the study of theoretical issues, his choice does not mean the neglect of the problems of praxis. *Crisis* is the best proof that we may not interpret the spirit of Husserl's transcendental phenomenology merely as a premise for a new point of departure for metaphysical reflections.

Husserl's philosophy could be defined as a theory of the meanings which constitute conscious human life. All his attention is concentrated on the meaningful complexity of the conscious, thinking ego. In Western philosophy, to investigate consciousness as a living '*I think*,' involves naturally a critical reexamination, beginning with Descartes, of the meaning of this expression. The Cartesian cogito establishes immediately, in the world of meaning constituted by the logos, living as thought, the affirmation of the existence of an I. That is, in the meaning (idea) corresponding to the word, this self-perceiving logos. The same deontological demonstration of the existence of God shows how Descartes linked his conception of the 'cogito' to a parallel between the significate and the significant of given linguistic structures; language is meant as existing for itself, independent of words, to use Saussure's terms. In the use of significate-significant parallelism in given linguistic structures that are independent of thought and living discourse, the distinction between *res cogitans* and *res extensa*, intellect and sensation, soul and body, are linked both to objectivism alienated from modern scientism and to the subjectivism of romantic idealism. This was the procedure of Pythagoreanism, and Descartes, starting from the earliest philosophical tradition, introduced it into philosophy and the modern natural sciences.

Husserl's investigation of the Cartesian cogito hinges on the crux of problems concerning, the reconciliation between symbolic communication and existence, life, and words, proceeding from the concrete biological experience of the human being. In the *Crisis* this investigation achieves satisfactorily definitive results.

In point of fact, in the first part of the work we cannot fail to see in an apparently static givenness of the structures of meaning in mathematical language an historico-temporal relationship. The criticism of Galileo is actually a criticism of Platonic Pythagoreanism through the Renaissance Neoplatonism in which the philosophy of the Italian scientist is contained.

Indeed, the work of a geometer or mathematician involves planning in some way. The way in which Husserl passes by means of planning to the discovery of diachronicity, of the historicity of the formation of mathematical languages, and of their even more peculiarly theoretical use is extremely significant:

> So alltäglich vertraut ist der Wechsel zwischen apriorischer Theorie und Empirie, dass wir gewöhnlich geneigt sind, Raum und Raumgestalten, über welche die Geometrie spricht, von Raum und Raumgestalten der Erfahrungswirklichkeit nicht zu scheiden, als ob es einerlei wäre. Soll aber die Geometrie als Sinnesfundamente der exakten Physik verstanden werden, müssen wir hierin und überhaupt sehr genau sein. . . . In der anschaulichen Umwelt erfahren wir in der abstrativen Blickrichtung auf die blossen raumzeitlichen Gestalten 'Korper' – nicht geometrisch – ideale Körper, sondern eben die Körper, die wir wirklich erfahren, und mit dem Inhalt, der wirklich Erfahrungsinhalt ist. Wie willkürlich wir sie in der Phantasie umdenken mögen: die freien, in gewissem Sinne 'idealen' Möglichkeiten, die wir so gewinnen, sind nichts weniger als die geometrisch-idealen Möglichkeiten, nicht die in den idealen Raum einzuzeichnenden geometrisch 'reinen' Gestalten – die 'reinen' Körper, die 'reinen' Geraden, die 'reinen' Ebenen, die 'reinen' Figuren sonst und die in 'reinen' Figuren Verlaufenden Bewegungen und Deformationen. . . . Die Phantasie kann sinnliche Gestalten nur wieder in sinnliche Gestalten verwandeln. . . . Die Dinge der anschaunlichen Umwelt stehen ja überhaupt und in allen ihren Eigenschaften im Schwanken des bloss Typischen; ihre Identität mit sich selbst, ihr Sich-selbst-Gleichsein und in Gleichheit zeitweilig Dauren ist ein bloss ungefähres, ebenso wie ihr Gleichsein mit anderem. Das greift in alle Veränderungen ein, und in *ihre* möglichen Gleichheiten und Veränderungen. Entsprechendes gilt also auch für die abstrakt gefassten Gestalten der empirisch anschaulichen Körper und ihrer Beziehungen. Diese Gradualität charakterisiert sich als eine solche grösserer oder geringerer Vollkommenheit. Praktisch gibt es wie sonst auch hier ein Vollkommenes schlechthin in dem Sinne, dass das Spezielle praktische Interesse dabei eben voll befriedigt ist.[7]

To recognize the practical origin of these ideals of rational perfection means that the ideas of geometry and mathematics are not relevant to the establishment of transcendental reality (one thinks of the importance of the thought of Cusanus in philosophical and scientific reflections of the Renaissance), but of a field of planning; they appear as forms of the logos of the temporality of planning which has at its foundations the concrete, living corporality of the thinking and acting human subject.

Of course, in order to retrieve the vital biological dimension of the logos is inconceivable in Husserl without his *epoché*. But that the *epoché* may lead to this recuperation and is, in fact, its true significance, is not determinable except each time when its vital dimension has been experienced. The path taken by Husserl is, then, difficult and tortuous. Its own foundations are continually checked with reference to philosophical enterprises. To suspend judgment, to withdraw from the natural acceptance of meanings already given, prepared, objectively present in one's own world, is indicative not only of a subjective capacity (for establishing such meanings) but also of the capacity (possibility, liberty) to change them, to reconstruct them in a plan, to accept or reject them. Being able to withdraw from the world with all that in it has meaning, to discover the foundation of such meaning, may be considered as the final and highest result of biological evolution. It is, of course, the negative side of the capacity to plan one's own world or to live. However, the human being, without being capable of it, would not be capable of being in a world of planned meaning, within a plan of his doing.

At the beginning of the *Crisis*, the contraposition between the practical sciences and philosophy seems a contraposition between answers to specific questions and the answer (whether sought or found) to the ultimate problems – i.e., insoluble through the concrete praxis of man. Consequently one finds that the contraposition is between a science conceived as the creative conquest of the object, that is, knowledge of the past conceived as a static and inert continuation of the past in the present, and a knowledge which is the rediscovery of the past and the present in plans for the future; between a thought and thinking, the lived and living. But obviously Husserl was deceived in a certain measure by the vision of the task with which he was concerned; this vision was different from the results the phenomenological method was capable of leading him to. In the first place he set out to establish the foundation of *knowledge*. The meaning of this word comes to us, as we have outlined, from a long and distant historical tradition, and bears the meaning of confrontation between subject and object, standing in front of each other. Once accepted in this sense, one could not avoid finding these two terms which confront each other in a typical, visual, and *theoretical* situation, whereas we should strive for a different formulation: not in the subject and object of traditional philosophy; not in the ego and nonego of Fichte, and not in the idea which becomes aware of itself in becoming. The ego self-objectifies itself in the act of self-awareness; as the origin of meaning it finds in itself *both* subject and object, i.e., a *cognitive* relationship. It is an arduous, strained conclusion which we may draw from Husserl's own dilatory question begging, a conclusion that underlines a *hic sunt leones* in his investigation. The self-awareness of the ego establishes in a "game of mirrors" a new form of parallelism between meaning and significate at the basis of the traditional split between *res cogitans* and *res extensa*, soul and body, God and the world.

Rome TRANSLATED BY TOM DAY

NOTES

[1] Ludwig Wittgenstein, *Tractatus Logico-philosophicus*, I, II, Surkamp, Frankfurt am Main, 1973, p. 11.
[2] *Ibid.*, 1.13, 2.012 (pp. 11, 12).
[3] Wahrig, *Deutsches Wörterbuch*, Berteilsmann Lexikon, Berlin, 1968–1971 (see under "Tatsache").
[4] G. F. W. Hegel, *Phänomenologie des Geistes*, Surkamp, Frankfurt am Main, 1970, p. 584.
[5] See page 1 of this article, and note 1.
[6] Karl Marx, *Thesen über Feuerbach, No. 2*, Marx-Engels Werke, vol. 3, Diez Verlag, Berlin, 1962, p. 5.
[7] Edmund Husserl, *Die Krisis der Europäischen Wissenschaften und die Transzendentale Phänomenologie; Husserliana*, vol. 6, Martinus Nijhoff, The Hague, 1962.

PART II

THE TELIC PRINCIPLES

A. TELOS AND THE CONSTITUTIVE CONSCIOUSNESS

RUDOLF BERNET

PERCEPTION AS A TELEOLOGICAL PROCESS OF COGNITION

Edmund Husserl's phenomenology is essentially dominated by teleological concepts. This is true of his philosophical analyses of the development of biological organisms as well as of his philosophical considerations of cultural phenomena. Both animal and human (transcendental) life has a goal. The history of interpersonal relationships as well as the tradition of philosophical thought are to be understood in terms of a teleologically organized meaningful context. This extensive use of teleological concepts is encountered with bewilderment or an ironical reaction of a Nietzschean variety on the part of most contemporary thinkers. They consider Husserl to be a belated representative of an outdated rationalism, and his teleologically determined theories appear, at best, as the expression of a naive faith. One of the objectives of the following considerations concerning the *teleological constitution of cognitive life* is to contribute to a revision of this widespread misunderstanding. I would like to show that Husserl's use of a teleologically functioning ideal of absolute truth can only be understood on the basis of his radical awareness of the finitude of human knowledge. The basic fact from which Husserl proceeds is the possibility of error, and the tension between the desire for absolute cognition and the necessary frustration of this desire. Husserl does not intend to resolve this tension by means of the teleologically anticipated ideal of cognition. On the contrary, a completely responsible form of cognitional praxis consists in fully accepting this tension and the resulting infinite philosophical task.

Of significance in connection with the tension between that which is cognitively sought after and that which is cognitively attainable, is the phenomenon of *perspective*, i.e., the contrast between one-sided perceptual appearance and the fully determined, genuine thing-in-itself. It may be asked whether it is justifiable to limit the consideration of the problem of the teleological structure of cognition to the abstract case of the *visual perception* of a thing. In order to justify narrowing the field of investigation in this way, it should be noted that within Husserl's phenomenological epistemology visual perception functions as the fundamental form of cognition. Furthermore, I am of the opinion that the fact that every concrete visual perception is embedded within a complex structure of other sensuous experiences is not essentially relevant for our problem.

Tymieniecka (ed.), Analecta Husserliana, Vol IX, 119–132.

The following considerations are divided into three sections. In the first, I would like to establish, by means of some general observations, the concept of that ideal referred to by the knowledge of what is transcendent ("*transiente*" *Erkenntnis*). Here, the difficulty consists in the fact that although this kind of knowledge strives toward adequate self-givenness of its objects, these objects, in principle, exclude such givenness. In the second section, the teleological function of this cognitive ideal will be illustrated using the concrete example of visual perception. I would like to show that the teleological anticipation of the thing-in-itself, phenomenologically understood, forms the fundamental, dynamic principle of organization for these sensual cognitive acts. In the final section, I would like to call attention to the practical motivation of this cognitive goal. The meaning of an unobtainable goal; i.e., of an endless process of cognition, is finally only intelligible within the context of striving for an ethically responsible life.

I

In this first section we are confronted with an extremely difficult task. On the one hand, inadequate cognition, i.e., the merely partial intuitive givenness of transcendent objects, is unthinkable apart from the concept of adequate cognition of an entirely determined object. On the other hand, the reality of the thing, as understood within the natural attitude, is excluded by the phenomenological reduction. Furthermore, this thing-in-itself should not be thought of as a "noumenon" in the Kantian sense. That which the appearing object is must be determined exclusively upon the basis of its phenomenologically pure givenness. However, the task of a pure phenomenological determination of the thing-in-itself received essentially different solutions in the *Logical Investigations* and *Ideas I*. In *Ideas I* elements of an epistemological image-theory (*Bildertheorie*), which still can be found in the *Logical Investigations*, are eliminated by Husserl's radical departure from all forms of epistemological realism. However, this essential improvement in the phenomenological formulation of the ideal to which the knowledge of what is transcendent refers, implies at the same time, as we shall see, a certain disintegration of this cognitive ideal. The ultimate goal of cognition is no longer the perfect cognition of the *object*, but rather the endless progression, as it were, of *cognition itself*.

In the *Logical Investigations* the connection between the appearance and that which appears is designated by the term "representation" (*Repräsentation*). This concept earned its specific phenomenological importance in its opposition to the epistemological image-theory. Among Husserl's contemporaries, notably Twardowski had returned to this scholastic theory under the influence of Brentano.[1] He thus hoped to find a solution to the well-known problem within the framework of an intentional theory of consciousness, of so-called objectifications having no existent object (*gegenstandslose Vorstellungen*). According to Twardowski, every conscious act has an intentional, i.e., mental, content which possibly corresponds to some external reality. Husserl's critique of this theory is well-known. According to him, consciousness of an image presupposes the possible self-givenness of its original. To distinguish, within the structure of intentional objectivity, between the mental content and the real (external) object leads to an infinite regress.[2] Intuitive representation, in Husserl's sense then, never means that the image becomes a representative, within consciousness, for the object which exists in itself. The represented object is the intentional object, whose reality can only be analyzed in terms of its phenomenological givenness, i.e., the intuitively fulfilled intention.

When an appearance brings the fully determined object to absolute self-givenness, this is an adequate representation. Here the intuitive intention is completely fulfilled: that which represents and that which is represented coincide.[3] On the other hand, inadequate representation, e.g., the actual adumbration of a three-dimensional object, is not purely intuitive givenness. The appearance does not bring the intentionally meant object to full self-givenness. The object is only partially self-given, and this self-givenness is necessarily accompanied by an empty intentional content. In the *Logical Investigations*, Husserl designates this empty content by the term "signitive" (*signitive*), in order to distinguish it from the pure intuitive component of the intentional act.[4] The appearance of the physical object is a "mixed" (*gemischte*) form of representation.[5] Certain parts of the intentional object are intuitively represented, others are merely signitively represented. The model which Husserl uses for the signitive form of representation, however, is the extrinsic connection between the intentional meaning and its linguistic expression. Consequently, the connection in perception between intuitive and signitive intentions is also extrinsic, i.e., a mere contiguity.[6] The transference of a specifically linguistic use of signs into a purely sensuous experiential framework is questionable. And, furthermore, the connection between intuitively fulfilled and empty (unfulfilled) perceptual intentions is hardly as extrinsic as that between sense-giving acts and their linguistic expressions. Finally, it is even more problematic to use this model in connection with the conscious relationship between the appearance and the thing itself.

The emptily apperceived reverse side of a thing is not represented in perceptual consciousness by a sign. The empty intention belongs essentially and inseparably to the perception itself, which is not a consciousness of a sign. Rather, the empty intention, which is essentially bound up with partial self-givenness, is an anticipatory consciousness of possible self-givenness.

Finally, this theory of inadequate perceptual representation as a mixed, signitively mediated form of representation is contradictory. Husserl himself becomes entangled in the absurdity of the epistemological image-theory which he criticized. This is shown by the fact that in the *Logical Investigations* Husserl posited a relation of "similarity" (*Aehnlichkeit*) or "analogy" (*Analogie*) between the appearance of the object and the object itself.[7] A relationship of similarity, however, can only hold where there exists the possibility of comparison, e.g., between a landscape and its pictoral representation. The phenomenologist, however, cannot compare the fully determined object and its modes of partial self-givenness. With the phenomenological reduction the reality of the thing-in-itself is excluded. What the thing is and its determinations are to be exclusively determined on the basis of its appearance. This is possible because the appearance is the *self*-givenness of the appearing object and not simply its image or sign in consciousness. However, if we posit a relationship of similarity between the appearance and the object, we will have presupposed that the object that is referred to by this similarity is something existing in itself, independently of consciousness. This phenomenological theory of representation thus contradicts itself by using a concept of objective reality which it intended to overcome with its critical new formulation of the being and determination of transcendent objects. A foundation of this sort is circular and contradictory.

Even more important in the present context is that the relationship of similarity used in connection with inadequate representation employs a concept of adequate givenness that is unacceptable within a phenomenological theory of "external" perception. To say that the appearance is only "similar" to the object, implies the possibility of a coincidence between the appearance and that which appears. The way in which inadequate representation is characterized as a "mixed" form of representation in the *Logical Investigations*, presupposes the model of adequate representation as it is found in so-called "inner" perception.[8] Thus, the ideal of adequacy implicated in this concept of inadequate representation is that of a possible identity between that which represents and that which is represented. This ideal, however, could only be realized if the three-dimensional spatial form of the intentional object could be transposed in its entirety into the essentially two-dimensional field of appearance. This is not only impossible, it is absurd. The absurdity consists in constructing the ideal of adequate givenness, and thus the specifically phenomenological concept of the spatial thing-in-itself as well, in accordance with a model of something which presents itself without perspectives.

Summarizing Husserl's position in the *Logical Investigations*, we find that it is certainly correct that the determination of inadequate self-givenness implies the concept of adequate givenness. For example, the one-sidedness or incompleteness of each adumbration of an object presupposes the idea of the thing-in-itself, or in pure phenomenological terms, the ideal of adequate givenness.[9] As we shall see later, the teleological anticipation of this cognitive ideal essentially determines the concrete structure of the perceptual process. Now, however, we still have to deal with the problem of a correct phenomenological determination of the thing-in-itself. Behind the problematic formulation in the *Logical Investigations*, the position found in *Ideas I* already appears in outline. If the thing-in-itself is not to be thought of as a real object belonging to the natural world, then it must be understood in terms of a phenomenological description of the intuitive givenness of the thing in consciousness. Thus, if the appearances of a thing are essentially one-sided modes of givenness, then this one-sidedness must be included in the concept of the thing-in-itself. To propose a form of givenness in which a spatial object could be given without perspective is unacceptable, even if only as an ideal. Still, the thing-in-itself is something other than the one-sided givenness of a thing, or the plurality of its appearances. Thus, the difficult task is that of accounting for the essentially perspectival nature of spatial objects, without identifying these objects with the plurality of their perspectival modes of givenness.

Let us now consider the presentation of this problem in *Ideas I*. The essential improvement of this presentation over that found in the *Logical Investigations* is its uncompromising criticism of the possibility for "realizing" adequate givenness of spatial objects.[10] Husserl here utilized the results of his so-called "Dingvorlesung" of 1907, in which he rejected the possibility of adequate knowledge of a spatial object even with respect to a single side, and emphasized the absurdity implicit in the idea of an "intuitus originarius".[11] In *Ideas I*, the partial character of each of the object's appearances refers primarily not to a fully determined thing-in-itself, but rather to the multiplicity of appearances. Thus, to say that every appearance is inadequate means that every appearance always refers to other possible appearances of the same object. Even a continuous examination of the object in a harmoniously connected series of appearances will never yield adequate knowledge of the object, because the possibility of a new or different determination can, in principle, never be excluded. Accordingly, the inadequacy of the cognition of transcendent objects is primarily understood in *Ideas I* in terms of the essentially incomplete nature of this cognitive process itself, i.e., its infinite nature.[12] Thus, Husserl concludes that the thing-in-itself must be phenomenologically defined taking into account this infinite character of every perceptual process. It appears to be very difficult, however, to satisfy this requirement. If the thing-in-itself can only be grasped through the concrete performance of an endless experiential process, two difficulties arise. First, since the experiential process is always infinite, the thing-in-itself will never be brought to fully evident, i.e., completed, givenness. Second, if the thing-in-itself forms the progressively constituted objective unity in the always incomplete experiential process, and as such is never adequately given, then its teleological function within the perceptual process becomes unintelligible. Husserl's solution to these problems consists: (1) in sharply differentiating between the thing-in-itself and the objective unity progressively constituted in an infinite experiential process; and, (2) in interpreting the thing-in-itself as a mere *idea*, not realizable in experience, whose teleological function in the experiential process is that of a regulating principle.[13]

What essential insights, then, does this interpretation of adequate givenness as a regulative idea – "idea in the Kantian sense"[14] – yield? First, the ingenious insight that although the necessarily endless manner in which transcendent objects are given precludes the possibility of their adequate givenness, still the *idea* of this endlessness can be grasped in absolute evidence. If the adequate givenness of a thing, qua *idea*, is fully evident givenness, then it becomes clear how this idea simultaneously regulates the dynamic structure of the experiential process and is nevertheless incapable of being realized in this process, i.e., that this idea is not a datum in experience. In the second section of this paper, I would like to elucidate the regulative function of this idea of adequate givenness within the context of visual perception. I want to show that the *teleological* anticipation of the adequate givenness of a thing is the fundamental principle regulating the structure of a continuous perceptual experience.

II

The fundamental concept of a phenomenological analysis of perception is that of *appearance*. Taken as pure phenomenological givenness, "appearance" should neither be understood as an image in consciousness nor as a sign which is the causal result of the appearing reality. As *self-givenness*, the appearance is the appearing objectivity; and yet, as *partial* self-givenness the appearance is never identical with that which appears in it. Husserl accounts for this two-sided nature of the appearance of an object by defining it as a necessary and inseparable compound of "authentic" and "inauthentic" appearances.[15] Formulated in noetic terms, this means that each successive appearance of a thing is a partially fulfilled intentional act, a combination of "pure perceptual content"[16] and apperceptive surplus. Put into an equivalent noematic formulation,[17] that which appears-as-such is an intuitive givenness, necessarily surrounded by a horizon of *possible* intuitive givennesses of the same object. The noematic concept of appearance has systematic priority insofar as it does not rely upon a specifically categorial performance, i.e., the apperceptive interpretation of a representing content, and thus can give a better account of the specifically sensuous character of perception.

The distinction between that which is empty and that which is fulfilled within the givenness of a single appearance, however, only becomes apparent within the *continuous synthesis of appearances.*[18] The fundamental structure of the perceptual process is the temporal stream of consciousness. Husserl consequently analyzes this process's duration as a continuum of always new "now-points" with continually new phenomenal content. A continuous synthetic process takes place within this stream of appearances, which binds together the multiplicity of ever-new appearances. For the sake of simplicity let us limit ourselves to the perception of a single object, neglecting its insertion within the world horizon. In this example, the continuous synthesis is a synthesis of "coalescence" ("*Verschmelzung*"), a "*Deckungssynthese*,"[19] which relates the multiplicity of appearances to the unity of that object which appears in all of them. This synthesis, a purely sensuous and passive accomplishment, is possible because every actual appearance, by means of its (inner) horizon, necessarily points to other possible appearances of the object. The occurrence of each new appearance is, to a greater or lesser extent, already indicated, i.e., "motivated," by the previous appearances. This motivation is not by any means arbitrary. Rather, possible appearances are structurally interconnected with the present appearance by relations of compatibility and contiguity. Thus, for example, in the present appearance of the table top the motivation for the possible givenness of a complementary appearance of the *same* surface is more coercive than the motivation for a possible appearance of the underside of the table. The most important principle organizing the horizon of possible appearances, however, is the teleological anticipation of the thing-in-itself, which is a unity given in all of those actual and possible appearances. This will become even clearer in the following discussion, as we attempt to characterize the continuity of appearances as a process of fulfillment.

The transition in perception from one appearance to the next, i.e., the actualization of just this or that possible givenness of the thing, is not only determined by the requirement for harmonization or for the uninterrupted continuation of the already past continuum of appearances. Rather, the new appearance should enhance the intuitive givenness of the appearing object, and thereby improve our knowledge of it. Thus, the continuous synthesis of appearances does not simply unify a multiplicity of appearances with respect to the appearing object. The performance of this synthesis is essentially co-determined by the interest in a continually better and ultimately perfect knowledge of the perceived object. In other words, *the synthesis of appearances is always at the same time a synthesis of fulfillment.*[20] The dynamics of this process of fulfillment, the actual cognitive interest within this process, is essentially determined by the teleological anticipation of absolute cognition, i.e., of adequate givenness of the appearing object.[21] This cognitive goal not only regulates the formation and structuring of the horizon of possible givennesses of the thing, but also constitutes the selective criterion, so to speak, for the actualization of one of these possibilities. The interest, then, affecting the concrete course of the perception of a thing, is directed toward those appearances which make possible an approximation of this ideal of the complete givenness of the object. Accordingly, for Husserl the perception of a thing is essentially a cognitive process in which the perceived object is continuously confirmed and more completely determined, i.e., pre-predicatively "explicated."

This is not the place to discuss the specifically sensuous form of this perceptual cognition and "explication," or to contrast it with categorial cognition which is built upon it. More important for us is the observation that the perception of a thing is an *endless* process of fulfillment. This can be phenomenally demonstrated in that every actual appearance is necessarily surrounded by an empty horizon. With each synthesis of fulfillment, i.e., with every intuitive actualization of that which is purely emptily intended, a new empty horizon arises.[22] And, in principle, the possibility can never be excluded – however long the preceding experience of confirmation might be – that a new appearance might conflict with this empty anticipation, i.e., "disappoint" it.[23]

Finally, as a process of fulfillment, continuous perception is simultaneously a *process of constitution.* That is to say, for Husserl, the process of taking cognizance of the object is not a progressive discovery of an objective reality which exists independently of consciousness and is, in itself, completely determined. Rather, the "object" is actually first built up in the progressing process of cognition in its phenomenological determination and existential validity (*Seinsgeltung*). The real "object" (*Gegenstand-in-Anführungszeichen*), which is being progressively constituted, is the intentional correlate of that unitary consciousness which encompasses the multiplicity of appearances in a continuous synthesis of fulfillment. Just as the multiplicity of appearances is teleologically oriented toward adequate appearance, so too, the object being constituted is a continuous approximation of the teleologically anticipated ideal of a fully determined thing-in-itself. The meaning of idealism within Husserlian transcendental philosophy cannot be understood without reference to the very close relationship between the cognitive and the constitutional processes. I cannot here go into the problem any further. It should be stressed, however, in view of continual misunderstandings, that in any case Husserlian idealism is not a "creative" idealism. The spatial object *qua* constituted, is not identical with constituting consciousness. It is not at all a structure of consciousness – it is not an object *of* consciousness, but an object *for* consciousness.

If one is interested in the teleological structure of the perceptual process, then the question as to whether this process can be described in purely noematic terms is very important. That is, a pure noematic description of the process of fulfillment implies a new determination of the teleologically structured cognitive interest, a determination in which this interest can no longer be understood in terms of an egological accomplishment.

Husserl's still unpublished contributions toward a pure noematic analysis of perception[24] focus upon the concept of the field of appearances. Thus, they are closely related to Gurwitsch's phenomenology of the transcendental field of consciousness.[25] Gurwitsch and Husserl first emphasize the noematic concept of appearance and devote particular attention to the horizon of possible appearances implicated therein. This horizon is a monadological system of interdependent perceptual possibilities. With the actualization of each new possibility, the entire system is modified. The transition of an appearance from pure possibility to actual givenness is described by Husserl in terms of its movement from the margins of this field to its center. Cognitive interest is responsible for this transition, or, in noematic terms, the associative demand for supplementary intuitive givenness. Here I cannot further pursue this, for the most part, merely fragmentary theory. I believe that its weaknesses are to be found not so much in the purely noematic description of the synthesis of fulfillment, and also not primarily in the fact that the description of the noematic field of appearance uncritically employs objective spatial structures. Rather, the most important question is whether a pure noematic theory of perception is capable of accounting for the kinaesthetic motivation of the "I can." The teleologically structured cognitive interest in the perceptual context is always realized kinaesthetically. Interest in a complementary, intuitive givenness of the thing is not sufficient to actually bring about this appearance. Kinaesthesias are also required, e.g., a movement of the head, which function in the service of this interest. However, if kinaesthesias originally possess the form of the "I can," then a nonegological formulation of the teleology operative in the perceptual process is questionable, at least within Husserlian phenomenology.

Nonetheless, I do not wish to contest the previously mentioned advantages of the noematic concept of appearance. In my opinion it is here that the essential merit of Gurwitsch's development of Husserl's theory of perception is to be found. The gestalt psychologist's interpretation of the field of appearance,[26] and in particular of the connection between the actual appearance and its horizon of possible appearances,[27] discloses entirely new possibilities for a phenomenological theory of perception. Primarily, the critique of intellectualism implied by the noetic concept of appearance (the dualism of a formative element and formless representational material) has decisive consequences for the understanding of the relationship between sensuously structured experience and (categorial) judgments. In connection with the question concerning the teleologically anticipated thing-in-itself, however, I find Gurwitsch's theory to be unconvincing. Gurwitsch identifies this thing-in-itself with the total system of its possible appearances.[28] Thus, the thing-in-itself disintegrates into the infinite number of its appearances. Correctly understood, however, the thing-in-itself has the form of an intentional unity constituted in this infinite multiplicity of appearances. The difference here is decisive. That is to say, if the idea of the thing-in-itself is identified with the idea of the totality of its one-sided (partial) modes of conscious givenness, then the thing-in-itself is a form of consciousness, even if an infinite one. In this way, however, Gurwitsch lapses into an absolute phenomenalism, and into the previously criticized "creative" idealism. If, on the other hand, the thing-in-itself is the unity constituted in an infinite experiential process, then the radical difference between the three-dimensional "object" and the two-dimensionality of its givenness within constituting consciousness is guaranteed.

III

In this, the third part of my presentation, I would like to begin with a summary of the Husserlian doctrine of the teleologically anticipated ideal of adequate givenness of a thing. With this as our basis, then, we can finally conclude with an attempt to understand the basic motivation of Husserlian phenomenology.

Husserl's determination of the teleological anticipation of adequate givenness of a thing is the solution to a paradox. This paradox consists, on the one hand, in the fact that adequate givenness of a thing can essentially never be realized; and, on the other hand, in the fact that nonetheless every appearance, every continuous multiplicity of appearances teleologically anticipates this adequate givenness, i.e., the thing-in-itself. Husserl solves this paradox by interpreting the adequate givenness of a thing as an "idea in the Kantian sense." That is to say: (1) Adequate givenness of a thing is not something "real," but rather a *teleologically anticipated idea.* (2) This idea is the idea of an infinite experiential process. The idea of infinite experience is, in contradistinction to the endless process of experience itself, adequately given. The concept of adequate givenness does not, therefore, imply the contradictory claim of completing that which cannot be completed. There is also no longer the assumption, as in the *Logical Investigations*, of the existence of a spatially transcendent object, offered without perspectives. The thing-in-itself is understood in pure phenomenological terms with reference to the infinite multiplicity of its appearances. It is a unity constituted by, but not to be identified with, this multiplicity. (3) The teleological anticipation of the idea of the thing-in-itself essentially structures every perceptual process; in other words, this teleologically anticipated idea has a *regulative function.*

A specifically phenomenological formulation of the thing-in-itself consists primarily in not presupposing its "reality," but rather in determining it solely upon the basis of the infinite multiplicity of phenomenologically pure, partially intuitive appearances. This is the meaning of a phenomenological reduction of the "natural" thing-in-itself. That which the thing-in-itself is, is to be determined by cognition and not the other way around. Since cognition of a spatial object is an infinite, i.e., incompleteable, task, one can never definitely determine exactly *what* and *whether* the perceived object is. In the transition from a "natural" to a transcendental epistemology interest is diverted from the *object* of cognition to the subjective *activity* of cognitive inquiry. The *telos* of this subjective activity is adequate givenness of the thing, which can never be realized within experience. In other words, the goal of the perceptual process is nothing else than this very process itself, i.e., the infinite progression within a harmonious series of experiences of ever more complete determination and confirmation of the object. It may be objected that this formulation is somewhat excessive, since Husserl himself in connection with the perception of a thing speaks of the satisfaction of the cognitive interest through *optimal*, but not adequate givenness.[29] As a rule, however, this cognitive interest and its goal of optimal givenness is something which changes from one cognizing subject to another. And even when, within different perceptual experiences, the interest and the perceptual circumstances are identical, we find that the same optimal givenness of the perceptual object does not equally satisfy the experiencing subjects. In any event, the concept of optimal givenness cannot simply be substituted for the concept of adequate givenness and its teleological function. It is true that the cognizing subject normally does not expressly strive after the infinite and thus incompleteable, progressive improvement of his cognition. But it is equally true that inadequate cognition (every incomplete process of cognition) structurally implies the idea of adequate givenness and that the adequate givenness of a perceptual thing necessarily points to an endless experiential process.

If the *telos* of perception is adequate cognition of the thing-in-itself, then perception becomes an infinite and thus unrealizable task. This thesis corresponds to Husserl's statements concerning the teleological nature of the history of philosophy.[30] Since its establishment by Plato, philosophy has essentially been characterized by the theoretical ideal of striving after knowledge for its own sake. The history of philosophy can be understood as a centuries-long endeavor to realize this ideal. Phenomenology, with its rigorous scientific point of departure and its scientifically regulated step-by-step procedure, is, however, first able to take this task on seriously. The systematic accomplishment of phenomenological science, however, will require the cooperation of generations of investigators, for phenomenology's task is an infinite one. What was true of the perception of a thing also holds for phenomenological science as such. In both cases, the teleological aspiration for absolute cognition of objective being becomes a pursuit of the infinite advancement of the process of cognition.

It can legitimately be asked, however, whether consciousness of the infinite nature of the theoretical task can actually serve as a goal or motive for theoretical activity. Is it not more natural to resign oneself than to hopelessly pursue an unending, essentially uncompleteable task? This question makes it clear, that the (essentially Cartesian) ideal of adequate cognition, taken apart from its practical motivation, is finally unintelligible. If the infinite character of the theoretical task does not deter, but rather encourages, the phenomenological investigator, this implies a practical motivation for the pursuit of knowledge for its own sake. The practical motivation at work in Husserl's philosophical theory is the ethical demand for *absolute responsibility*; a responsibility which extends not only to theoretical statements and investigative activities, but also to human life as a whole. As envisioned by Husserl, the phenomenologist theorizing uncompromisingly, i.e., in a completely responsible manner, faces an endless task. He accepts this task, finally, not out of interest in the object of his cognitive activities, but rather out of an interest in the theoretical research itself. Also, (endless) theoretical investigation interests him not only as an attempt to escape death, but because he believes his activity to be the best and worthiest form of human life. The highest form of human life is, according to Husserl, a search for absolute self-responsibility.

University of Louvain

NOTES

[1] K. Twardowski, *Zur Lehre von Inhalt und Gegenstand der Vorstellungen*, Wien, 1894 (trans. Grossmann, *On the Content and Object of Presentations*, M. Nijhoff, The Hague, 1977). For Husserl's discussion of this book cf. *unpublished manuscript at the Husserl-Archives Louvain* (to be henceforth referred to as MS.) *K I 56* (1894);*MS. A I 7* (1896); *Logical Investigations*, trans. Findlay, Routledge and Kegan Paul, London, 1970 (hereafter cited as *LI*), pp. 290, 657–58; *Ideen zu einer reinen Phänomenologie und phänomenologischen Philosophie*, Bk. I. M. Niemeyer, Halle, 1913 (hereafter cited as *Ideen I*), §129, p. 267 (the original pagination can be found in the margins of the Husserliana edition).

[2] *LI*, pp. 593–96. *Ideen I*, §52, pp. 99–102; §90, pp. 186–87.

[3] *LI*, pp. 542–43, 713, 761–62, 866.

[4] *LI*, pp. 713, 717–18, 731–32.

[5] *LI*, pp. 713, 632–33, 740, 742.

[6] *LI*, pp. 717, 732, 742.

[7] *LI*, pp. 310, 741, 861.

[8] *LI*, pp. 542, 544.

[9] *LI*, pp. 713–14.

[10] *Ideen I*, §§43, 138.

[11] *Ding und Raum*, Husserliana, XVI, §§33–34, 48–49, 51–53, 65; *Beilage* V.

[12] *Ideen I*, §143, p. 297; §149, p. 311.

[13] *Ideen I*, §§143, 149.

[14] *Ideen I*, § 143, p. 297.

[15] *Ding und Raum,* §§16, 18.
[16] *LI*, p. 713.
[17] For the noematic concept of appearance cf. *The Idea of Phenomenology*, trans. Alston and Nakhnikian, M. Nijhoff, The Hague, 1970, pp. 8–9, 11. *Zur Phänomenologie des inneren Zeitbewussiseins (1893–1917)*, Husserliana, X, pp. 336, 339, 348. *Ideen I,* §88, p. 183; §98, p. 208; §99, p. 209; §133, pp. 274–75; §135, p. 279; §138, pp. 286–87; §150, p. 314.
[18] For the general structure of the continuous synthesis of appearances cf. esp. *LI*, pp. 721, 736, 746, 789–91; *Ding und Raum,* §§19, 26, 28–30; *Ideen I,* §§118, 131.
[19] *LI*, pp. 746, 790, Cf. *Ding und Raum,* §§26, 29–30; *Ideen I,* §§118, 131.
[20] Cf. esp. *LI*, Investig. VI, §§16, 24, 29, 47; *Ding und Raum,* §§29–30, 32f.; *Ideen I,* §§86, 135, 138, 144, 149–50.
[21] *LI*, pp. 713–14, 720; *Ding und Raum,* §32, pp. 107–8; §33, pp. 112, 115; §35, pp. 124–25; *Ideen I,* §143; §149, p. 311.
[22] *LI*, pp. 721, 736; *MS. M III 2 II 2*, p. 47 (1913, revision of Investig. VI).
[23] *MS. M III 2 I 3*, pp. 55f. (1913, revision Investig. VI).
[24] Cf. esp. *MS. A VII 7*, p. 5 (1933); *MS. A VII 13*, pp. 6a (ca. 1925), 20a, 27af., 37af. (1921); *MS. A VII 15*, pp. 8bf. (1932); *MS. A VII 26*, p. 2a (1925); *MS. C 16*, p. 76b (1932); *MS. D 1*, pp. 5b, 9 (1932); *MS. D 3*, p. 11a (1920); *MS. D 13 I*, pp. 47a (1920–1930), 52af., 69af. (1921).
[25] Cf. esp. A. Gurwitsch, *The Field of Consciousness*, Duquesne University Press, Pittsburgh, 1964 (hereafter cited as *Field*) and A. Gurwitsch, *Studies in Phenomenology and Psychology*, Northwestern University Press, Evanston, 1966 (hereafter cited as *Studies*), pp. 332–49.
[26] *Field*, pp. 217–18; *Studies*, p. 348.
[27] *Field*, pp. 184, 216–20, 222–23, 277–78; *Studies*, pp. 347–48, 341, 146.
[28] *Field*, pp. 184, 219, 222, 227; *Studies*, p. 146.
[29] *Ding und Raum,* §§32, 36f.
[30] Cf. *Die Krisis der europäischen Wissenschaften und die Phänomenologie*, Husserliana, VI and *Erste Philosophie (1923/24) I*, Husserliana, VII.

DAVID CARR

INTERPRETATION AND SELF-EVIDENCE

Hermeneutic philosophy, understood as a general theory of human understanding, is associated with the names of Paul Ricoeur and Hans-Georg Gadamer. Ricoeur and Gadamer developed their hermeneutic philosophies in relative independence of one another, and their theories first emerged in rather different forms: Gadamer's in 1960 in *Wahrheit und Methode*, a historically presented conception applied primarily to the practice of the humanistic disciplines; Ricoeur's in 1965 in *De l'interpretation*, an essay on Freud.[1] But both philosophers take their cue from Heidegger in *Being and Time* and this lends their efforts a common provenance and a common accent.

Through Heidegger, and in Ricoeur's case more directly, hermeneutical theory also descends from Husserl. But while Husserl exercises some positive influence, he also provides a foil for hermeneutics. In many respects, Husserl's thought, in its overall structure and basic themes, seems to represent to both Gadamer and Ricoeur the most elaborately articulated version of the philosophical tendency to which hermeneutics is most fundamentally opposed.

In the following essay I would like to examine some of the basic ideas of the hermeneutical philosophers in the light of this opposition to Husserl's classical phenomenological position. I think that some hermeneutical criticisms of Husserlian phenomenology are justified and some are not. But my purpose is not to adjudicate between two schools of thought; nor is it to produce some hybrid between the two. Rather, I think that an examination of hermeneutical ideas from this point of view can bring to light some of their underlying presuppositions which have not yet been seen. In particular, it will be seen that hermeneutics shares with phenomenology, and indeed most modern philosophy, an unexamined basic concept that it has not yet acknowledged.

It is primarily in its capacity as heir to the Cartesian tradition that phenomenology comes under the hermeneutical attack. Though Husserl's criticisms of Descartes are many and very important, according to this view, phenomenology is still at bottom a form of Cartesianism. This characterization of phenomenology centers on two closely related doctrinal emphases: one is a certain elevated, even pretentious conception of the capacity of human knowledge, expressed in the notion of philosophy as rigorous science; the

Tymieniecka (ed.), Analecta Husserliana, Vol IX, 133–147.

other is its focus on subjectivity as the beginning and end point of all knowledge. Against this view, hermeneutics stresses above all the finitude of the human condition and the resultant finitude of human knowledge, even self-knowledge.

With its stress on finitude and its antisubjectivism, hermeneutical philosophy participates in the general anti-Cartesianism of some of the most influential twentieth-century thought. We find the same notions stressed in Wittgenstein's language games as forms of life and in Quine's ontological relativity. Everywhere, it seems, it is thought to be important, as it was for the early figures of eighteenth-century empiricism, to curb the pretensions of human knowledge, at least philosophical knowledge; and it matters little to the tenor of the message whether the source of man's limitations lies in his role as language-user or his embeddedness in a historical tradition. In fact, these two notions at bottom come to the same thing.

Confining ourselves here, however, to the hermeneutical critique of phenomenology, let us seek out those key elements of Husserl's classical position, which, in the eyes of these critics, seem central and vulnerable. I have already suggested that there are two such elements. First, Husserl presents a theory of the essence of human knowledge or understanding whose centerpiece is the theory of *Evidenz* – which I prefer to translate as "self-evidence." The measure and aim of knowledge in its preeminent form – *Wissenschaft* – is grounded judgment, and self-evidence is what grounds judgment in the preeminent sense. In a case of self-evidence "the affair [*die Sache*], the complex (or state) of affairs, instead of being merely meant 'from afar,' is present [*gegenwärtig*] as the affair 'itself,' the affair complex or state-of-affairs 'itself' . . . "[2] Self-evidence, says Husserl, is "an '*experiencing*' of something that is, and is thus; it is precisely a mental seeing of something itself".[3] The object which is intended is also given; and furthermore, it is given just *as* it is intended. To encounter something in this way is furthermore to acquire an "abiding possession" to which one can return again and again.[4]

The second element consists of a doctrine concerning where the self-evidence appropriate to *philosophical* science is to be found. The ideal of self-evidence is never fulfilled, for example, in the external world. There and in other spheres of knowledge it is burdened by a one-sidedness, incompleteness, and uncertainty that makes it a mere presumption, not an actuality. Above all, it is burdened by unclarified presuppositions or prejudices. Only in the sphere of reflective self-awareness are such defects not *in principle* to be found. Only here is the ideal of self-evidence *in principle* fulfillable. In this sphere, furthermore, the structure of all other spheres, and indeed the reason for their defectiveness with respect to self-evidence, can be self-evidently recognized.

Hermeneutic theory opposes each of these two elements in turn. Against Husserl's phenomenological theory of knowledge is ranged a theory of human comprehension derived from Heidegger's notion of *Verstehen*, a term which now takes over the central place occupied by consciousness in Husserl's thought. Understanding is not a human faculty or a particular sort of act but a fundamental dimension or *Existential* of *Dasein*. And *Dasein*'s essential finitude is exemplified in his understanding. It is a mistake to suppose that the one-sidedness, incompleteness, and lack of certainty that is characteristic of our understanding can be overcome. These features, which Husserl correctly saw and identified, are not accidental but quite necessary characteristics of our encounter with the world. Above all, it is a mistake to think that the presuppositional or prejudice-laden character of our comprehension is something to be put aside. It is the very nature of our understanding to be ahead of itself, to approach its objects and its world with a prior structure of comprehension, a prejudice in the literal sense.[5] Husserl even recognized the source of this structure of comprehensive when he insisted on the temporality and historicity of consciousness.[6] Yet he failed, on this view, to draw the proper consequences of what he himself saw. The theory of self-evidence is incompatible with this view of human understanding.

And against the second element of Husserl's classical conception, hermeneutical theory makes the point that self-understanding is *no exception* to the conditions of human comprehension. In fact, according to Heidegger's theory, all understanding is self-understanding, the projection of one's own possibilities. The supposed Cartesian self-coincidence or self-transparency is only self-comprehension reduced and abstracted into an empty and meaningless identity.[7] Genuine self-understanding is possible, but it is no less historical and finite an understanding than any other.

Let us now turn to the theory of the structure of human understanding that lies at the base of this hermeneutical critique and serves as the source of the notion of finitude. One of the first things that the expositors of this theory usually say is that the structure of human understanding is by no means to be construed as a kind of barrier or obstruction which stands between the knower and the known. This is already found in Heidegger's theory of *Verstehen* ("understanding") and the closely related concept of *Auslegung* ("interpretation"). "In interpreting," says Heidegger, "we do not, so to speak, throw a 'signification' over some naked thing which is present-at-hand . . . "[8] The process of interpretive understanding, in which something is understood *as something*, is not a process in which the thing is hidden from us, but precisely the process through which it is revealed.[9] "In the projecting of the understanding, entities are disclosed in their possibility".[10] "When entities within-the-world are discovered along with the Being of *Dasein* – that is, when they have come to be understood – we say that they have *meaning* [*Sinn*]. But that which is understood, taken strictly, is not the meaning but the entity . . ."[11] Thus meaning is not some intermediate domain (*Zwischenreich*)[12] standing between me and the object of my understanding: rather, the latter is accessible to me only as having meaning and, without meaning, would be hidden.

Here Heidegger is essentially repeating a Hegelian point. One recalls the masterful way in which Hegel, in the first paragraph of the introduction to the *Phenomenology*, reduces to absurdity one of the guiding assumptions of modern philosophy since Descartes. If we conceive of the faculty of knowledge as a kind of instrument for grasping the object or as a medium through which it is discerned, doubts arise about the ability of this faculty to do its job. For an instrument transforms what it works upon, and a medium may distort the image of what appears. Once these doubts have been raised, we wish we could somehow dispense with the instrument or medium, or at least subtract from the final product its distorting effects. But, Hegel reminds us, the instrument or medium we are talking about is precisely knowledge, and if we dispense with it or subtract from its effects, we are worse off than before: for now the object is unknown, and rather than a direct, undisturbed grasp of it we are left with no grasp at all. What is wrong according to Hegel is the original metaphor through which we conceive of knowledge: "what calls itself fear of error reveals itself as fear of the truth".[13]

Something like this theme has been taken up by post-Heideggerian expositors of hermeneutical theory, such as Ricoeur. Prior to the notion of finitude, which is a negative concept, he says, is the "entirely positive" notion of the "ontological conditions of comprehension".[14] This condition is best expressed, according to Ricoeur, in the notion of belonging (*appartenance, Zugehörigkeit*) a notion he takes over from Gadamer, who in turn adapted it from Count York von Wartenburg.[15] York was speaking of our knowledge of history in particular, and referring to the fact that we are interested in and can know about history because we ourselves are historical. This, of course, is the point on which Heidegger draws in his own discussion of history, historicity, and historiography. But in the hands of Heidegger, Gadamer, and Ricoeur, the notion of belonging as the basis of knowledge is not limited to the sphere of history. Or rather, historical knowledge becomes the paradigm of all knowledge. A striking reversal takes place: natural-scientific knowledge, which for most modern philosophy has served as the model, is now seen as a mere "subspecies [*Abart*] of understanding".[16]

As Ricoeur points out, the concept of belonging does not refer merely to a relation between subject and object.[17] Rather, it means that subject and object belong to the same sphere, that for the knower the object to be known belongs, in advance, within a horizon of familiarity. It is in virtue of this horizon, which constitutes the *Vorgriff* with which the knower enters the situation and partakes of it, that the object is rendered accessible to him; without this it would be utterly strange, would be strictly speaking meaningless. Because it has meaning for him in advance, he can ask meaningful questions about it in the effort to come to understand it.

This, then, is the positive hermeneutical concept of understanding, of which the negative notion of finitude is only the reverse side. But here we might ask, how do we pass from the obverse to the reverse? Does the conception of human understanding we have just been sketching really imply, in any sense, the idea of finitude normally associated with it by the hermeneutical philosophers? Linking this conception with Hegel, as we just did, causes us to wonder. Indeed, if we examine more closely the idea of belonging, and that of the horizon of familiarity, as sketched above, we find a few more historical precedents which not only make these ideas seem less innovative than they originally appeared, but also raise doubts as to the link between this and the notion of finitude.

Consider, for example, Plato's theory of knowledge. Here it is precisely because man's soul belongs within the same sphere as the forms and is not merely confined to the temporal world, that he can come to know them. If there is finitude or limitation here to man's ability to know true reality, it derives not from his belonging to the world of the forms, but rather from the extent to which he is *not* entirely of that world. The horizon of familiarity, too, has its counterpart in Plato's theory of recollection. Man can come to know the forms fully because he is already somehow familiar with them in advance.

This conception of belonging, entirely without skeptical or finitist consequences, has had its echos throughout the history of philosophy. In Vico it is applied directly to the problem of our knowledge of history. History for Vico is the only thing that we can fully know, and the reason is that we have made it ourselves. Thus, unlike nature, it belongs fully to the human sphere, and its inner workings are accessible to us in a way that nature's are not.

In modern philosophy, of course, the notion of the horizon of familiarity finds its primary counterpart, as Heidegger acknowledges in one passage, in the concept of the *a priori.*[18] Kant even comes close to saying about nature what Vico says about history: that we can know it because we make it. This does not mean, of course, that we create it in the sense of giving it its existence, but we do determine, through *a priori* concepts, what counts as nature, and thus we make sense of it.[19] Again, while human finitude plays an important part in Kant's philosophy, it is not derived from the horizon of familiarity. On the contrary, what is within the horizon is precisely what we can know fully, in the form of mathematics and natural science, whereas it is that which lies totally outside such a horizon which escapes our ability to know and defines the limits of our reason.

These observations suggest that the transition from the notion of belonging, and that of the horizon of familiarity, to the concept of human finitude, is not a necessary one. They make it clear, in fact, that other considerations must be involved, in addition to the ones we have mentioned, if this transition is to be made in the manner of hermeneutical philosophy. It may be that the object – a person, a social event, a text, even a natural event – is accessible to us because it belongs to the same sphere, because it is within the same horizon with us in some sense. But if our grasp is limited, one-sided, incomplete, this is because there is *another* possible sphere, another horizon, to which the object *also* belongs while we do not. The term "horizon" suggests perspective, of course, and we could say there is another possible perspective on the object. Put simply, in the language of hermeneutics, *another interpretation* is possible. Furthermore, in order to make the hermeneutical point significant, it must be said that this alternative interpretation is somehow closed to us. The *Vorgriff* which *we* bring to the situation rules out an alternative framework which may be brought to bear on the object by others or under other circumstances. Thus our situation limits our perspective, rules out our being simultaneously in another situation which is nevertheless considered possible.

The idea of the finitude of human comprehension, then, derives not merely from that of belonging, or of the horizon of familiarity, but rather from the notion of *alternative* – and perhaps mutually exclusive – spheres of belonging or horizons. But more than this. If the notion of finitude is to make any sense, it must be the *same object* which is subject to different interpretations or perspectives. It is our grasp of *this object* that is shown to be limited by the consideration of another possible grasp of the same object. If the alternative perspective in question were totally unrelated to my own, without a common object, then it would make no sense to say that its possibility is what limits my understanding of this object. A common object, then, with a possible variety of different but equally correct perspectives on it, is required if we are to make the transition in question.

What does this, in turn, commit us to? A certain notion of the *object*. As we said before, this object may be another person or his actions, a historical or social event or entity, a text, possibly even a natural object or event. But in any case we are saying that the object is not exhausted by the understanding we have of it. It may be accessible to us in virtue of belonging within the horizon to which we ourselves belong. But it has an identity and a set of possible determinations which transcend this horizon, determinations that may be accessible to others under other circumstances. What these circumstances are, of course, and what it is about the object that is known, we cannot say. But the object in this sense serves, as in Kant's notion of the limiting concept, to curb the pretensions of our own understanding.

As in Kant's case, the notion of object as limiting concept goes hand in hand with a concept of the alternative perspective. In Kant's case, of course, the latter is the *intellectus archetypus*, a mind of intellectual intuition that knows the object as it is in itself, that is, presumably, in all its possible determinations, including those that are known to us. The finite intellect is measured against the infinite intellect. For the hermeneutical philosopher the alternative is more likely to be but another finite perspective, conceived in the same terms as ours – that is, as being likewise historically situated and laden with prejudices of the same sort as ours, but different in content.

But how different is this, really, from the Kantian conception? For the finitist, presumably, no single perspective is able to exhaust the object. But this suggests two possibilities. Either the object is exhaustibly determinable by a certain number of perspectives, which, taken together, would constitute complete knowledge of it; or, alternatively, the object is inexhaustible – i.e., is open to infinitely many interpretations. But it could be said that these infinitely many interpretations are precisely what define the infinite intellect Kant was speaking about. To be sure, the ideal completion of interpretations which would take in the whole object would not be brought under the unity of one intellect, i.e., synthesized in the manner in which we unify the various strands of our own knowledge. But the point is this: the finitist thesis of the hermeneutical philosophers cannot even be formulated without tacit reference to an object which lies in some respects beyond the grasp of the finite intellect, which corresponds in some way to a totality – *Inbegriff* – of interpretations, and against which any given interpretation is measured in order to be declared finite.

There is some resistance to admitting this among the hermeneutical philosophers. Gadamer, for example, writes:

Whereas the object of the natural sciences can probably be defined *idealiter* as what would be known in a completed knowledge of nature, it is senseless to speak of a completed knowledge of history; and precisely for this reason the talk of an object in itself, corresponding to this [historical] inquiry, can in the last analysis not be cashed in.[20]

But Gadamer must see that without such a conception, in order to be consistent, hermeneutical theory would have to give up such notions as "incomplete," "one-sided," "relative," "finite." There would simply be different grasps or interpretations, ranged alongside one another; and with nothing against which to measure them, we might as well call them complete as incomplete.

These considerations allow us to see that hermeneutical philosophy shares with Husserl's phenomenology and indeed most of modern philosophy a fundamental conception of human understanding that lies so deep and is so all-determining that we ought to accord it the status of a basic metaphor or myth. It is more important than the concept of the intervening medium criticized by Hegel, and is in fact shared by Hegel. Sellars has spoken of the myth of the given in empiricism, but this in my view is only one version of a deeper myth we might call "the myth of the measure" or of "the final appeal." The measure may stand at the beginning of experience, as it does for the empiricists and rationalists, as an unmediated datum or clear and distinct perception; or it may stand at the end of a long process of experience, whether individually or historically conceived, as in the case of full-blown Hegelian idealism – what Gadamer calls "total mediation." To mention another, perhaps more important division, that between the gnostics and the skeptics of modern philosophy, the measure may be attainable and actually attained, or it may be unattainable. But running through all these struggles and divisions is the shared presupposition which underlies each position and enables it to make each of the assertions it makes. The hermeneutical position we are discussing, in the context of these struggles, is perhaps the most sophisticated of all. The very thing that renders the object accessible to us – interpretation – is also the thing that keeps it from being totally accessible to us. Mediation is not a barrier but a means to the object, as the term *Vermittlung* implies. Yet, mediation cannot be totalized. But the *it* which is rendered thus imperfectly accessible to us remains as the Archimedian point for the very articulation of this position. To be sure, the dependence of the position and its articulation upon this *it* is for the most part unrecognized or unacknowledged.

But it does come through. In Heidegger it can be found in the key phrase of this theory of *Auslegung*, the phrase *etwas als etwas verstehen*. Attention is understandably directed by Heidegger and his followers to the second *etwas* and, even more important, to the *als*. The "as-structure" of understanding becomes the focus. But the first *etwas* is just as important; without it the whole notion of interpretation would be senseless. It is this *etwas* to which we turn with our interpretation, which we try to disclose and illuminate when we understand it *as* something; it is what we try to bring close to us and render comprehensible. In this sense it is prior to our understanding, it is the pregiven focus of our interpretative endeavor. At the same time, it is this *etwas*, somehow identical, that can be subject to another interpretation and thus rendered accessible in a different way, perhaps to others at another time.

To be sure, as lying somehow both prior to and beyond our interpretation, this *etwas* seems to be without content and perhaps even meaning for us. This is the case if we accept the principle that something is meaningful only *as* something. This raises a paradox which turns out to be a version of Plato's in the *Meno*: if this *etwas* has no meaning for us, how can it be anything at all for us, and thus how can it be something that we turn to and try to understand? Perhaps we can avoid this paradox if we say that nothing ever comes to us uninterpreted in some way. It would be in keeping with Heidegger's analysis, I believe, to say that the object is first passively encountered as already interpreted, with its received sense and significance, and that the individual's understanding consists in his either accepting and appropriating this interpretation or forging a new one. In the latter case, however, the *etwas* remains, in the sense of our discussion earlier, as that which is identical between the new and the old, that which the new interpretation discloses for us in a way that is more appropriate or authentic than the old.

Thus this first *etwas* – to speak with Kant we might call it an *etwas* = *x* – remains as a sort of indispensable placeholder, and it seems as impossible to get rid of it as it is difficult to say anything else positive about it.

What are we to say about the myth of the measure, or the myth of the *etwas* = *x*? We have suggested that it explicitly or implicitly underlies modern philosophy and that for hermeneutical philosophy it is maintained as an unarticulated and unacknowledged principle. If this is true, this gives it the status of an unadmitted presupposition, a *Vorurteil*. Is this all we can say about it? Or is this a prejudice that, once recognized, can be examined with regard to its legitimacy?

The concept of the object we have just been developing was entailed, we claimed, by the simple assertion that, for any given object, another interpretation is possible, besides our own, which is somehow equally valid and which is not collapsible into our own. Perhaps it is this claim which should first be examined. When the hermeneutical philosopher asserts this, how does he know it, since the other interpretation in question is supposedly beyond his reach? Is this claim merely an unjustified dogmatic assertion?

Yet it seems absurd to characterize as an unjustified dogmatic assertion the innocuous claim that, for a given object, "another interpretation is possible." The reason for this is, of course, that this possibility is evinced constantly in our everyday experience. Perhaps there is something like a completely naive and straightforward grasp of an object in our experience, such that no question of alternative interpretations arises, or indeed of interpretation in any sense, i.e., where the object is simply there for us with a univocal and unquestioned sense. But in most cases, the fact that another interpretation is possible is not something we need to be told by philosophers. We recognize it every time the apparent sense of a perceived event, the remark of a friend, or an item in the newspaper becomes questionable and has to be reassessed. Or, such an object can present itself right from the start as ambiguous, carrying its alternative interpretations, so to speak, along with it as it enters our experience. If we encounter such alternatives within the confines of our own experience, we encounter them all the more in social interaction. The meaning of an event is not always agreed upon by the member of even the closest community, and here my actual encounter with alternative interpretations may involve rational discussion, persuasion, complex political maneuvers, or even violence.

Thus, meeting up with other interpretations is in a way one of the most common and pervasive occurrences in our encounter with ourselves, with our world, and with others. There is, of course, a difference between the situation I have just been describing and the thesis of the hermeneutical philosophers. While the "other interpretations" we have just been describing are encountered within our own experience, hermeneutics seems to be speaking of another interpretation which lies somehow beyond my capacities even for encounter. I can only encounter the interpretations of others because I can encounter *them* (the others) and can communicate with or at least understand them. This communication presupposes some common ground, some shared *Vorstruktur* as a condition for recognizing our very disagreement. Still, the hermeneutical thesis can legitimately draw upon this experience of disagreement for support. Just as there *are* other interpretations that I actually encounter, so there *may be* still others that I do not and indeed cannot encounter in virtue of my limited historical situation. For example, the interpretations which future generations will put on the events of the present day are in principle closed to me. And the hermeneutical thesis requires not the existence of different interpretations beyond my experience, but only their *possibility*.

But there is an aspect of the situation we have been describing that is just as important and to which the hermeneutical philosophers seem to pay too little attention. The encounter with another interpretation sets up a tension, a certain problem to be resolved. In a way the matter is simplified if I just reject the other's interpretation and maintain that I am simply right. But if I grant a certain validity to the other's interpretations, I am at the same time acknowledging that my *own* is "just" another interpretation, and this is not a recognition with which I can remain comfortable. In some ways this is like the hermeneutical situation that Gadamer speaks of, where the object in its strangeness calls for understanding, or Ricoeur's point of departure with the double meaning or plurivocity of the text. But with the actual encounter, not just with the object to be interpreted, but with another interpretation of the object, a special sort of problem is created. A certain critical distance has opened up between the object-for-me, to put it in Hegelian terms, and the object-in-itself. The very act of according even a preliminary validity to the other interpretation initiates a kind of comparison between the object and my interpretation of it. This comparison has the purpose of bringing me back into accord with the object, of restoring the straightforward givenness of the object before it was torn away from me, so to speak, by the intrusion of the other. This purpose may be achieved, either by rejecting my interpretation in favor of the other's, rejecting his and reinstating my own, rejecting both in favor of a third, or combining the two into a third.

It is during this process that one's prejudices come in for critical scrutiny. Even the hermeneutical philosophers, starting with Heidegger, admit that there is a difference between naively assumed and unexamined prejudices and critically evaluated presuppositions. In Heidegger we find this expressed as follows: speaking of the "positive possibility of the most primordial kind of knowing," he says:

> To be sure, we genuinely take hold of this possibility when, in our interpretation, we have understood that our first, last, and constant task is never to allow our fore-knowing, foresight, and foreconception to be presented to us by fancies [*Einfälle*] and popular conceptions, but rather to make the scientific theme secure by working out these fore-structures in terms of the things themselves [*aus den Sachen selbst her*].[21]

In this passage, if I understand it correctly, Heidegger seems to envisage just the sort of critical comparison I was speaking of where my interpretation is examined in light of the *Sache selbst*. Gadamer echos this in commenting on Heidegger:

Even for Heidegger, historical knowledge is not planning projection, not the extrapolation of goals of the will, not the arrangement of things according to wishes or prejudices or suggestions of those in power; rather it remains something measured against the things [*Anmessung an die Sache*], *mensuratio ad rem.*[22]

And he speaks in similar terms himself when he writes of distinguishing legitimate from illegitimate prejudices,[23] and even "*true* prejudices, through which we *understand*, from *false* prejudices, through which we *misunderstand*."[24] The problem with Gadamer's theory, and Heidegger's too, is that they tell us precious little about how these distinctions are to be made and how the *Anmessung an die Sache* is to be achieved.

But even putting this crucial problem aside, they also fail to recognize the *point* of all these maneuvers. What is the point? Surely it is not to escape the "mediation via interpretation" altogether. But it is to reach an interpretation of the object that cannot be superceded – the *true* interpretation, if you will, one of which we can no longer say, "another interpretation is possible." In other terms, the point is to reach a conclusion of the process in which there is no longer a gap between the object for me and the object as it is in itself. Or, in still other terms, to reach that point where the object is given just as it is intended.

It is obvious that we have now returned, via the theory of interpretation and some Hegelian terminology, to something like the Husserlian theory of self-evidence. We have not claimed, of course, that such a point of coincidence (*Deckung*) of intention and the given is ever actually reached, or even that it can be reached. We have described it, rather, as a conception or an idea that draws us on in our encounter with the object, as a guiding or regulative principle of the understanding. But this is how Husserl described it too, though he took different positions throughout his life on whether this ideal could ever be achieved in particular cases or in particular domains. But even the description of the ideal has its variations in his writings. Early in his career, he was content simply to state the ideal, to define that endpoint or *telos* of our cognitive endeavors, to tell what it would be like if it ever were reached. But later, when he became more interested in consciousness as a dynamic and progressive process, he stressed not only this ideal endpoint but also the manner in which it functions in the life of consciousness and how it structures the variations and maneuvers through which the understanding is drawn in its endeavor to reach it. This is so to the extent that, in the later work, the "theory of self-evidence" becomes the description of this process itself. Indeed, in the *Crisis*, as is well known, Husserl tries to come to terms with the process of interpretation in precisely the context of tradition and history with which hermeneutical theory is so concerned.

But let us recall that we are inquiring here into the validity of the hermeneutical thesis that "another interpretation is possible." We have pointed out how this thesis is legitimated in part by the actual encounter, within our everyday experience, with other interpretations of the events and actions around us. But we have also suggested that the thesis has further implications that are not spelled out by the hermeneutical philosophers, implications that we tried to articulate as the commitment to the idea of the object and of the full determination of the object. Can this commitment be similarly justified or validated by an appeal to experience?

In one sense, of course, it cannot. While we do encounter other interpretations in our experience, we do not, perhaps, ever encounter the fully determined object as long as we are engaged in the process of interpretation itself. And, if the hermeneutical thesis is correct, we never come to an end of this process in our experience, i.e., to that place where all other interpretations would be *a priori* excluded.

In another sense, however, we do encounter the fully determined object – or rather the *idea* of the fully determined object – *in* the very process of understanding itself. But we encounter it precisely as the regulative principle which drives our inquiry on its way. The fully determined object is not, strictly speaking, something we encounter *in* experience, but is rather a structural feature *of* our experience itself which gives sense and direction to what we do. The legitimation derives, then, not merely from engaging in the process of interpretation, but from reflecting on its structure. And this means that such legitimation is not empirical but eidetic and transcendental in precisely Husserl's sense. It is eidetic because it is the recognition of a structure common to all interpretive processes; and it is transcendental in that what is recognized – the presence of the fully determined object as regulative principle – is a condition of the possibility of any interpretive process.

What I have been saying amounts to the claim that something like the Husserlian notion of self-evidence is after all – though perhaps not admittedly – involved in hermeneutical theory. But in one sense I would claim that Husserl is more sophisticated. For while hermeneutical theory seems tacitly committed in a straightforward way to a kind of in-itself lying beyond the grasp of our interpretations, Husserl sees this in-itself as an ideal and transcendental structure. Operating as he does within the *epoché*, Husserl does not commit himself ontologically on this in-itself beyond all understanding. Does this make his theory more idealistic? Or simply less dogmatic?

I would like to turn now to the second phase of the hermeneutical critique of phenomenology, which I think can be discussed briefly in the light of the foregoing response to the first phase. This second critique is directed against Husserl's notion that, given the *theory* of self-evidence, reflective self-awareness, our apprehension of our own subjectivity is the only place where genuine self-evidence is to be found. While my remarks on the first phase of the hermeneutical critique have been critical of the hermeneutical approach, in this case I am more in agreement with it. There is some justification for saying that Husserl moved away from this Cartesian position in his later works, as many commentators have claimed. But the real problem is that Husserl tended to confuse, even in his latest works, the essentially Cartesian motive of finding unassailable self-evidences in self-apprehension, with the essentially Kantian motive of finding the transcendental structure which makes experience possible. The true accomplishment of the phenomenological *epoché* is not the turn inward, but, as Ricoeur says, the turn to meaning. What is genuinely discovered is not the solitary ego but intentionality in its various forms and functions. And to take intentionality seriously is not merely to describe consciousness but also to describe what it is *of* and how this *of*ness is achieved. Thus phenomenology must be as much about the world as about consciousness. Further, in the transcendental turn, there is no reason to suppose that the self-evidences achieved are any more infallible or unassailable than any other sort.

University of Ottawa

NOTES

[1] Hans Georg Gadamer, *Wahrheit und Methode*, 2nd edition, J. C. B. Mohr, Tübingen, 1965; Paul Ricoeur, *De l'interprétation*, Editions du Seuil, Paris, 1965. The present essay is an expanded and revised version of a contribution to a symposium on "Phenomenology and Hermeneutics" held at the Western Division Meeting of the American Philosophical Association in Chicago, April, 1975.
[2] Edmund Husserl, *Cartesian Meditations* (transl. Dorian Cairns), Martinus Nijhoff, The Hague, 1960, p. 10.
[3] *Ibid.*, p. 12.
[4] *Ibid.*, p. 10.
[5] Martin Heidegger, *Being and Time* (transl. John Macquarrie and Edward Robinson), Harper and Row, New York, 1962, p. H. 150. (Pages preceded by "H." refer to the

original German pagination in the margins.)
[6] *Wahrheit und Methode*, p. 500.
[7] *Being and Time*, p. H. 144.
[8] *Ibid.*, p. H. 150.
[9] *Ibid.*, p. H. 149.
[10] *Ibid.*, p. H. 151.
[11] *Ibid.*
[12] *Ibid.*
[13] G. W. F. Hegel, *Phänomenologie des Geistes* (ed. J. Hoffmeister), Felix Meiner, Hamburg, 1952, p. 65.
[14] Raul Ricoeur, "Phenomenology and Hermeneutics," *Nous*, **9**, March, 1975, p. 88.
[15] *Wahrheit und Methode*, p. 247 and note.
[16] *Ibid.*, p. 245. Cf. *Being and Time*, p. H. 153.
[17] "Phenomenology and Hermeneutics," p. 88.
[18] *Being and Time*, p. H. 150.
[19] Cf. *The Critique of Pure Reason*, A 92, B 125.
[20] *Wahrheit und Methode*, p. 269.
[21] *Being and Time*, p. H. 153.
[22] *Wahrheit und Methode*, p. 247.
[23] *Ibid.*, p. 261.
[24] *Ibid.*, p. 282.

BERNARD P. DAUENHAUER

THE TELEOLOGY OF CONSCIOUSNESS: HUSSERL AND MERLEAU-PONTY

In his *Phenomenology of Perception*, Merleau-Ponty details a number of substantive disagreements which he has with Husserl. These disagreements can be taken as descriptive of the difference between what can be called existential phenomenology on the one hand and eidetic phenomenology on the other. In this paper I wish to focus on one of these disagreements which I take to be of capital importance, namely, the difference between the accounts of the teleology of consciousness found in *Phenomenology of Perception* and in Husserl's *Formal and Transcendental Logic.* [1]

First, I will clarify the difference between these two notions of the teleology of consciousness. Then I will argue that Husserl's and Merleau-Ponty's notions of the teleology of consciousness are intrinsically bound up with their respective claims concerning the character of intentional performances. Specifically, for Husserl, there are three distinct sorts of intentional performances, namely, intuitive, pictorial, and signitive performances. For Merleau-Ponty, however, all consciousness is in some measure perceptual and all intentional performances are signitive. Third, I will assess these two positions in the light of the phenomena they claim to describe. I will argue that these positions are not mutually exclusive but that neither, if taken alone, is sufficient to describe fully the entire range of phenomena. I will suggest, but not argue, that this fact both points to a need for an ontology to complete the phenomenological program and provides some minimal conditions which an acceptable ontology must satisfy.

I. HUSSERL'S AND MERLEAU-PONTY'S NOTIONS OF THE TELEOLOGY OF CONSCIOUSNESS

A. Husserl's notion

For Husserl, intentional performances are constitutive or "sense-bestowing" performances. The teleological function of intentionality is nothing other than this bestowal of sense (*FTL*, 245). To uncover the full history of a sense it is necessary to trace its development both to its beginnings and to its

Tymieniecka (ed.), Analecta Husserliana, Vol. IX, 149–168.

culmination in eidetic intuition. This entire history of the development of a sense is, for Husserl, teleological.

On the basis of *Formal and Transcendental Logic*, there are three aspects of this teleological history which can be distinguished. First, there is the orientation to the *goal* of intentional performances. From this standpoint, an analysis of the stream of constitutive performances reveals the following. On the one hand, the passive performances of consciousness are oriented to the constitution of abiding objects as grasped pre-predicatively. The pre-predicative grasping of these objects, in turn, is oriented to the active performances which constitute the sense of objects, i.e., objectivities of the understanding, in judgment taken in the broad sense. On the other hand, constitutional analysis reveals that the active performances which straightforwardly constitute categorical objectivities themselves point back both to an essential form of the objectivity in question and to an essential mode of conscious activity essentially correlated to this sort of objectivity. In short, consciousness in all of its intentional performances, both passive and active, is oriented toward eidetic necessities.

But constitutional analysis reveals even more concerning the goal of intentional performances. Every single pulse of psychic life is teleologically connected in a synthetic unity. Husserl writes:

> The "Objectivating" performance – to which all single intentional processes at a multiplicity of levels and in relation to various objects (which nevertheless combine sensefully to make up "worlds") are subservient – makes it ultimately necessary to have the whole universality of psychic life in view, as correlated with the ontic universality (the universality of the intrinsically All of the objects). This teleological structure of intentional life, as a universally Objectivating life, is indicated by the fact that object and judgment (in the widest sense) belong together, and by the universality with which we can freely submit any already-given object to our categorical actions. (*FTL*, 262–63)

Thus every particular intentional performance is teleologically structured in such a way that, by taking it as a clue, one can bring to light both the *a priori* unitary structure of the ego-pole which governs all intentional performances of whatever sort and the ontic *a priori* of the object-pole, or world, according to which objects of any sort can be constituted. Likewise, the essential correlation between ego-pole and object-pole is revealed. But the object-pole, as the world or the totality of all actual or possible objects in general, is itself constituted: "Accordingly, this constituted world is preceded by my ego, as the ultimately constitutive subjectivity" (*FTL*, 251).

Therefore, by virtue of the teleology of consciousness, the Ego as ultimate constitutive subjectivity is seen to exist for itself with apodictic necessity. By contrast, the world as the constituted totality of all actual and possible objects is seen to be dependent upon the constitutional style of the Ego (*FTL*, 251–52). The world, then, as totality of objects in general, is founded upon the Ego. This does not, of course, render the world contingent. Rather, it belongs to the very teleology of consciousness that it constitutes not merely objects within the world, but the world itself. This is the goal of consciousness, the *terminus ad quem* of its entire stream of intentional performances.

But *Formal and Transcendental Logic* teaches us that it is not enough only to discover the *terminus ad quem* of the history of sense. One must also disclose the beginnings, the *terminus a quo*, of the genesis of sense. Active genesis of sense presupposes the passive genesis of sense in pre-predicative experience. And this passive genesis itself has a history. Pre-predicative experience or encounter (*Erfahrung*)[2] is "the performance in which for me, the experiencer, experienced being 'is there,' and is there as what it is . . ." (*FTL*, 233). This encounter as an intentional performance is itself constitutive. But it constitutes the objects as transcendent objects. Though the object draws the ontic sense peculiar to it from the mental processes of encounter, this ontic sense of transcendence belongs to the proper essence of the encounter itself. Thus encounter is "the primal instituting of the being-for-us of objects as having their objective sense" (*FTL*, 164). Though the transcendent object is indeed constituted, nonetheless

> the object claims an existence extending beyond the multiplicities of actual present experience. . . . The object is evidently not itself the actual and openly possible experiential processes constituting it; nor is it the evident possibility, connected with this process, the possibility, namely, of repetitive synthesis (as a possibility pertaining to 'I can') (*FTL*, 165).

To reduce the object to such processes is to fall into psychologism. Though it is within the encounter that the object is revealed, the object is there revealed as an encountered transcendent object.

But what is involved in encountering a transcendent object? I take it that we may consider the Introduction in *Experience and Judgment* to be a faithful account of Husserl's position during the period I am here concerned with. If that is so, then we find that even the pre-predicative encounter with individual objects is not, in all respects, the ultimate starting point for the teleological process of consciousness in the genesis of sense.

Individual objects encountered pre-predicatively are pregiven in a passive *doxa* which is prior to and presupposed by all judicative activity or cognition properly so called. These individual objects are simply presumed in the certainty of a belief. The "preliminary presence" of these pregiven "presumed" objects stimulates and sets cognitive activity in motion: "Before every movement of cognition the object of cognition is already present as a *dynamis* which is to turn into an *entelecheia*."[3] These objects are said to affect us. That is, they stand out from an always copresent environment and thus attract our interest to them. This environment, the domain of what is always already there without awakening any specific attention on our part, is presupposed for any object to affect us. This domain, as that field which is correlative to a unity of passive *doxa*, can appropriately be spoken of as an actual world. This world is the universal ground prior to and presupposed by cognitive activity.

> It is this universal ground of belief in a world which all praxis presupposes, not only the praxis of life but also the theoretical praxis of cognition. . . . Consciousness of the world is consciousness in the mode of certainty of belief; it is not acquired by a specific act which breaks into the continuity of life as an act which posits being or grasps the existent or even the act of judgment which predicates existence. All these acts already presuppose consciousness of the world in the certainty of belief. (*EJ*, 30)

Thus at bottom all conscious performances take their point of departure from a universal passive belief. And the correlate of this belief is the world as universal ground or environment whence particular objects can affect consciousness. This *Urdoxa* with its correlative world is of course uncovered only through genetic analysis. It could not be noticed except through the encounter of consciousness with particular objects. But when it is uncovered, it is uncovered as that which is presupposed for such an encounter.

In this regressive analysis back to beginnings of the teleology of the constitutive performances of consciousness, consciousness in the mode of certainty of belief is, as far as I can see, strictly correlative to world as universal ground presupposed by all praxis. There is, at the level of encounter, no basis for assigning a priority to consciousness comparable to that assigned to the Ego as constituting subjectivity over the world as the constituted totality of all actual and possible objects. But at the same time it is countersensical to consider this world as universal ground to be prior to the *Urdoxa* of consciousness. The primal encounter of consciousness with this universal ground as transcendent is always and necessarily an encounter. And every encounter is an intentional performance.

In point of fact, Husserl argues that the subjectivity encountering the world in this primary *Urdoxa* is itself constituted by what he calls the "livingly flowing present." But for present purposes I need do no more than mention this fact.[4] It is enough to emphasize that, for Husserl, the entire history of the development of a sense is teleological. When fully analyzed, a sense is seen to originate in what can be called an incorrigible encounter with the world understood as ground of passive belief in being and to culminate in the indubitable grasping of the world as totality of objects of experience in general.[5]

But there is a third aspect of Husserl's notion of the teleology of consciousness which deserves notice in this context. This aspect concerns the manner in which a specific sense is progressively constituted. If we consider the manner in which a sense, or any "segment" of a sense, of any sort is developed, we find that consciousness moves in an alternating pattern. That is, a sense is built up in an alternation between encounter and an abiding grasping, both taken in a maximally broad sense. Using terminology usually associated with the predicative, I would put it this way: "S is p" expresses the fact that consciousness, while abidingly grasping the already constituted sense of S, now encounters S as likewise p. This encounter leads to the constitution of the abiding sense Sp, which then serves as the basis for a new encounter expressible as Sp is q. And so forth. This alternating movement between encounter and abiding grasping takes place at all levels of both passive and active genesis of sense. This pattern of alternation in the development of sense is not a contingent feature of consciousness. Rather, it belongs of teleological necessity to consciousness.[6] It can be discerned at play not only in the constitution of the world and objects in the world, but also in the self-constitution of the Ego. It is at play in all constitutive performances of any sort and at any level.

In summary, then, by focusing on the goal, the beginnings, and the alternating pattern of progression of constitutive performances, one can see the scope and meaning of Husserl's concept of the teleology of consciousness. The teleologically governed constitutive performances of consciousness take their point of departure from a consciousness of the world in the mode of certainty of belief, proceed to elaborate senses according to a rule-governed pattern of alternation, and tend to culminate in the intuition of the ultimate subjectivity constituting in eidetic necessity the world as rule-governed "totality of objects of experience in general" (*EJ*, 39).

B. *Merleau-Ponty's notion*

Though the place of the notion of the teleology of consciousness in *Phenomenology of Perception* is not as obvious as it is in *Formal and Transcendental Logic*, it is nonetheless of capital importance there. For Merleau-Ponty, the teleology of consciousness is the intrinsic orientation of consciousness toward the world and truths concerning the world. But this world not only contains ontic contingency within it. It is itself radically, ontologically contingent. The ontological contingency of the world is what "forms the basis once and for all of our ideas of truth. The world is that reality of which the necessary and the possible are merely provinces" (*PP*, 398). Truth is not absolute knowledge. Correspondingly, there is no absolute falsity. Thus the teleology of consciousness consists in that orientation to progress from our first possession of true ideas to more perfect ideas. And this forging of ever more perfect ideas goes on, in principle, endlessly (*PP*, 396).

This understanding of consciousness and its teleological function is involved in Merleau-Ponty's claim that all consciousness is, in some measure, perceptual consciousness (*PP*, 395). Several other theses of Merleau-Ponty belong together with this claim: (1) As Husserl has shown, the exploration of perceptual objects is in principle incompleteable (*FTL*, 246). (2) All thematic intentionality, whether pre-predicative or predicative, is made possible by an underlying nonthematic, living, operative intentionality which predelineates my thematic performances (*PP*, 418; *FTL*, 235). (3) The world as the temporal synthesis of horizons has a logic or permanent style which is not constituted but which is pregiven as an individual being, but which is nonetheless an *omnitudo realitatis* to be deciphered. The possibility of deciphering it is definitively established by our first sensory experience. It is to this logic or style of the world that "my body in its entirety conforms, and through which things of intersensory significance become possible for us" (*PP*, 326).

In light of these theses the place of the concept of the teleology of consciousness in *Phenomenology of Perception* can be brought to light. As in part I A, I will consider this notion under the aspects of the goal, the beginnings, and the manner of progression of the teleological function. Admittedly, the distinction between goal and beginnings, when Merleau-Ponty's notion of the teleology of consciousness is under discussion, is somewhat artificial. But it does facilitate the explication of his position.

The fundamental goal of consciousness as *ek-static*, or of subjectivity's basic project, is to inhabit the world. This world has a logic of its own which all vision assumes. Empirical perceptions endow this world project with specific form, but they do not give rise to it (*PP*, 405). All sense-bestowal, then, is centripetal as well as centrifugal. In reflection upon perception, structures are discerned. But these structures are derivative from, rather than regulative of, the living significance which arises in the subject's dwelling in the world.[7] Consciousness is indeed teleologically oriented to achieve abiding acquisitions; but these acquisitions are nonetheless temporal. Thus there is no radical distinction between the ideal and the factual. As Merleau-Ponty puts it, "there is not one rational truth which does not retain its coefficient of facticity. . . . Thus every factual truth is a rational truth and vice versa" (*PP*, 394).

But there is nonetheless a primacy of the factical over the eidetic. As Merleau-Ponty makes clear in the preface to *Phenomenology of Perception*, essences are not the object of philosophy, but rather are the means whereby we become acquainted with and prevail over our facticity (*PP*, xiv–xv). The world, then, and the I, as unitary fields of experience, cannot, in principle, be brought to intuitive presence. Nor can there be an intuition of the eidetic necessities governing the correlation between these unities. In intentional performances the unity of both I and world is invoked, rather than grasped in direct experience (*PP*, 406).

These unities, world and I, are polarized in the direction of one another. "The world is inseparable from the subject, but from a subject which is nothing but a project of the world, and the subject is inseparable from the world, but from a world which it projects itself" (*PP*, 430). Nonetheless, the world is the cradle of meaning, the native ground of rationality. We can be present to ourselves only because we are present to the world. Consciousness, then, in its personal existence, in its active performances, in reflection, rests on the foundation of an acquired and stabilized, prepersonal existence. In this sense then, the goal of intentional performances is, on the one hand, to reveal the world as that transcendent, open field within which significance can arise, but whose unity can only be invoked rather than experienced. On the other hand, these performances reveal a subjectivity without which there would be no world but nonetheless one which at bottom has an acquired existence and whose unity likewise is no more than invoked. In brief, then, what the teleology of consciousness leads intentional performances to disclose is (1) that subjectivity is both dependent and yet irremovable, and (2) that both world and subjectivity, rather than being rule-governed structures, are open, correlative fields of experience. There is a "logic" to these fields, but that "logic" does not rule; rather, it enables.

Shifting now to that aspect of the teleology of consciousness which bears upon the beginnings of sense-bestowals, we find that, for Merleau-Ponty as for Husserl, there is a primary belief, an *Urdoxa*, in a world which, as antepredicative, is as yet not differentiated into real objects and momentary phantasms. This primary belief in a world underlies all positional acts including perception properly so called. Perception and perceptual consciousness do not yield apodictic certainty because they are never necessary and because "perception entails a process of making explicit which could be pursued to infinity and which ... *could not gain in one direction without losing in another, and without being exposed to the risks of time*" (*PP*, 343; italics mine). Perceptions correct one another against the background of this world of primary belief. But perception itself is opaque. It brings into play my sensory fields "which are my primitive alliance with the world ..." (*PP*, 424). Ultimate consciousness is nothing other than consciousness of the present, that is, consciousness of existing toward and belonging to the antepredicative world. This is simply the experience of presence in the world.

From another standpoint, this primary belief in an antepredicative world involves a recognition of what Merleau-Ponty calls a primary past which is beyond thematization and on the basis of which any thematic present becomes possible (*PP*, 423–24). This primary past is not itself a moment in empirical time. Rather, it is that fundamental aspect of the subject-world correlation which reveals the all-encompassing temporality of the world. It is the horizon on the basis of which the mutual presence of self and world is recognized, and on the basis of which past, present, and future can be distinguished.

Nonetheless, this primary past is the horizon of the present in which we are centered: "Time exists for me only because I am situated in it, that is, because I become aware of myself as already committed to it ..." (*PP*, 423). Thus it is the present whence our view of time emanates, but this present alludes to both the infinitely open future and the primary past.

The passage of time, then, is not something to which I am submitted. It is something I perform. Subjectivity is itself temporality. It is not infratemporal.

> "In" my present, if I grasp it while it is still living and with all that it implies, there is an *ek-stase* towards the future and towards the past which reveals the dimensions of time not as conflicting, but as inseparable. ... Subjectivity is not in time because it takes up or lives through time, and merges with the cohesion of life. (*PP*, 422)

Nonetheless, this subjectivity is not the absolute source to which all else is relative. I do not create time. "I am not the initiator of the process of temporalization; I did not choose to come into the world, yet once I am born, time flows through me, whatever I do" (*PP*, 427). Temporality both prevents my absolute coincidence with myself and yet is the means of establishing my own reality as myself (*PP*, 427).

In effect, then, a consideration of the teleology of consciousness reveals a twofold horizon. On the one hand, there are the world and subjectivity understood as fields whose unity is invoked in intentional performances. On the other hand, there is the primary past whence subjectivity in its present temporalizing arises. Within this twofold horizon, a third aspect of the teleology of consciousness can be detected. This aspect is that which deals with the manner in which a particular sense is progressively achieved.

As mentioned above, for Merleau-Ponty all consciousness is in some measure perceptual consciousness. Further, all consciousness is active transcendence. Thus all the acts of the *I* outstrip themselves and leave no residue of an interiority of consciousness. Taking vision as exemplary, we find that vision is something I actually effect.

> I reassure myself that I see by seeing this or that, or at least by bringing to life around me a visual surrounding, a visible world that is ultimately vouched for only by the sight of a particular thing. Vision is an action . . . an operation which fulfills more than it promises, which constantly outruns its premises and is inwardly prepared only by my primordial opening upon a field of transcendence, that is, once again, by an *ek-stase*. Sight is achieved and fulfills itself in the thing seen . . . but it is nonetheless of its essence to take a hold upon itself in a kind of ambiguous and obscure way, since it is not in possession of itself and indeed escapes from itself into the thing seen. . . . Thus the perceptual "synthesis" has to be incomplete; it cannot present me with a "reality" otherwise than by running the risk of error. (*PP*, 376–77)

To the extent that all consciousness is perceptual, what holds for vision holds for all intentional performances. What is packed into this passage is at least the following set of claims: (1) In perceiving an object, whatever consciousness contributes, the effective perception always contains more than consciousness could contribute. (2) But there is no perception without the active performances of consciousness. (3) In perceiving, consciousness does not come into self-possession but rather distances itself into the object seen. (4) Yet the object perceived cannot be perceived without running the risk of error, which error can only be rectified by a further perception.

These claims can be reformulated by saying that all perception, on the part of both the perceiver and the perceived, involves both presence of significance and absence of significance. Neither the perceived nor the perceiver is empty of significance, and yet both are polarized in the direction of a significance which they either necessarily lack or necessarily await. It is not the case that either of them awaits some well-defined, predelineated significance. Rather, as fields, they await the emergence of significances. These emerging significances always point to the fields whence they both arise and maintain or change their validity. All significance, then, is temporal and refers to a correlative, and likewise temporal, absence of significance.

On such a view, there can be no essences, in Husserl's sense, holding either for conscious performances or for objects. Consciousness, then, is teleologically oriented always to outstrip its own performances, under the motivation of the very presence-of-significance – absence-of-significance structure of its own prior performances and under the guidance of the "logic" of the world which it inhabits. For Merleau-Ponty, then, in contrast to Husserl, the manner of progression of a stream of intentional performances is not one of alternative abiding grasping and encounter. Rather, it is a steady stream of gearing into the world one always already inhabits. For Merleau-Ponty, the gears always slip somewhat, whatever the level or type of performance being considered.

II. THE CORRELATION BETWEEN THESE TWO NOTIONS OF THE TELEOLOGY OF CONSCIOUSNESS AND THE CORRESPONDING CLAIMS CONCERNING THE CHARACTER OF INTENTIONAL PERFORMANCES

In preparation for an assessment of Husserl's and Merleau-Ponty's respective notions of the teleology of consciousness I want to show now how each of these notions is correlated to important claims about the character of intentional performances. Husserl's notion is intrinsically correlated to his distinction between intuitive, pictorial, and signitive performances. By contrast, Marleau-Ponty's notion is intrinsically correlated to his claim that all intentional performances are at bottom signitive.

If Husserl's account of the teleology of consciousness is to be consistent, there must be some intentional performances which are not signitive, but rather are intuitive. That is, there must be some performances in which the object is, strictly speaking, directly and immediately present to consciousness and not merely intended in its absence through some sign, anticipation, or representation. This is obviously the case if the *terminus ad quem* of all intentional performances is the achievement of universal rigorous science[8] or, in other terms, is the constitution of the structurally complex transcendental Ego with its correlative world understood as the totality of objects of experience in general. For unless the *a priori* structural necessities which make all experience possible can be brought to intuitive presence, then there cannot be a science in Husserl's sense of the term.

Similarly, the distinction between intuitive and signitive performances is required, for Husserl, if there are to be signs. In experiencing signs as signs one intends both a presently intuited object, e.g., a stone slab, and a presently absent object, e.g., a historical person or event. The ways in which one intends these two objects, the present stone and the absent person or event, must, according to Husserl, be kept distinct if one is to make sense of what is involved in experiencing signs. Further, it should be noted that in the experience of a sign the signitive intention, here the intending of the historical person or event, presupposes the intuitive intention, the grasping of the stone slab. Thus, for Husserl, without intuitive intentions signitive intentions are not only impossible but are countersensical.

A second and, for present purposes, more instructive consideration shows that even the genesis of a particular sense, if it is to proceed by way of the Husserlian pattern of alternation between abiding grasping and encounter, requires that there be both intuitive and signitive performances. That is, among the multiplicity of performances required to constitute any particular object either pre-predicatively or predicatively there must be some performances which are intuitive and some which are signitive.

Two examples will make this clear. First, consider the constitution of the sense of a particular physical object, a rock for example. In principle, the exploration of the rock can go on *ad infinitum*. But in the process of exploration, abiding acquisitions are achieved, are brought to intuitive presence. The rock is extended, divisible, subject to causal influences, etc. All subsequent explorations, e.g., concerning its chemical properties, its relations to other physical objects, its suitability for conversion into cultural objects, etc., presupposes these abiding, intuitive achievements. In short, though the rock can be signitively intended in an indefinite number of ways, all these signitive intentions presuppose an intuitive holding of the rock as selfsame object. Further, all these signitive intentions themselves are teleologically oriented to give rise to intuitive performances.

Second, consider a particular subset of odd prime numbers. Again, in principle the exploration of this subset can go on *ad infinitum*. But again, abiding acquisitions are achieved. Every member of this subset is divisible only by itself and 1, is either of the forms $2k + 1$ or $2k + 3$, etc. All subsequent elaborations presuppose these abiding, intuitive achievements. And all subsequent exploratory, signitive acts both presuppose abiding achievements and are themselves teleologically oriented to give rise to intuitive performances.

In summary, for Husserl, if there is to be the possibility of progressive exploration and multiple encounter and the possibility of unfulfilled anticipation, one must indeed acknowledge that there are signitive intentions. But if there is to be something determinate to be progressively explored and repeatedly encountered and concerning which one has anticipations, there must also be intuitive intentions.

By contrast, Merleau-Ponty's understanding of the teleology of consciousness as the intrinsic orientation to a world which is both radically contingent and yet which has a "logic" of its own precludes any fundamental distinction between intuitive, pictorial, and signitive intentional performances. Both the world and subjectivity have their respective "styles" in accord with which subjectivity lives out its projects. But these styles are not manifestations of *a priori* necessities. Rather, they are pregiven as actual fields always permeated by facticity.

Whatever is constituted within these fields always retains this coefficient of facticity. Explorations of these fields can indeed go on indefinitely. But unlike Husserl, Merleau-Ponty maintains that all acquisitions achieved *en route* remain historical and thus are in principle open to being superceded. The teleology of consciousness, therefore, is the ordination of consciousness to the attainment of acquisitions which contain within themselves the more or less explicit call for supercession.

If, as Merleau-Ponty claims, the constitution of sense is both genetic and centripetal as well as centrifugal, then all intentional performances must in the last analysis be signitive. Each performance must point both to the presence of what is thematically intended and to the absence of that which is correlative to what is intended but is not itself thematically intended. That is, each intentional performance, whatever its object and however it intends that object, always at the same time invokes both the world and the I as unitary fields of experience which themselves cannot be fully thematized. As a consequence, for Merleau-Ponty, one must admit that all intentional performances are radically signitive in that they always allude to that which can never be brought to intuitive presence, namely, the world and subjectivity as indefinitely open fields of experience.

Likewise, and perhaps more instructively, if the way in which any particular sense is progressively constituted does not proceed by the way of alternation between abiding grasping and encounter, but rather by way of a process which cannot gain in one direction without losing in another (*PP*, 343), then all intentional performances must be signitive. This is so because each sense has a history whose successive phases involve both gains and losses. The sense at any particular moment is what it is only in likewise signifying the presently absent gains and losses of both its past and future moments.

The contrast between Husserl's and Merleau-Ponty's notions of the teleology of consciousness, then, can be summarized in this way. For Husserl, in principle there is a well-defined *terminus ad quem* to the stream of intentional performances. The stream of intentional performances is teleologically oriented to bring to intuitive presence the transcendental Ego as that rule-governed conscious life which constitutes, according to *a priori* necessities, the world as the rule-governed totality of all actual and possible objects of experience. Within this life, at least in principle, specific essences are brought to intuitive presence precisely as eidetic necessities. This universal *terminus ad quem* stands in polar opposition to the *terminus a quo* of all conscious performances, namely, the passive belief in the world as universal ground or environment within which objects can affect consciousness.

For Merleau-Ponty, on the other hand, there is in principle no well-defined *terminus ad quem* to the stream of intentional performances. Rather, the teleology of consciousness consists in the orientation of the stream of performances to inhabit the world. This world indeed has a "logic" or style. But it is also radically contingent and historical. Though consciousness is irremovable from the world, it is nonetheless dependent upon the world as that in which it arises and lives as temporality. All conscious performances, then, retain their reference back to that "primary past" which lies at the basis of all discriminations between past, present, and future. Thus there can be, strictly speaking, no well-defined polar opposition for the stream of intentional performances between a *terminus a quo* and a *terminus ad quem*.

Bound up with these different views of the goal of the entire stream of intentional performances is the contrast in the claims of Husserl and Merleau-Ponty concerning that which particular performances can achieve. For Husserl, not only at the *terminus ad quem* but also *en route* there are performances which are so full, which so intuit their object, that they require nothing beyond themselves for their completion. Though not all performances are of this sort, those which are not – those which are either signitive or pictorial – are intrinsically oriented to lead to intuitive performances. But for Merleau-Ponty it is impossible for any performance to achieve the self-sufficiency implied in Husserl's concept of an intuitive performance. Thus all intentional performances are signitive.

For Husserl, then, consciousness is teleologically oriented to the achievement of the intuitive presence both of particular objects and of the world as the totality of objects of experience. In principle this goal can be realized, though of course this constitution of the world is an infinite task. But for Merleau-Ponty consciousness cannot be oriented to such a goal, for such a goal is in principle impossible. Consciousness, however, is not absurd. It is not oriented to that which is in principle impossible. Thus each of its performances, since it in principle calls for supercession, is signitive.

III. AN ASSESSMENT OF THESE TWO NOTIONS OF THE TELEOLOGY OF CONSCIOUSNESS

The ultimate test for these two notions of the teleology of consciousness is their fidelity to the full range of phenomena which are to be described. I will argue here that both the claims of Husserl and those of Merleau-Ponty, from different perspectives, have merit. Though I doubt that there is any phenomenon which can be satisfactorily described exclusively in the terms of either Husserl's or Merleau-Ponty's characterizations of intentional performances, I will not here argue for so broad a claim. It is sufficient for me to claim that there is at least one set of phenomena, namely, that composed of expression and silence, which cannot be satisfactorily described exclusively by either of their terms, but rather that the full description of this set of phenomena requires the use of both of their characterizations. My thesis, then, is limited to the claim that these two positions are not in principle mutually exclusive, but neither is one reducible to the other without distortion of the description of at least some phenomena. This thesis is nontrivial in that it throws into sharp relief a need within the programs of both eiditic and existential phenomenology for an ontology which can accommodate the requirements arising from the experientially supported but different notions of the teleology of consciousness proposed by Husserl and Merleau-Ponty.

In defense of this thesis, I want first to point out that Husserl and Merleau-Ponty mean different things when they characterize intentional performances as signitive. For Husserl, as intentional performance is signitive if the object of the performance is somehow absent but is meant *through* some other object, word, sign, etc. That is, what makes a performance signitive is the character of the connection between a particular performance and its specifically thematized object, namely, the relation of absence between consciousness and its object. For Merleau-Ponty, by contrast, every intentional performance is signitive because each performance both is called for by a previous performance and calls for a subsequent performance. That is, what makes a performance signitive is that it is intrinsically a member of a stream of performances, which stream has no ideal terminus.

In a similar vein, it is noteworthy that though Husserl and Merleau-Ponty both speak of active intentional performances, they do not mean the same thing by the term "active performance." Husserl distinguishes between passive or spontaneous performances and active performances. Active performances stand in correlative opposition to passive performances. They are those performances which I freely perform when I submit pregiven objects of any sort to the requirements of my praxis, whether this praxis be the praxis of life or the theoretical praxis of cognition. Thus all active performances are founded upon passive or spontaneous performances. Some of these active performances are signitive. But in principle no passive performance can be signitive.

For Merleau-Ponty, on the other hand, all conscious performances are responses to the environing world which is always already there with its own "logic" or style and which solicits our responses. But since both the world which calls and the I which answers are open fields of experience rather than *a priori* rule-governed structures, every conscious performance is contingent and not necessitated. Since no performance is necessitated, none are strictly spontaneous or passive. Thus, for Merleau-Ponty, all performances are active.

Given these clarifications, one finds that the characterizations of intentional performances presented by Husserl and Merleau-Ponty are not mutually exclusive. Since they are not, then the respective notions of the teleology of consiousness with which these characterizations are intrinsically correlated are not mutually exclusive either. But the question which remains is: Are both of these characterizations of intentional performances needed for a faithful description of the full range of phenomena? My thesis is that both are needed, since there is at least one realm of phenomena whose description cannot be completed exclusively in terms of either Husserl's characterizations or those of Merleau-Ponty.

Let me turn now to the realm of expression, which embraces all verbal, musical, depicting, and gestural performances, and its connection with the phenomenon of silence. The entire realm of expression is predicative in the broad sense. An intentional analysis of this realm shows that it is founded upon a realm of pre-predicative experience and is inaugurated by a motivated interruption or silencing of the realm of pre-predicative experience. Further, the realm of predicative experience is itself complex. It contains several levels, and the transition from level to level is effected by a motivated interruption or silencing of the stream of performances at one level, thus opening the way for performances at another level. Finally, the realm of expression taken as a whole itself founds a realm of postpredicative experience, experience of the nonexpressible, which realm is itself inaugurated by a motivated silencing of the stream of expressive performances. It is important to notice that these several performances of silence are all active, in Husserl's sense, but yet are not in any proper sense predicative. The performances of silence and those of expression are mutually dependent upon one another for their full sense. That is, performances of silence have and retain sense only in conjunction with predicative performances. But it is likewise the case that predicative performances require performances of silence to achieve their full sense.[9]

The realms of both pre-predicative experience and postpredicative experience contain performances which fit Husserl's concept of intuitive performances. Unless some object or set of objects were brought to intuitive presence, there could be no motivation for the silence which inaugurates the shift to the predicative realm. Performances in the predicative realm clearly endeavor to express, directly or mediately, something about objects experienced pre-predicatively. But the realm of expression is that realm where signitive performances are found. Since the predicative realm as a whole is founded upon the pre-predicative realm, and since all signitive performances are predicative, then there is a legitimate basis for Husserl's distinction between those intentional performances which are signitive and those which are not. In other words, the distinctiveness of each realm, with the predicative founded upon the pre-predicative, provides legitimacy for Husserl's distinction between intuitive and signitive performances.

However, the same intentional analysis of the phenomenon of silence shows that the predicative realm itself motivates a silencing which inaugurates a post-predicative realm. The predicative realm indeed founds the postpredicative realm, but it does so in virture of a recognized intrinsic insufficiency of performances within the predicative realm to deal comprehensively with all that is experienced. This analysis likewise shows that the predicative realm itself arises on the basis of a recognized intrinsic insufficiency of pre-predicative performances to yield the full sense of what is experienced within the realm. Thus, though there is a founding-founded relationship among these three realms, the realm which founds does so at least in part in virtue of its own intrinsic insufficiency. And yet the founded realm must always refer back to the founding realm for both the completion and maintenance of its own sense. In short, no realm, and no performance in any realm, is in the final analysis self-sufficient. From this standpoint there is legitimacy in the claim that, in Merleau-Ponty's sense of the term, all intentional performances are signitive.

The intentional analysis of the phenomenon of silence also throws light upon the question of whether all intentional performances are active or whether one must distinguish between active performances and passive or spontaneous ones. The silence which interrupts the stream of pre-predicative performances and inaugurates the "predicative turn" as well as the stream of predicative performances does involve a motivated but deliberate distancing away from the stream of pre-predicative performances. The pre-predicative performances are not themselves founded upon such a deliberate distancing and thus can be appropriately designated, in Husserl's sense, as passive or spontaneous.

But each pre-predicative performance explores a world given initially in a primary *doxa*. Predicative performances continue this exploration of one and the same world. Each particular performance, whether pre-predicative or predicative, then, is a response to that which is primordially given as the world that is always already there. Thus there is justification for the claim that all intentional performances in every realm are, in Merleau-Ponty's sense, active.

The upshot of these considerations is that there is at least one set of phenomena whose full intentional analysis requires that one recognize both the features emphasized by Husserl and those emphasized by Merleau-Ponty in their respective characterizations of intentional performances and their correlated notions of the teleology of consciousness. This set contains those phenomena which belong to the realm of expression as intercalated with silence.

Specifically, without a distinction like that of Husserl's between signitive and intuitive performances, there is no way satisfactorily to distinguish linguistic and other expressive performances from extraexpressive performances. In the absence of such a satisfactory distinction, no growth in precision can take place concerning either of these sorts of performance. That is, without a distinction of this sort, the distinctions between the expression and the expressed on the one hand and between expression and silence on the other collapse with a concomitant loss of intelligibility in each case.

Further, without a distinction like that of Husserl's between active and passive or spontaneous performances, there is no way satisfactorily to account for the manifest differences between those performances which are predicative and those which are not. In the absence of a distinction of this sort there is no way satisfactorily to deal with the differences between those performances which are perceptual, in the broad sense, and those which are judgmental, again in the broad sense. Thus without a distinction of this sort, the uncovering of the sense both of particular performances and of types of performances, especially in this instance performances of silence, is stymied.

On the other hand, Merleau-Ponty's claim that all intentional performances are signitive calls attention to the openness of the correlative fields of experience, namely, the fields of world and subjectivity. These fields and the multiple actual and possible experiences which occur within them perpetually allude to one another horizonally. The indefinitely explorable connections between any particular intentional performance and these fields as well as the experienced inexhaustibility of each field itself reveals that it is in principle impossible to *exhaustively* distinguish one realm from another. The intentional analysis of silence confirms both the inexhaustibility of every realm and the intrinsic intercalation of each realm with every other realm. Thus to acknowledge the inextricable interconnection of all fields and of all performances within them it is appropriate to say that, in Merleau-Ponty's sense, every intentional performance is at bottom signitive.[10]

Similar considerations lend weight to Merleau-Ponty's claim that all intentional performances are active. Perceptual performances, performances of silence, and judgmental performances, in the concrete stream of lived experience, all mutually call for and respond to further actual and possible performances of each of these sorts. No individual performance of any sort can stand alone. Each is senseful precisely as a performance only insofar as it is intercalated with other performances of each sort.

This set of considerations, based on the analysis of the set of phenomena which belong to the realm of expression as intercalated with silence, shows that both the characterizations of intentional performances offered by Husserl and those offered by Merleau-Ponty have legitimate bases in experience.[11]

If I am correct in claiming that each of their characterizations of intentional performances is intrinsically bound up with their respective notions of the teleology of consciousness, then it follows that each of these notions too has a legitimate basis in experience. Their notions of the teleology of consciousness are not incompatible. Rather, they are complementary. Any comprehensive, faithful description of experience must embrace that which each of their positions brings to light.

The way in which their respective notions of the teleology of consciousness complement one another can be put in the following way. Husserl's notion of the teleology of consciousness is faithful to the experience of the idea of science in what can be called the Platonic sense. This idea involves the teleological orientation to eidetic intuitions, universality, and the transcending of all contingency. It is experienced as an infinite ideal. On the other hand, Merleau-Ponty's notion of the teleology of consciousness is faithful to the experience of the historicity, finitude, and contingency of all conscious life, including those performances in this life which intend the Platonic idea of science. In short, Husserl's position accounts for the sensefulness of the ideal of science in the Platonic sense. Merleau-Ponty's position accounts for the historicity of all sense, including that of the idea of science.

But however complementary these two position are, they remain nonetheless two positions. No easy synthesis of them can be performed. And since each is legitimately drawn from the description of experience, neither of them can be either abandoned or reduced to the other.

This situation points to a need within each position, and these positions are representative of eidetic phenomenology on the one hand and existential phenomenology on the other, for an ontology which is sufficiently subtle to accommodate the legitimate demands of both positions. This paper makes no direct contribution to the development of the requisite ontology. But by showing that a full description of experience must recognize both the legitimacy and the insufficiency of both of these positions, it contributes to the clarification of the minimal conditions which any acceptable ontology must satisfy.[12]

NOTES

[1] Merleau-Ponty explicitly borrows from *Formal and Transcendental Logic* the term "teleology of consciousness." For the most part my discussion here will be confined to *Phenomenology of Perception* (transl. by Colin Smith), Routledge & Kegan Paul, London, 1962, hereafter cited as *PP*; and *Formal and Transcendental Logic* (transl. by Dorion Cairns), Martinus Nijhoff, The Hague, 1969, hereafter cited as *FTL*.

[2] I will hereafter, except when quoting from Cairns' translation which renders *Erfahrung* by "experience," follow Robert Sokolowski in translating *Erfahrung* as "encounter." Sokolowski's treatment of pre-predicative constitution, in his *The Formation of Husserl's Concept of Constitution*, Martinus Nijhoff, The Hague, 1964, pp. 173–77, is quite helpful.

[3] Husserl, *Experience and Judgment* (ed. by Ludwig Landgrebe; transl by James S. Churchill and Karl Ameriks), Northwestern University Press, Evanston, 1973, p. 29; hereafter cited as *EJ*.

[4] For a good summary account of Husserl's position on this point, see Sokolowski, *op. cit.*, pp. 199–201.

[5] I have developed this point in detail in a different context in my 'Husserl's Phenomelogical Justification of a Universal Rigorous Science, ' *International Philosophical Quarterly* 16 (March 1976), 63–80. Let me add that it is worth considering whether the primal encounter with the world is a peculiar operative intention which can never be made fully thematic. This question is prompted by J. N. Mohanty's thesis that the concept of 'intentionality' is an analogical rather than a generic concept. See his *The Concept of Intentionality*, Warren H. Green, St. Louis, 1972, pp. 178–82.

[6] I am extending here the account of this pattern of alternation given by Sokolowski, *op. cit.*, p. 171.

[7] *PP*, 429–50. See also, Merleau-Ponty, *The Structure of Behavior* (transl. by Alden

L. Fisher), Beacon Press, Boston, 1967, pp. 220–24.

[8] I take it that, in the works of Husserl under consideration here, the aim is still to achieve universal rigorous science.

[9] The intentional analysis upon which these remarks rest is made in my 'Silence: An Intentional Analysis,' *Research in Phenomenology* 6 (1976), pp. 63–83.

[10] I would suggest that Merleau-Ponty's position at bottom, though not explicitly, embodies an elaborate, expanded, ontological version of the logical principle that the greater the extension, the lesser the intention and vice versa. Thus he holds that the greater the precision, the lesser the richness, the greater the temporal immediacy, the lesser the expressibility, etc. This is what I take is packed into his claim that gain in one direction is intrinsically conjoined to loss in another.

[11] I believe, but am not now prepared to show, that analyses of memory and imagination would add confirmation to the thesis I am defending here.

[12] Promising hints toward such an ontology have been provided by Robert Sokolowski. See his *Husserlian Meditations*, Northwestern University Press, Evanston, 1974, and especially his 'Ontological Possibilities in Phenomenology: The Dyad and the One,' *The Review of Metaphysics* 29 (June, 1976), 691–701.

SERGE VALDINOCI

PHENOMENOLOGIE ET TELEOLOGIE

Reprise: des questions de fond

Depuis 1907, data à laquelle Husserl prononça les leçons formant le texte principal de l'*Idée de la Phénoménologie*, jusqu'en 1934, où le devenir de l'humanité européenne le convain d'écrire la *Krisis*, la préoccupation centrale de l'écrivain n'a pas changé. En 1907, la possibilité d'unifier l'expérience est étendue à l'infini: non seulement l'immanence réelle, mais encore l'immanence intentionnelle des objets idéaux trouvent unitairement leur fondement dans la conscience, conscience dans les lois de laquelle se joue toute la philosophie. De même en 1934, Husserl appelle les européens à l'unité d'une pensée qui assure le fondement de tout logos philosophique. Entre temps, l'auteur condamne dans *Logique formelle et Logique transcendantale* la désarticulation de la philosophie contemporaine: il y a en effet autant de philosophies que de philosophes. Contre la disparité des recherches locales Husserl réclame l'unité d'une science universelle qui respecte le télos de la vraie pensée née en Gréce. Unitaire de droit selon Husserl, la phénoménologie porte normativement un seul télos: réaliser *la* philosophie. Notre propos vise essentiellement à tenter d'assumer la vigueur des exigences husserliennes en confrontant ces dernières avec les réalités conceptuelles enfermées dans la phénoménologie. Cette analyse nous aidera alors à situer la phénoménologie dans le temps présent.

Par rapport au temps présent une constation immédiate s'impose: les recherches phénoménologiques ont certes inspiré d'innombrables efforts; l'existentialisme, l'esthétique et plus généralement une certaine terminologie véhiculent le grand précepte husserlien: respecter le sens et l'amener à l'évidence. Cela n'empêche point que ces pratiques se distinguent nettement de l'idéal d'apodicticité auquel se réfère constamment Husserl. Le télos phénoménologique est donc subverti. Précisément, la postérité a réutilisé certains concepts – la variation eidétique, l'intentionalité etc. – pour les employer à d'autres fins. Sans crainte de se tromper, on peut même affirmer que Husserl n'a pas trouvé, en toute rigueur, un seul continuateur. Ainsi, la téléologie phénoménologique est morcelée *de fait*: paradoxalement le télos s'éclate en des télos partiels. C'est dans ces termes qu'il est permis de dire que la phénoménologie *a* un télos aujourd'hui. De surcroît la pensée husserlienne *est* un télos par rapport aux philosophies qui précèdent ce qu'on a appelé le

Tymieniecka (ed.), Analecta Husserliana, Vol. IX, 169–182.

radicalisme phénoménologique: il faudrait en effet s'interroger profondément sur la signification à accorder à la radicalité de la nouvelle pensée. Si Husserl affirme avec force que son propos continue l'idée essentielle de la philosophie, il reste néanmoins que certains concepts sont repris à l'histoire empirique de la philosophie: pour preuve admettons un instant que la conscience, le sujet transcendental, l'intentionalité illustrent des thèses apodictiques, bien qu'elles soient extraites de l'histoire – *Historie* – de la philosophie. Mais si c'était vraiment le cas, pourquoi donc Husserl se livrerait-il à une réadaptation et à une transformation du sens des concepts? Par exemple, le sujet transcendantal n'est pas le sujet kantien, ni celui de Natorp etc. Tout est donc réévalué. Ceci signifie que ce redressement des concepts, qui n'est pas même contemporain de leur insertion dans la phénoménologie – puisque le sujet, pour reprendre le même exemple, change de contenu au fil des années – illustre combien la pensée de Husserl *est* par certains points fondamentaux le *télos* de modes de penser antérieurs. Ecartelée entre la téléologie qu'elle a et celle qu'elle est, la recherche husserlienne est en quête d'une unité qui règle de droit sa position face au passé et face à l'avenir.

Il existe pourtant une situation phénoménologique dans laquelle le regard du passé sur la nouvelle philosophie et le regard de celle-ci sur l'avenir se confondent; c'est celle où les deux téléologies se développent à partir d'une véritable auto-fondation phénoménologique. Lorsque le point de départ de droit est bien posé, la reprise de ce qu'il y a d'essentiel et de vrai dans le passé se coordonne avec l'institution d'une loi qui décide également de la vérité de l'avenir. A cette loi, Husserl donne le nom de 'réflexion philosophique'; la prise de conscience réflexive opère la continuité téléologique: la phénoménologie *a* le télos qu'elle *est*. Une conséquence s'impose immédiatement, capitale pour la suite de notre démarche: étant donné que la pensée de Husserl dépend de l'exigence que nous venons de découvrir, alors la continuité phénoménologique est soumise à une nécessité de cohérence; avant tout, les rapports téléologiques avec l'histoire en général sont déterminés par ceux qu'elle entretient avec elle-même: à clairement parler, la fondation scientifique de la phénoménologie impose à cette dernière une cohésion interne sans laquelle la question de l'unicité du télos phénoménologique face à l'histoire ne se pose même pas. En fait, il doit exister un rapport d'*expression* entre le télos de la recherche husserlienne et la téléologie immanente à la phénoménologie; ainsi le télos s'exprime nécessairement, et unitairement, dans les enchaînements conceptuels à l'intérieur de la science. Dans ces conditions, la phénoménologie est la voie royale qui conduit à la rationalité universelle.

Nous sommes gouvernés de l'intérieur par cette règle fixe, la seule qui conserve à la pensée de Husserl un destin en continuité avec le projet phénoménologique; mais il nous semble possible d'isoler un schéma conceptuel qui détermine la continuité téléologique. Si cette possibilité était bien exprimée, alors le télos de la phénoménologie trouverait un correspondant *dans* la phénoménologie. Dans cette perspective, retenons les implications du concept d'intentionalité. L'intentionalité est le principe unificateur de la conscience (*Ideen I*: p. 168 de l'édition allemande), le principe sans lequel un télos interne n'existe pas. Par ailleurs la conscience réfléchissante est le dévelopment même de la phénoménologie, en somme, le véhicule du télos externe et historique de la phénoménologie. Ainsi le rapport d'*expression* est respecté. Il est profondément reconnu que la conscience intentionnelle vise un objet: l'intentionalité institue la relation de connaissance, et l'object est donné face à moi. Toutefois, l'acte de connaissance ne déborde pas la sphère d'immanence: c'est ce qui explique que l'objet est plus un fil directeur qu'une donnée absolue. En d'autres termes, le mouvement par lequel la visée intentionnelle se dirige vers l'objet intentionnel appelle un mouvement inverse qui renvoie à la conscience la nécessité de constituer l'object. Cette situation complexe doit être êclaircie en procédant à partir de l'exigence de téléologie interne. Celle-ci est possible en général dans la mesure où le développement de la connaissance s'intègre dans le *système* des opérations de l'ego. Ce point est particulièrement mis en valeur dans les *Mèditations cartèsiennes* où Husserl insiste sur l'existence d'une clôture phénoménologique qui détermine le champ télélogique. Par ailleurs, le mouvement d'aller-retour de l'intentionalité, que nous venons de noter, est lui aussi profondément réglé: aller à l'objet signifie rencontrer ce qui en lui est essentiel: le fait de l'objet vrai exprime l'essence de ce dernier. Inversement, le retour intentionnel sur le sujet est également prescrit par l'essence de l'objet qui motive la connaissance. Par exemple, telle essence individuelle d'objet fait partie de l'ontologie régionale nature et celle-ci renvoie essentiellement à son mode de formation dans le sujet. Mais le mouvement de retour, appuyé sur l'essence, renferme une signification de plus que l'aller intentionnel: comme nous l'avons simplement indiqué l'essence *motive* la connaissance (ainsi s'exprime Husserl dans *Ideen I* p. 284 édition allemande). Ceci semble montrer que la barrière de l'essence, qui renforce l'aspect systématique du processus de connaissance, est encore subordonnée à une finalité intérieure au système. En effet, le concept de motivation introduit nécessairement une finalité interne (*Ideen I*, p. 89 même édition, note 1). Si l'essence déclenche la motivation, quel est le télos? Pour Husserl, c'est sans conteste le sujet qui se laisse motiver (*Ideen II* in Husserlianan II, p. 223); l'auteur affirme encore que le "concept de motivation" se rapporte aux actes du "je" (Husserliana XI, p. 386). L'ouvrage de B. Rang, *Kausalität und Motivation* renferme des analyses qui vont dans le même sens (p. 138 note). La motivation rend possible, en fait, la jonction entre l'intentionalité objectivante et l'intentionalité constituante qui retourne l'objet à ses conditions de possibilité dans le sujet. En définitive, le concept d'intentionalité autorise la prise en charge du système de la connaissance, fondé sur la référence à l'essence, par un télos interne qui rattache l'objectivité universelle à la subjectivité universelle. Tel est le résultat auquel nous parvenons après avoir analysé les conditions générales d'une théorie de la connaissance chez Husserl: la téléologie interne à la phénoménologie montre combien les concepts doivent *exprimer* le contenu du concept de conscience intentionnelle afin que soient évitées les téléologies individuelles qui désarticuleraient l'unité phénoménologique. Dès lors, l'expression a une fonction à deux niveaux: le télos de la phénoménologie exprime le télos dans la phénoménologie qui exprime le regroupement (la polarisation des concepts autour) vers celui de conscience intentionnelle.

C'est donc de ce dernier point qu'il faut partir pour suivre la construction progressive de la téléologie phénoménologique. Il est remarquable à cet égard que la phénoménologie renferme très nettement une téléologie immanente. La première phénoménologie, celle qui s'annonce dans les *Recherches logiques* et se trouve dans *Ideen I* est toute entière une philosophie de l'objectivation. La métaphore de la vue dirige toutes les analyses consacrées à la connaissance. Nous reconnaissons ici l'aller intentionanel vers l'objet. Dans ces limites, l'*eidos* est le terme du processus de connaissance. Au cas où la perception ne peut être immédiate – comme dans la perception sensible, ou dans l'appréhension du cours du temps par exemple – Husserl substitue le gouvernement de l'Idée à celui de l'*eidos*. Toutefois, rien ne change puisque l'idée ne fait que différer à l'Infini la perception reportée et engage alors la phénoménologie dans la nécessité de vérifier à l'infini des thèses provisoires mais prédéterminées. Par contre la situation est transformée à partir des années vingt et surtout dans la *Krisis*. Le modèle de connaissance par le regard noétique animant une matière non intentionnelle perd sa prépondérance. C'est le monde temporalisé de la vie – *gezeitigte Lebenswelt* – qui constitue l'unité sémantique primordiale. La clôture de l'objet-objectivé fait place à l'ouverture du monde. Parallèlement, le monde est l'unité motivante à partir de laquelle les actes du sujet prennent sens. C'est pourquoi l'intentionalité d'acte est fondée dans une intentionalité d'horizon. Ainsi, le monde reprend le rôle joué par l'essence dans la première phénoménologie: il est l'essence elle-même déployée dans un horizon universel. La nouvelle dimension phénoménologique reproduit bien la téléologie que nous avons mise à jour en élucidant les conditions générales de l'unité de la phénoménologie à partir de ses éléments internes: la motivation répond à l'objectivation, la constitution temporelle approfondit l'intuition objectivante. L'ambiguïté de l'intentionalité est respectée. La dynamique conceptuelle ainsi engendrée nous conduit au premier niveau d'expression où la conscience occupe une place centrale: la conscience simple objective en thématisant; la réflexion de la conscience, c'est-à-dire l'insertion de la problématique du monde dans la conscience, est le terrain où s'effectue une interprétation à l'infini réglée par l'essence du monde. De surcroît, ce télos dans la phénoménologie se transforme facilement en télos de la phénoménologie. En effet, la véritable pratique philosophique est pour Husserl la prise de conscience – ou réflexion – universelle. Dès lors, la dimension intersubjective effectue la jonction entre la partie – individuelle – et le tout – universel, L'intersubjectivité transcendantale s'enracine dans une conscience universelle dont la réflexion produit la philosophie transcendantale. A ce second niveau, définitif, on retrouve le schéma intentionnel déjà élucidé: dans la *Krisis* Husserl montre abondamment que le thème de réflexion est l'histoire de la pensée. La phénoménologie apprend tout d'une bonne lecture du passé qui livre l'*objet* vrai de la philosophie. Ainsi, la réflexion porte sur la Raison comme totalité unifiée, et non parcellarisée dans des pratiques scientifiques naïves et hétéroclites. Inversement l'Idée de la philosophie, qui livre l'essence de la vérité, *motive* la phénoménologie à constituer la philosophie rigoureuse par une critique de la Raison (*Kritik der Vernunft*). L'intentionalité est aussi bien objectivante que motivante. En somme le télos de la pensée de Husserl exprime le télos dans la philosophie husserlienne. L'exigence d'expression est respectée; la philosophie actuelle, incluse dans l'universalité de l'autodéveloppement de la phénoménologie, devrait être téléologiquement unifiée dans et par la pratique assidue de la phénoménologie.

Pourtant, notre évocation de la philosophie du temps présent a ruiné une telle perspective. Cette constation, que nous réitérons, suscite deux remarques: soit notre philosophie est, comme Husserl l'a beaucoup craint, retombée dans l'empirisme le plus total. Alors l'humanité n'as pas saisi la chance de salut que lui offrait l'auteur dans la *Krisis*. Soit, au contraire, c'est la phénoménologie elle-même qui n'a pas réussi à établir une véritable cohérence dans ses propositions; cette lacune aurait fait manquer à Husserl la réalité d'une autofondation scientifique, qui est la prémisse d'un autodéploiement universel. Par souci de prudence, approfondissons cette dernière possibilité avant d'évoquer la première. Le passage de la finalité interne de la phénoménologie à sa finalité externe – pour reprendre une terminologie kantienne – exprime certes le rapport d'expression qui relie intrinsèquement les deux téléologies. Toutefois, ce faisant, nous avons respecté la définition husserlienne de la téléologie comme "Forme de toutes les formes" (E III9 p. 65); de la finalité interne à la finalité externe, nous avons laissé se déployer "la forme la plus générale (. . .) la dernière mais en soi la première" (E III 9 p. 65) étant donné que la relation d'expression entre les deux niveaux téléologiques recouvre le même schéma. Ceci considéré, il y a risque de cercle vicieux puisque nous utilisons une proposition husserlienne, qui fait partie de l'intérieur de la phénoménologie, à des fins universelles, alors que nous cherchons justement à justifier l'universalité de la phénoménologie. En toute rigueur, la définition de la téléologie par Husserl ne peut être exportée avant que ne soit fondée universellement la philosophie de Husserl. Ainsi la finalité externe de la phénoménologie considérée dans sa *forme* peut très bien se trouver en opposition avec la finalité de certains *contenus* phénoménologiques – par exemple les concepts qui sont réutilisés sans que soit repris le contexte – employés par la philosophie de notre présent. En retour, il est possible que la finalité interne de la phénoménologie considérée comme *forme* entre en contradiction avec la finalité de certains concepts si on examine sérieusement le *contenu* de ceux-ci. Dans ces conditions, nous sommes contraints de nous pencher, à l'intérieur de la pensée phénoménologique déjà, sur les rapports entre la forme téléologique de la conscience intentionnelle et la téléologie des contenus conceptuels impliqués par l'intentionalité.

Reprenons alors le schéma initial qui a servi de fil directeur, schéma qui a révélé deux dimensions de l'intentionalité, une qui éclaire le mode de donnée essentiel de l'objet, l'autre qui constitue l'object-dans-le-monde. Cette ambiguïté est fort importante car elle retrouve, en les réunissant, les deux définitions principales de l'intentionalité constituante que la postérité a reconnues chez Husserl. Dans les deux cas, cependant, la conscience demeure conscience-de-quelque chose. C'est donc à partir du champ d'objectivité – qui réunit tous les "quelque chose" – qu'il nous faut procéder. Or, la notion d'objet s'enrichit constamment au fur et à mesure que Husserl approfondit son travail. Dans les *Recherches*, l'objet est donné, dans *Idées I* il est constitué, plus tard il est présupposé dans une genèse hylétique. La complexité du statut de l'objet retentit sur celui de la constitution dont dépend l'objet constitué. Fondamentalement, la constitution est d'abord une production d'actes constituants – *dans Idées I* – . Ensuite la constitution comme acte ne sera plus que la reprise patente d'un constitué latent: la véritable opération se confond avec le devenir passif du temps. Et, comme le thème de la constitution intentionnelle structure la phénoménologie, ce que nous avons vu, c'est toute la phénoménologie qui verse d'un point de vue dans un autre. Dès lors, l'harmonie entre intentionnalité objectivante et intentionalité motivante, qui a semblé évidente, cache un problème. Plus précisément, il ne faut pas confondre la motivation par l'essence dans *Idées I*, qui renvoie à une constitution de cette dernière par le sujet, et la motivation par les données hylétiques prises dans la genèse temporelle, qui éveillent et stimulent le sujet. Si, certes, la motivation est toujours rapportée à un sujet, il est clair que la phénoménologie a changé de sens avec la constitution. L'essence n'est plus le complément nécessaire du fait, mais elle fait partie intégrante du monde, et ce dernier se constitue dans le temps, temps dans lequel se constitue le sujet. L'insertion des *data* hylétiques dans le temps introduit alors la dimension du corps constitué, mais aussi celle du corps temporel constituant, étrangère à *Idées I*. Le concept de conscience kinesthétique prend dorénavant une signification centrale, comme l'a montré U. Claesges. En résumé, ce qui était constitué – l'essence, le corps – devient en un sens constituant. L'évolution du concept de constitution a donc dramatisé la phénoménologie qui se restructure en fonction de la transformation du concept. La constitution impose sa téléologie à la phénoménologie, c'est le point important.

Cependant le schéma que nous venons de décomposer offre une autre possibilité d'analyse. Le premier point fixe apparent était le quelque chose objectif. Le second se trouve dans l'unité de la conscience d'où procèdent les espèces de l'intentionalité. La région de la conscience est en effet le terrain d'investigation privilégié du phénoménologue. Dans la cinquième des *Recherches logiques*, Husserl développe l'extension du concept de conscience, la conscience étant le socle et l'origine de tous les actes. Cette attitude de Husserl est encore totalement présente dans son cours de 1907 sur *chose et espace*: "la conscience elle-même n'a besoin d'aucun porteur", dit-il (Husserliana XVI p. 40). Exprimée autrement, la situation révèle que les vécus "n'appartiennent à personne" – *niemandes* – (Ibid. p. 41). Progressivement, pourtant, le centre d'intentionalité se métamorphose. Dans *Idées I*, l'auteur introduit le je dans la conscience, en tant que "transcendance immanente" (*Idées I*, p. 97–98, texte allemand). Ce je est d'abord absolument vide de contenus; il ne vit que de la relation intentionnelle à l'objet; il est le pôle d'unité auquel sont référés tous les actes intentionnels. Puis, emporté par son mouvement, Husserl concrétise le je: le pôle subjectif n'est rien sans ses habitus (Manuscrit E III 2 – p. 5a 1920–21). Cette nouvelle donnée épistémologique entraîne la phénoménologie dans un renouvellement continué de ses fondements: la domination thématique du sujet concret signifie que le sujet prend une dimension temporelle puisque les lois de formation de l'habitus sont passives. Dès lors, le sujet est constitué dans le temps avant que de constituer les objets. Enfin, une dernière réduction emprisonne définitivement la sphère de la conscience – qui, coextensive à la temporalité, semblait déborder le sujet – dans une activité primordiale du sujet qui dépasse même la constitution du temps et conditionne celle-ci: en tant que *Nunc stans*, le sujet domine sa propre constitution par le temps (Manuscrit C 17 IV, p. 1–2–3–4). Ainsi, l'indétermination du pôle subjectif de l'intentionalité, qui est celui auquel il est nécessaire de se référer constamment pour comprendre que c'est toujours le même objet qui est là, face à moi et constitué, engage en fait la phénoménologie dans un univers mouvant. Au fur et à mesure que la pensée de Husserl s'enfonce dans l'infini de la découverte et substitue des formes différentes du *Ich* à la forme épistémologique primordiale de la conscience, la constellation conceptuelle de la phénoménologie change de figure. Bornons-nous à quelques exemples pour élucider cette assertion. Dans la dérive phénoménologique apparaît la fonction nouvelle du corps qui, à l'époque des *Recherches* et de *Idées I* était simplement représenté dans la conscience par l'hylé. Au lieu de cela, la perspective corporelle croise dorénavant celle de l'intentionalité: le concept 'd'instinct transcendental' impose en effet sa connotation corporelle à l'activité du je. Dès lors, l'extension du sujet, comme *nunc stans*, dans des zones extraordinaires correspond à la nécessité de penser, dans son ordinaire, une réévalution définitive de la phénoménologie que Merleau Ponty a seulement tenté d'envisager. Une nouvelle détermination de la conscience semble s'imposer aussi: la conscience devient un phénomène intersubjectif qui caractérise plutôt l'humanité dans son ensemble que l'individu dans ses fonctions de connaissance. L'humanité est ainsi appelée à une prise de conscience théorique de l'Idée de la philosophie, incitée aussi à la conscience morale de sa Responsabilité. Ce déplacement de l'accent, qui ne renie certes pas les analyses antérieures de la conscience, prépare toutefois un rééquilibrage de l'ensemble conceptuel de la phénoménologie au profit d'une perspective socialisante et historique; la temporalité individuelle s'intègre dans l'histoire collective du monde. Mais cette transformation conceptuelle ne fait que souligner la complexité des variations du concept de Ich: successivement transcendant, immanent puis transcendant comme *Nunc stans*, individuel puis à tendance plutôt universelle, actif, passif et actif à nouveau, le je est le thème instable d'une quasi rhapsodie philosophique. C'est pourquoi les dernières tentatives de Husserl prennent un sens inouï dans la phénoménologie: le flottement de la thématique de l'intentionalité et la difficulté d'une problématique univoque de la constitution l'amènent à donner un contenu de plus en plus fort au concept de *reconstruction*. La phénoménologie est à la recherche de la clarification de sa genèse, qui est en même temps la genèse du monde. Ce concept de reconstruction intervient de plus en plus souvent, et à tous les niveaux, dans la théorisation du dernier Husserl; en général, la reconstruction philosophique caractérise le passage du *latent* au *patent* (Manuscrit AV 22, p. 243, années 30). Sont ainsi évoquées la "construction indirecte" du premier flux, du commencement absolu de l'ego, de la naissance, de la mort etc. De surcroît l'autoclarification de la phénoménologie est bien solidaire de l'autoélucidation du devenir du monde: les analyses sociologiques de Husserl sont souvent des reconstitutions historiques; l'origine de la société humaine, la naissance de la philosophie, la genèse de la société à partir de la famille, l'évolution monadique universelle (cf. Toulemont: *L'essence de la société selon Husserl*) occupent longuement sa réflexion. En somme, la reconstruction prend le relai de la constitution. Ce point nous semble très important: deux voies sont possibles à partir du schéma intentionnel conscience-de-quelque chose. La décomposition de ce schéma en privilégiant l'objet nous amène à concevoir la phénoménologie à travers la constitution: la téléologie de la phénoménologie est fondée dans l'approfondissement de ce concept. Dans ce cas, toutes les questions se règlent en rapport avec le fait que la constitution s'effectue dans le *présent vivant.* Jansen montre bien, par exemple, dans *Geschichte und Lebenswelt* que le sens passé de la philosophie est constitué dans le présent (p. 123). Il faut done revenir à la constitution actuelle. Au contraire, la mise au premier plan de la conscience, dans le schéma intentionnel, nous contraint de reconnaître la mise en place progressive du concept de *reconstruction*: la reconstruction découvre le latent pour rendre compte du patent. L'examen des contenus conceptuels de la forme générale de l'intentionalité, laquelle unifie le télos de la phénoménologie, montre que les contenus dominants – la constitution et la reconstruction – imposent deux téléologies distinctes. De plus la dés-unification téléologique semble définitive puisque la reconstruction implique toujours une constitution primordiale d'une part, et que la constitution primordiale est à ressaisir dans une reconstruction, d'autre part.

Cependant, si la téléologie des contenus conceptuels impliqués dans l'intentionalité coupe le télos *dans* la phénoménologie en deux, la loi d'*expression* qui établissait la continuité entre le télos unitaire dans la phénoménologie et celui *de* la phénoménologie est démentie. Elle s'applique à la *forme* téléologique et non à la téléologie demandée par les *contenus.* Plus précisément, il semble que la philosophie de Husserl doive s'éclater si on la considère dans sa dimension historique. L'accent peut être porté d'abord sur la notion de constitution rapportée au présent. En effet, il est important de souligner que la phénoménologie place alors le hiatus qui sépare la science de la naïveté dans le présent. La tradition philosophique l'avait surtout rattaché à l'éternité d'une Idée (Platon) d'un Dieu (St Augustin – Descartes), ou d'une fonction (la normativité des catégories de Kant). Hegel le place dans une maturation temporelle qui rapporte l'immédiateté à elle-même par la médiation du temps. Chez Husserl le hiatus est dans le présent et dans son mode de structuration: l'actualité décide de la vérité; dans le présent coexistent condition et conditionné. Cette idée a un télos dans la mesure où elle se retrouve dans une pratique philosophique contemporaine qui met au premier plan une analyse des structures. Il y a une identité de fonction entre le présent vivant de Husserl et l'épistémé structurante selon M. Foucault: l'épistémé conjoint aussi condition et conditionné dans le présent d'une socle épistémologique. Plus généralement, toutes les analyses structuralistes se fondent sur cette différence qui intervient hors de l'histoire. Le télos s'affirme encore étant donné que le modèle structural est souvent confondu avec la réalité – la structure est l'essence de la réalité, malgré les appels à la prudence de Lévi Strauss – et que le présent vivant structurant de Husserl est également la réalité en acte. Toutefois la structure, selon les structuralistes, n'est pas subjective, ce qui fonde toute la discontinuité entre Husserl et cette philosophie. Ce sont donc bien certains contenus phénoménologiques, seulement, qui *ont* un télos.

L'accent peut être également porté sur les connotations du concept de reconstruction, qui implique aussi une téléologie. Toutefois, nous voyons que la phénoménologie *est* le télos d'une pratique philosophique consacrée par la tradition. Le télos de la phénoménologie est Dieu comme forme des formes, finalité des finalités. La phénoménologie reprend donc en elle une procédure fondée sur la supériorité de la cause finale qui inspire classiquement nombre de philosophes depuis Aristote. Le télos de la reconstruction phénoménologique est donc issu d'un style philosophique provenant de la toute puissance de la cause finale. Dans la phénoménologie se *reconstruit* le schéma directeur de ces philosophies: parvenir à Dieu par la raison en se fondant sur l'estimation de la bonne finalité. La phénoménologie *s'identifie* même à ce modèle car Dieu s'annonce comme télos dans la subjectivité transcendantale. Dans la philosophie du moyen âge, au contraire, les voies philosophiques vers Dieu sont des raisonnements extérieurs à la révélation. Certes, chez Spinoza et Hegel, la philosophie s'identifie également à la théophanie, mais la finalité n'intervient pas puisque nous partons alors de Dieu, et non pas ni du monde ni du sujet individuel. En ce sens, Spinoza et Hegel diffèrent des philosophies fondées sur l'idée d'une relation de finalité entre le monde et Dieu et ne préfigurent pas la reconstruction de Husserl. Par opposition à la dialectique de Hegel, la phénoménologie est le télos d'une conception *analytique* de la raison qui sépare les contraires pour les réunir après examen de la cause finale. Par opposition à Spinoza, la phénoménologie est le télos d'une philosophie qui s'appuie sur l'idée de création, où la créature est rattachée à Dieu par une cause finale également. En ce sens la pensée de Husserl est le terme d'une certaine Idée de la philosophie qui se reconstruit en elle. La désintégration de la phénoménologie est donc opérée: la pensée de Husserl est la reprise d'un modèle passé; elle n'est pas autofondatrice.

Ecartelée de droit, et non plus seulement de fait – ce que nous avions déjà remarqué au départ de notre recherche – entre le télos qu'elle a et le télos qu'elle est, la phénoménologie ne laisse pas de place à la loi d'expression qui règle la relation entre la téléologie dans la phénoménologie et la téléologie de la phénoménologie. En effet il manque une unité de sens qui s'exprime de l'intérieur vers l'extérieur. Toutefois, une remarque s'impose. L'analyse a porté sur les deux pôles intentionnels: l'objet qui est constitué, et la conscience-sujet, qui est reconstruite. Or l'objet est une réalité fixée par l'attitude naturelle, idéalisée. De même, la conscience, à l'instar de tout ce qui est dans le temps, est idéalisée (*Husserliana* VIII, p. 398). Il en va de même de ma vie en général, qui est une "idée limite" (*Philosophie première 2*, p. 225 traduction francaise). Dans ces conditions, il manque à notre recherche la thématisation de l'opération intentionnelle qui porte les deux idéalités: celle-ci produit certes les idéalisations, mais elle est surtout animée par l'opérer qui rend possible toute idéalisation; la *fungierende Intentionalität* est la fonction ultime à laquelle se réfèrent la formation du sens et la mise en forme idéalisée du sens. Pourtant, les travaux de Husserl basculent inéluctablement de l'analyse de la conscience à celle de l'objet et inversement; en somme, la phénoménologie oscille entre sa dimension génétique et sa dimension statique, conscience et objet étant deux mises en forme – ou configurations – sémantiques. Un problème apparaît donc: il faut pouvoir concilier l'exigence originaire de la formation du sens et les formes conceptuelles dans lesquelles le sens *s'exprime*. C'est peut-être ici que se situe la possibité de la véritable *expression* phénoménologique.

Il semble que cette question puisse être éclaircie par l'examen de la situation existentielle de Husserl. Dans *Idées I* (p. 201 du texte allemand), l'auteur se décrit comme un explorateur qui détermine les "configurations phénoménologiques". Dans cette mesure, le sens est délimité, objectivé; le principe des principes, ou *l'originäre Anschauung* (*Ideen I*, § 4), se donne également dans des "limites". L'explorateur reconnaît donc les formes nouvelles plus que les lois de formation. Cette attitude, malgré les analyses génétiques, caractérise toute la pensée de Husserl qui culmine dans une phénoménologie eidétique (cf. *Méditations cartésiennes*, fin du § 37), c'est-à-dire dans une doctrine qui met au premier plan l'élucidation des limites – en vertu de lois essentielles – dans lesquelles se donnent les choses. Cette politique de la découverte recouvre, malgré sa prudence, un *a priori*: la mise en place des différentes limites dans lesquelles se donne le sens impose l'idée d'une hiérarchie conceptuelle. Les essences individuelles ont les limites les plus étroites; c'est pourquoi elles dépendent des essences moyennes – les ontologies régionales – qui sont universelles; enfin, ces dernières sont elles-mêmes rattachées à une essence supérieure, celle d'ego, qui est à la fois individuelle et universelle. La hiérarchie ainsi décrite fonde les deux télos que nous avons analysés: vers le haut, nous trouvons le télos de la conscience, vers le bas de la hiérarchie s'annonce le télos prescrit par l'objet. Le destin des téléologies de la phénoménologie est ainsi relié à celui de la signification de la hiérarchie conceptuelle, que nous allons analyser.

La fonction de l'essence comme limite est de pré-voir le tout dans lequel s'intègrent les éléments à venir: l'essence de rouge recouvre par exemple tous les moments rouges. L'essence d'ego, ou essence supérieure, pré-voit un espace qui est le tout des touts universels régionaux. Ainsi, au fur et à mesure que les éléments sont découverts ils sont dès l'abord mis à la bonne place dans le tout qui les accueille. Ce processus cache une énorme difficulté. En effet, dans die *Krise des Apriori* L. Eley a bien montré que l'essence est vide de contenu (p. 51; 65). Ce que Eley appelle: *Entsachlichung des Wesens*, correspond au statut de l'essence; cette dernière est traitée, dans le chapitre I de *Idées I*, comme le complément du fait: tout fait se rattache à une essence (§ 2), subordonnée elle-même à une hiérarchie d'essence (§ 12). L'essence est donc le corrélat du fait, le statut déterminant du fait. C'est pourquoi l'essence est une *référence* épistémologique; elle n'est pas thématisée comme *signification* avec un contenu: détermination du fait, elle est elle-même indéterminable. Nous retrouvons, avec l'enchaînement vide de contenu des essences, l'unité téléologique formelle de la phénoménologie qui *exprimait* la continuité de la finalité interne à la phénoménologie et de la finalité externe de celle-ci. Cette forme des formes est à présent démasquée. Elle se situe aux antipodes de la nécessité d'approfondir la *fungierende Intentionalität*, puisqu'elle relie des formes et non des modes de formation.

Notre critique a pourtant un sens positif: la véritable expression phénoménologique ne peut que se chercher sous les hiérarchies conceptuelles et formelles. Nombreux sont d'ailleurs les textes de Husserl qui distinguent au coeur de la sphére phénoménologique – obtenue après *époché* –, entre une naïveté transcendantale et la véritable apodicticité transcendantale. Lorsqu'il affirme par exemple "qu'une critique de la transcendance exige au préalable, pour devenir radicale, la critique de l'immanence" (*Husserliana* VIII, p. 378), l'auteur met au premier plan la nécessité d'une autocritique. L'autocritique retrouve une exigence de purification qui nous rapproche de la *fungierende Intentionalität* phénoménologique. C'est dans une autocritique complète que la phénoménologie évite le risque de se laisser imposer des fausses téléologies internes: cela suppose notamment que la fonction de l'époché soit renouvelée: d'une part l'époché n'exclut pas ce qu'elle place entre parenthèses, comme l'indique lui-même l'auteur (cf. *Husserliana* VIII, p. 432), qui révise ainsi ses positions de 1907 et 1913. Dès lors, le sujet n'est plus la forme fondamentale qui implique la forme objet, et ces deux formes ne déterminent plus deux téléologies qui partagent la phénoménologie: l'élément premier est le sujet-monde d'où tout est dérivé. D'autre part, l'autocritique implique que l'*époché* soit appliquée à "l'immanence" de la phénoménologie, c'est-à-dire à la phénoménologie elle-même. La phénoménologie est alors mise à distance d'elle-même. Nous trouvons ici, peut-être, une voie d'accès à la phénoménologie qui est en même temps une voie d'accès à la philosophie en général. Située à distance d'elle-même, la phénoménologie universelle a en effet quelque chance de se re-trouver, dans l'ailleurs apparent des autres philosophies. Il y a ici un programme de recherche qui vérifierait notre hypothèse: d'abord, repérer les véritables moments critiques dans la pensée de Husserl, ceux qui engagent *toute* la phénoménologie et son télos; ensuite, procéder à partir des philosophies antérieures et postérieures pour détecter les continuités phénoménologiques et rassembler ensuite le grand corps parcellaire du Sens. Telle serait la démarche qui donnerait vie à l'hypothèse selon laquelle loin de la phénoménologie husserlienne, nous sommes dans la *vraie* philosophie phénoménologique.

Si l'hypothèse avancée peut être approfondie, elle signifie en dernier lieu que la distinction entre avoir un télos (dans l'avenir) et être un télos (par rapport au passé) est dépassée en ce qui concerne la phénoménologie. Cette dernière est alors désindividualisée, à distance d'elle-même dans le passé-futur. Husserl a clairement montré que la philosophie – c.est-à-dire l'essence phénoménologique de la philosophie – est un phénomène de civilisation qui brasse les époques historiques. Dans ces conditions, la pensée de Husserl n'est qu'un index ou un révélateur: elle joue dans l'histoire le même rôle que tient l'objet dans la phénoménologie; la vraie position de l'objet révèle la distance pertinente, ou la vraie différence – appelée encore constitution – qui le sépare du sujet. Le bon accès à la phénoménologie met à jour la bonne perspective qui révèle la vérité d'une civilisation: nous retrouvons ainsi la nécessité de l'autocritique. En transformant la phénoménologie en un index qui laisse se déployer l'autocritique, nous respectons encore une loi de la phénoménologie: en effet, les lois portant sur l'être trouvées à partir des prémisses phénoménologiques peuvent en retour être appliquées à la phénoménologie (*Idées I*, p. 123, édition allemande). Ceci montre à quel point cette dernière accepte d'être intégrée dans ce qu'elle intègre et combien elle suscite la distance en elle-même. Ellen s'alimente à ce qu'elle constitue et brise très nettement, en retour, les fausses distances. En conséquence la phénoménologie n'est aucunement un être conceptuel, une entité mais c'est la mise en place du bon écart qui déplace les fausses évidences dans l'histoire, fait surgir le sens, et accepte de se voir métamorphoser en retour. C'est pourquoi notre tâche est de radicaliser la "mauvaise conscience" de la pensée occidentale en retournant d'abord au révélateur des problèmes, c'est-à-dire à la phénoménologie. Il s'agit d'instituer en la phénoménologie la bonne perspective, en neutralisant – par *époché* – les éléments faibles et en accentuant les données fortes, et ainsi, en instaurant les écarts dans la pensée de Husserl. Husserl lui-même contribue à instituer les écarts: 1900 marque la coupure avec la psychologie qu'il a pratiquée en 1890; 1913 instaure la bonne distance entre l'objet et le sujet etc.

En définitive, la seule finalité qui s'impose est celle de l'unité téléologique de la civilisation occidentale: ceci suppose la suppression du télos propre à la Phénoménologie car celle-ci n'est pas un Etat dans l'Etat de la philosophie. En outre, cette démarche annonce la tâche d'autofondation de la philosophie universelle, tâche qui tente d'éclairer la perspective pertinente qui mettrait la philosophie à distance de la naïveté. La distance serait alors enfin à l'extérieur du domaine philosophique et l'autocritique – qui suscite cette distance critique – serait également menée à bien. Tels sont les éléments d'une téléologie de la phénoménologie.

University of Nancy

B. TELEOLOGY OF THE PERSON AND OF HUMAN EXISTENCE

PAOLO VALORI

MORAL EXPERIENCE AND TELEOLOGY

Is the Phenomenon of Moral Experience Open to a Teleological Perspective?

I. THE PROBLEM

Out of the confusion produced by the numerous interpretations of Husserl's phenomenology, one of the problems that arises more frequently and is seen as one of the most decisive is the following. Husserl's phenomenology, in particular his early work, is based on the eidetic insight, that is, it searches for the meaning or the "logos" of phenomena.

His procedure appears to a certain extent rather static, methodological, neutral, disinterested, visual, theoretical, contemplative, analytical, in a word – if we can use a word that seems to presuppose an interpretation – "Platonizing." It is well known, however, that Husserl rejected this accusation of Platonism, either true or disguised. His *eidos*, he insisted, is not a metaphysical idea or an ontological reality, but a meaning intended by the constituent consciousness. But the label of Platonism has remained quite tenaciously attached to phenomenology, in spite of the protests of its founder, and has been curiously repeated even by Husserl's critics. Thus the Marxists of the *Voprosy Filosofii* generally consider phenomenology as a desperate attempt of bourgeois philosophy to restore, in a modern way, the theory of the hyperuranic world.

Scholastics, too (for example, J. Maréchal), oppose the "pure" philosophy of Husserl[1] – which they construe as a philosophy which is merely static, contemplative, representative – to the dynamism of "action" led by Blondel.

On the other hand, it is undeniable that, particularly in the work of Husserl's second period, the Husserl of the *Crisis*, there appears an historico-teleological dimension of reality understood as the life-world, the totality of human experience that develops in time in a dialectical process of "auto-consciousness." This new dimension of phenomenology would overcome the sphere of mere methodology and would open out to a true ontology of human existence.

What should we think of these two aspects of phenomenology? Should one see in these two attitudes an incurable *aporia*, an irremediable *caesura*,

Tymieniecka (ed.), Analecta Husserliana, Vol. IX, 185–191.

perhaps a break – as someone has said – between the first Husserl, the conscientious recorder and meticulous analyst of the intentional structures of consciousness, and the later Husserl, the anxious prophet of the imminent dissolution of European humanism (we remember that the *Crisis* belongs to the times of 1937)? In such a situation would this second phase be an unjustifiable intrusion into a closed system, into a well-defined methodology that would consistently reject it? This paper is an attempt to reply to this question.

II. THE REPLY

We do not accept that there is an insoluble *aporia* between the two dimensions of phenomenology, though of course we do not deny accentuation of different aspects in its theoretical structure and in its historical itinerary. The first dimension, more static and theoretical, opens, with no break in continuity, onto the second, more teleological and dynamic dimension. Now, one of the elements which solidly unites the two phases of phenomenology is the moral engagement.

In the early Husserl the disinterested search for truth is seen also as an essential ethical task which makes the philosopher a "servant of humanity." In the later Husserl, on the other hand, man's attempt to improve is represented in particular by his growing appreciation of his humanity itself, an appreciation which is often obscured by the practical, egoistic ends of technology and of particular sciences. And so the contemplative-intuitive aspect and practical dynamism, intellectual methodology, and teleology, are brought together in Husserl by the idea of "having to do" something. In other words, moral experience is not an accessory element of Husserl's philosophy but an essential element of it.

In phenomenology it is not necessary to choose between the primacy of the True and that of the Good – as Maréchal's article seems to indicate; in fact, phenomenology seems to open out to a teleological panorama, if not to a metaphysical-theological one, as suggested in certain unedited manuscripts quoted by Roth.[2]

Concretely we would like to demonstrate (1) that an ethical demand and an explicit thematization of it exist throughout the itinerary of Husserl; (2) that this phenomenology of moral experience opens out coherently and without forcing to a teleology; and (3) that this teleology has specific characteristics which distinguish it from Hegelian and Marxist historicism, and in general from immanentism, as well as from classical metaphysics. In short, the phenomenology of moral experience, present throughout Husserl's work, coherently assumes in its last phase a teleological dimension with certain particular characteristics (personalism, freedom of the individual, search for a moral value, existential concreteness, etc.), which distinguish it from the transcendentalist, historicist contemporary teleologies.

This does not mean that phenomenology is transformed into metaphysics. I remain of the opinion, already often defended,[3] that Husserl's phenomenology is substantially a methodology. This, however, does not exclude but rather opens out onto a kind of teleological ontology which is not easily defined. Perhaps it could be described as existential and personalistic vitalism?

III. THE PHENOMENOLOGY OF MORAL EXPERIENCE

There is no doubt that phenomenology, even though it accentuates the theoretical-doxical aspect of knowledge, has never omitted its axiological-ethical aspect. It is sufficient to recall the passage in the *Logical Investigations*[4] in which it is said that "the object [*Gegenstand*] of phenomenological research can be . . . a thing, an event, a mathematical connection, a being, a having-to-be [*Seinsollen*]." Further evidence for this view may be found in chapter 95 of the *Ideen.*[5]

The *Ethical Investigations* of A. Roth have brought together the numerous fragments of manuscripts concerned with ethics; and the justly famous works of M. Scheler have also shown the fruitfulness of the phenomenological method when it is applied to the dominion of moral research.

But the usefulness of this methodology is not only a question of historical interest which regards so-called Husserlian "orthodoxy." Theoretically it can be maintained that the phenomenological method shows its depth, even its vocation, in the moral sphere. In fact, this would constitute in a way a privileged domain for it.

In ethics that insufficiency of purely empirical factuality appears even more clearly – morality does not attend to facts but to values – as does the insufficiency of logic which is only discursive and formal – morality does not regard the merely abstract coherence of reasoning, but the normativity of the action to be performed: the *bonum faciendum*. Therefore, it is precisely in moral philosophy that the inadequacy of positivistic and logistic theories clearly appears.

The phenomenological method, on the other hand, satisfies a demand which is common among moralists and men of culture today, that morality should be founded not from above (*von oben*) but from below (*von unten*);

that is, not on a metaphysical structure already constructed and therefore *a priori*, but on an effectively verified description of the phenomena of consciousness. Such are, for example, the phenomena of axiologico-ethical judgment, of obligation, of choice, of regret, etc. on which the analyses of Husserl and Scheler have thrown much light. Paul Ricoeur's book, *Saggio su Freud,*[6] contains some interesting comparisons of the methods of psychoanalysis and phenomenology.

IV. PHENOMENOLOGY OF MORAL EXPERIENCE AND TELEOLOGY

From these necessarily brief remarks concerning the phenomena of moral experience it appears already quite clear that the description of such phenomena – that is, a value to pursue, an obligation to perform, a choice to make, a guilt to avoid, etc. – has to involve also a certain dynamism which breaks the static nature of pure speculation and points to a finalistic orientation, a "telos" to aim for. Phenomenological analysis is no longer only a theory of disinterested observation, but a praxis – even though guided by a theory; it is no longer an idea of action, but an action, no longer the search for the true alone, but also for the good. In other words, the cold light of intentional analysis no longer looks only at the process of consciousness, but takes part in it too; it thus seems to be animated by a new and original dialectic. This appears especially in the *Crisis* in which the idea of an *entelecheia* immanent in the development of European humanism is in a way the leitmotiv of the whole volume.

The thematic of action, on the other hand, persuades all the considerations of the *Ethical Investigations.*[7] Mention is made there of formal and material axiology, of the norm of duty,[8] of the categorical imperative,[9] of the relationship between virtue and happiness,[10] of the auto-determination of man as his central ethical task, etc.[11]

It is curious to note how Husserl does not despise the eudaemonistic aspect of existence. He writes that "happiness belongs necessarily and analytically to value".[12] He continues, "a value is not a value for me unless in the apprehension of it I find joy, and without joy the world itself would be without value".[13] Without speaking of real *eudaemonism* one cannot help noticing a notable difference from the rigid formalism of Kant!

Pure phenomenology, then, has an unsuspected dynamism of action in so far as it tends to a final end – in a certain sense even theological[14] – to be aimed at. In this way phenomenology has overcome its methodological limits, and the question arises, Is it to enter the more demanding sea of ontology?

On the other hand, it is difficult to deny that this has not been introduced almost surreptitiously and extrinsically, but is coherently connected with the whole structure of the thinking of Husserl, even if – without a doubt – it is more accentuated in its later phases.

V. BUT WHAT TELEOLOGY IS INVOLVED HERE?

The problem remains of determining more accurately the constitutive characteristics of this teleology which phenomenology reaches in the last stage of its itinerary. The history of ancient and modern philosophy is full of teleologies – religious, providentalist, evolutionist, progressist, historicist, idealistic, Marxist, etc. There is, accordingly, considerable opportunity for misunderstanding. On the other hand, the *Crisis*, the *Ethische Untersuchungen*, and the volume of Diemer entitled *Versuch einer systematischen Darstellung seiner Phänomenologie*[15] give us some fairly precise indications on this subject.

First, it is clear that Husserl's teleology is clearly distinguished from that of classical German Idealism, in particular of Fichte and Hegel. Transcendentalism of that type is explicitly mentioned and refuted in the *Crisis* as obscure and incomprehensible,[16] "metaphysical,"[17] not rigorously scientific,[18] mythical,[19] taken as folly by the common man;[20] substantially, not verified and verifiable, but "constructed" with a certain arbitrarity, etc.

Let us be content with one citation which sums up all these criticisms fairly lucidly: "How was it that to this style – that of Hegel's *Phenomenology of the spirit*, mentioned previously – there remained attached mythical conceptual structures, an interpretation of the world based on obscure metaphysical anticipations; that is, how was it that they – the idealistic philosophers – were unable to elaborate a rigorously scientific conceptuality and a corresponding method, and how was it that all the followers of Kant continued to envisage a philosophy in this style?"[21] Husserl is here opposed to the rationalism of those systematic constructions – a rationalism which must not be mistaken for authentic philosophical rationality.[22] In other works too Husserl's diffidence favors Idealism, and in general the great philosophical "edifices" "constructed" but not founded on experiential intuition.

Much less, in my opinion, can Husserl's teleology be assimilated to a dialectic of the Marxist kind which assumes, even more than Idealism, the character of a completed metaphysical system. Finally, phenomenological teleology is distinguished clearly too from that of classical metaphysics.

The telos of Husserl, it seems clear to me, is not the *esse subsistens* of the Thomists or Augustine's "greatest good." It is not a being of nature, but neither is it pure act understood as supreme perfection. The telos of the individual existences remains, as they live in intersubjectivity and acquire in history consciousness of their rationality. This phenomenological type of reflexion can, on the other hand, acquire, by admission of Husserl himself, the value of "an interconfessional way to God".[23]

VI. CONCLUSION

The preceding remarks open the way to new questions which, in my opinion, must remain unsolved. As always, Husserl's phenomenology leaves us hungry when we deal with ultimate and "metaphysical" problems.

Is Husserl's teleology perhaps a lived existential personalism which sees in the telos God understood in the traditional sense? Or is the telos of Husserl only the immanent Absolute which has to exist as the ultimate sense of history? Or is it the case that immanence and transcendence do not exclude but complete each other? These are difficult questions which the professor from Freiburg, out of intellectual honesty, did not wish, or was unable to settle with a definitive and clear answer.

Let it be enough that we have demonstrated how in the philosophy of Husserl phenomenology and teleology are linked not extrinsically but quite coherently. Phenomenology is not, therefore, quite as contemplative, speculative, and static as it seems at first sight.

Università Gregoriana

NOTES

[1] J. Marechal, *Phénoménologie Pure et Philosophie de l'Action*, Habbel, Regensburg, 1930, pp. 377–400.

[2] Alois Roth, *E. Husserls Ethische Untersuchungen*, Martinus Nijhoff, The Hague, 1960, pp. 167–168.

[3] P. Valori, *Il metodo fenomenologico*, Rome, 1959, p. 198.

[4] E. Husserl, *Logische Untersuchungen*, Niemeyer Verlag, Halle, 1928, pp. 229ff.

[5] E. Husserl, *Ideen*, 3d ed., Halle, 1928, I, 198ff.

[6] P. Ricoeur, *Della interpretazione, Saggio su Freud*, Il saggiatore, Milan, 1966.

[7] Cf. the above cited A. Roth, *E. Husserls Ethische Untersuchungen.*

[8] *Ibid.*, p. 128.

[9] *Ibid.*, p. 136.

[10] *Ibid.*, p. 158.

[11] *Ibid.*, p. 149.

[12] *Ibid.*, p. 158.

[13] *Ibid.*, p. 159.
[14] *Ibid.*, p. 168.
[15] Meisenheim on Glan, 1966; cf. especially pp. 379, 389.
[16] *Ibid.*, p. 224.
[17] *Ibid.*, p. 225.
[18] *Ibid.*
[19] *Ibid.*, p. 227.
[20] *Ibid.*, p. 225.
[21] *Ibid.*
[22] *Ibid.*, p. 350.
[23] Cf. Diemer, *op. cit.*, p. 378.

ARMANDO RIGOBELLO

THE PERSON AS THE ACCOMPLISHMENT OF INTENTIONAL ACTS

1. SETTING THE PROBLEM

Aiming in the present essay to show how the person may be conceived through its intentional genesis as its highest accomplishment, let us, to start, point out the difficulty inherent in the traditional philosophical perspective in a speculative way. The most coherent theoretical positions, like those of Spinoza, draw their conclusions by relegating the "person" to the psychological level.[1] The operative term here is "person." Spinoza, for instance, expresses his dismay about the impossibility of the speculative thinkers of his time to attribute a meaning to the term "person" at all. Comparably, in the present day philosophy, Giovanni Gentile, talking about the "I," identifies it with the totality of self-consciousness, attributing Spinoza's own failure to the "naturalistic" approach.[2]

In fact, whether we consider the pure rationalism or the *concrete logos* of transcendental idealism, we see that the person is not attributed a speculative capacity. The conditions for a theory of the person are not much more favorable within the frame of the classic phenomenological perspective, the more so because of the difficulties arising between the confrontation of the concrete description of personal life and its interpretation. This difficulty points out, on the one hand, that we should take into account the speculative meaning of "person", and, on the other, the structure of the relation between lived experience as an irreducible phenomenological datum and its interpretation. In the present study we intend to formulate a proposal of research, assuming, as a working hypothesis, a model in which the intertwining of the phenomenological datum of experience, of the hermeneutical activity, and of the personal presence is expected to clarify the problem we have announced to start with.

We can then formulate the proposed model in this way: "The intentional acts of consciousness, those which are phenomenologically relevant, reach their accomplishment in the personalistic horizon. In this accomplishment consciousness reaches its 'most proper' (one's very own) explication."

The domain to which this model applies is that of consciousness, understood as the point of tension between the datum and the meaning-bestowing

Tymieniecka (ed.), Analecta Husserliana, Vol. IX, 193–201.

pole, and therefore the realm of interpretation. Personal life should thus be defined by hermeneutical activity. With reference to the latter the personalistic horizon is circumscribed. The personalistic area constitutes, indeed, the "most proper" horizon of consciousness. We propose to investigate it by attempting a "reduction within the reduction"; we do not intend, however, to proceed in the direction of primordially given, as Husserl does in paragraph 44 of *Meditation V*, but to start with the direction of the complexity of the "superior ego."

We have to suspend the validity of every cultural construction and world values, to redirect our quest radically towards a center, towards an original point of the constitutive activity with its complete horizon of possibilities as a presupposition from which the entire spectrum of values comes forth.

2. "APPERTAINING" SPHERE OF THE "PROPER" (ONE'S OWN) AND THE SPHERE OF THE "MOST PROPER" (ONE'S UNIQUE)

On the one hand, Husserl worked out a "reduction within the reduction" in order to obtain a strict and exclusive sphere of the ego, explicating thus gradually the primitive experiencing of the intentional operative system down to the level of the corporeity. The world of values, of culture, of history, belongs on the other hand to the sphere of the ego which reveals itself in the relations with other men. In his reduction, this world is left out and the ego isolated from it. Therefore Husserl's concept of "appurtenance" of the "proper" is primordial but isolating the subject from the rest of the world.

Although Husserl made provisions to avoid solipsism at this level, yet the "proper" remains restricted to the sphere of the primordial. The inquiry we would like to attempt is meant to grasp the domain of the proper as it extends into the sphere of extraneity.

The sphere of the "proper" constitutes the superior world of human experience. The question arises whether it is possible to individuate the condition of becoming visible and operating proper to the subject. Such a question presupposes a concept of the "proper" which is not resolved by what Husserl attributes to it in the quoted paragraph 44. How is this concept reached?

If we take the way indicated above, a "reduction within the reduction" in the sense of the complexity of the ego, the reduction brought radically to this level leads us not to the private and primordial sphere characterized by its "appertaining nature," but to the original and elementary emergence of the relation between the ego and the other, or, to be more precise, to the elementary and original dualistic structure of the ego (the primodial "appurtentivity" and the transcendent "extraneity"), to the paradox concluding the afore-mentioned paragraph 44. Nevertheless this paradox does not stem from reflection on the data of the reduction, but it is itself a new datum resulting from the same reductive process.

It is not a secondary variant of the same method, but on the contrary it differs fundamentally from the whole setting, full of consequences. We need to pick up the relation ego-other, the appurtentivity and extraneity at their original state, so to say, retaining it as a primary cause which is at the base of the most complicated elaborations of human spirituality. Of this primary cause the ego has a direct experience which, in particular conditions of radical reduction, leads to the most elementary source of its own reflex life (which is different from the elementary quality of perceiving corporeity proper). In other words, if the reduction illustrated in paragraph 44 gives us the *genetic process* of the first emergence and shaping of the life of the ego, the reduction proposed here obviates an *atemporal genesis*, which constitutes and is the form of the very life of the ego.

Therefrom we may draw consequences of considerable significance with regard to the relations "proper" and "extraneous" and to their thematical counterposition. The elementary, primordial "proper", which has its typical form in "proper corporeity," is opposed to extraneity; but there is a "proper" on a superior, that is, more developed, level of consciousness in which the extraneity does not mean detachment but enters into the sphere of the "one's own" as a constitutive element. It is no more the radical negation of the proper, but a constitutive element of the self-consciousness of the ego. We would say, then, that the opposition of the proper and the extraneous vanishes since the proper exceeds its elementary manifestations; gaining in depth, and becoming present to itself, it becomes aware of still another presence, of a "secret guest" founding its very interiority.

Paraphrasing one of St. Augustine's concepts, one might say that deep inside every one of us there is a *proper, our very own sphere, which is more our intimately own than we are to ourselves*; this spheres manifests itself as an *interior extraneity*, "an other" that gives rise to the perception of the infinite and the awareness of our finitude. For Husserl the *other self* remains on the outside of my own being even if I stay with the other in a maieutics of mutual spiritual enrichment. By contrast, here we discover the other self in the recess of our own ego. What Husserl said about coporeity proper, extraneity as a source of cultural forms, retains its genetic validity. Whereas the results of our present "reduction within the reduction" moving towards interior depth of man can be considered givenness which is already interpreted.

3. THE "MOST PROPER" AS "AUTHENTIC," THE "VERIFICATION" AS "AUTHENTICATION"

Let us attempt to analyze what we experience in our consciousness on a superior level. We could isolate, through the reduction complex (which serves for individuation), certain states of consciousness and the corresponding horizons of experience. There is a state or condition of consciousness, which we would consider as *solipsism*. A state in which one is aware of oneself as being "*solus ipse*"; this awareness makes one realize how insignificant are the cultural structures, values, the ideals we strive for, the projections of hope. One becomes aware of their relativity when one finds oneself in a solitude – not to be interrupted – in which the phenomena of moral, spiritual, and cultural life in general appear as masks, illusions, deceit, and one realizes that admitting their insignificance is an act of absolute sincerity with oneself.

At another level of consciousness the breaking with solipsism and the retrieved communication with others generates an equilibrium of axiological formations and practical commitments. We could see this level as that of *pragmatic communication*. The phenomenological reduction, by putting within brackets its contents, sheds light on the functional significance of communication with the other. The other's extraneity is here of a constitutive nature, insofar as it allows our own subject to detach itself from us. Thereby the pragmatic communication, a function certainly useful for daily experience, but irrelevant for our quest, defines the sphere of functional alienation. At times, it causes even a delay in the movement towards fundamental problems.

But there is still another perspective or level to be distinguished, which we call the *communication of authenticity*. With it the "reduction within the reduction" comes to its best focus. Indeed, after we have already distinguished the horizon of a solipsistic sincerity, the sphere of intersubjective communication needs to be further radicalized. In the first reduction we separate sincerity from solipsism, in the second communication from pragmatism. Moving away from the psychological shades of the first reduction and from the practical finalities following from the second, we thus obtain a more restricted perspective where sincerity and communication are intertwined in a new constitutive form. We would have thus found with the other self a common constitutive level of consciousness, in the "authentic." Mere sincerity is a moral value, which cannot be established speculatively; on the contrary, authenticity corresponds to sincerity on theoretical grounds; it is the expression of what is "more proper," one's very own, constitutive in a radical sense.

We experience current reflexive life in the world of moral, spiritual, and cultural values. At times we are under the impression that we are really more ourselves when we are detached in solitude (in an absolute immanence of ourselves with ourselves alone), and we see values, ideals, and moral intentions falling down as elements lacking significance. Then we either fall in a state of sincere despair or in a state of aridity which we have grown accustomed to call 'realistic' and 'unprejudiced' (an attitude often taken by egocentric pragmatism).

At other times, on the contrary, we have clear awareness that we reach *authenticity* while breaking out of solipsism and when the other self penetrates the sphere of our experience. Not as an extraneous other who confronts us, not as an instrument for our own affirmation, but as an "alter ego," as a partner in a disinterested communication ("disinterestedness" could be seen as the moral equivalent of rigor). The disinterested communication purifies us so radically that it leaves in us what is more proper; it renders us *authentic*.

We could analyze these two contexts of experience by means of a phenomenological method, and we would find the rigor of the radicalized reduction only in the second, of the deepening which the second experience represents while bringing into a new configuration of the ego and the other, of the proper (one's own) and more proper (one's very own), of sincerity, functionality, and finally of authenticity.

What is the condition by which the other is the partner in disinterested communication? What values has he to share with me, besides the technical exercise of rationality? The foundation of disinterested communication, which at the limit is accomplished in the "charitable" dimension of "interpersonal communication," is that other who becomes "more proper to us than we are to ourselves." Indeed, the *original depth of ourselves*, that which is more proper, and which constitutes our authenticity, is that presence within our presence to ourselves, that interior extraneity.

The clarification of the status of the more proper than the proper, of this presence within presence, of this interior extraneity leads this research beyond the phenomenological grasp and requires an interpretation whose dialectical nature is implicit in the very terms used to indicate that singular alterity. The counterdistinction resides in the very expressions. What does the expression "proper more proper" mean in the light of a common logic?

What is the meaning of "presence within presence"? Above all, the expression 'interior extraneity' contains, in the light of common sense logic, an internal contradiction.

This contradiction can only be resolved dialectically. Inside the immanence of the ego the testimony of the transcendence of the other self gains evidence: an immanent testimony of a transcendental Other. The dialectical interpretation, following the Platonic pattern, leads us, thus, to the metaphysics of transcendence. That "interior extraneity" which makes me pass through a crisis with my solipsistic limits and with my egocentric reckonings, which urges me in my solitude to open myself in communication, but which at the same time makes me feel unsatisfied with a merely pragmatic communication and opens me to communication as an interpersonal communion, is the presence of God in us, the "secret guest" noticed by Socrates as "voice," and to whom Augustine had given a personal form. The relation with other man is an *incomplete image* of an intended relation with this Other towards whom it is intentionally directed. His presence breaks up the solipsism and is, at the same time, the principle of unification progressing from the merely pragmatic and functional communication to the communion between persons. This communication, understood as a communion, is not entirely consumed because of the limitations of the human condition, but it remains an incomplete image. Act of communication stretching out towards accomplishment delineates the space of personal life in its inalienable singularity and in the constitutive communicative opening. It signifies man's vocation of transcendental destiny, which he must not evade, and calls for his faithfulness to a commitment of communication from which it cannot be dispensed.

There is a lot to be gained from the above investigation of communication. The communication with the other self is not concerned simply as an *intersubjective verification* (compare Heidegger's memorative thought). Indeed we can say that, following Husserl, we have understood the efficiency of this intersubjective verification as revealing the objectivity of a world which is not only "for me," but which is "for everyone," of a world being "correlative to communication."

We could say that, developing further our reflection toward foundation of intersubjectivity which resides in the "interior extraneity" of the Other, we are more fully justified in speaking of *authentication* than of verification. Have we therefore passed from verification to *reality*? We would rather say that we have reached a level where verification and reality are engaged in one *constitutive effort*. The authentication and its concrete practice is nevertheless left to the risk of a personal initiative. The allusive, anticipative thought finds in the authenticity of interpersonal communication the *maieutics* of the proper content, but its accomplishment remains, in the last instance, gratuitous.

4. INTENTIONAL ACTS AND THE PERSON AS A METAPHYSICAL HYPOTHESIS

The results we have so far obtained can be summarized in two points: First, we have devised a method: the *reduction within the reduction*. Second, we have shown the existence of a *constitutive atemporal genesis*. These points appear in two perspectives: the awareness of *interior extraneity*, and the *communication through authenticity*. The two perspectives, organically tied together, are those of intentional acts. They concern acts which do not reach their conclusion within the empirical delimitation of the behavior, but are tending further. They do not therefore get exhausted in their representations. On the contrary, they cannot form the object of a complete representation.

The awareness of the presence of myself to myself – interior *par excellence* – and at the same time exterior – are structurally incomplete acts in the sense that they are not consumed in the psychological dynamism which carries them; neither are they completed in the formal structures that shape them. The methodologies and the structures that express them, as the reduction within the reduction and the atemporal genesis, cannot be psychologically explained and cannot be represented either: their nature is allusive; they contain phenomena in their eidetic matrix but not in their empirical configuration. All this clarifies the intentional character of the acts thus evidenced and of their structure, which is a condition of their becoming explicit.

At this point we are led back to the theme of person. At the beginning of the present communication we hinted at the difficulty of a speculative concept of the person; we hope that our preceding reflection will help toward the resolution of the problem set initially. By so doing, we will contribute to the explication of the proposed model.

The intentional acts, those indicated and partly described in our study, are pointing toward a horizon; they are oriented acts; although allusive, they are alluding in a common direction, and oriented toward something that has the semantic capacity to comprise them, to harmonize them and express them in unity. The plurality of intentional acts, while directing itself toward a common horizon, takes shape in the process of unification, which constitutes a center of interpretation. The center of interpretation is the person. "It is proper to the essence of person," as Max Scheler points out, "that he exists and lives only while *accomplishing intentional acts*".[3] At this point our discourse crosses that of Scheler, but our itinerary was different and the convergence of our views is vague.

In the quotation we have reproduced from Scheler there is an adverb which particularly holds our attention: "only." A person in fact cannot be made an object, he is not a metaphysical substantiality, he is simply an *accomplishment*, an end toward which substantial acts are directed. We have tried to describe these acts in some of their operations, such as the awareness of the Other in the sphere of consciousness (the interior extraneity) and the authentication of such awareness in communication. The person, within which these acts are articulated as if he were the pole granting them meaning, constitutes for the same reason the code of their interpretation. We have hinted at "reinterpretation" in the previous paragraph. Their qualification as intentional acts belongs to a first interpretation, while their qualification as intentional acts directed toward a personal horizon belongs to a second type and stage of interpretation. The emergence of the personalistic horizon is not only the end of the process, but it is a level at which the single acts acquire a convergent meaning; they express a unitary plexus and they give place to the operative image of what we call person.

'Person' remains a phenomenological connotation and belongs to an ontological limbo. The reductions, intentionalities, their intricacies, hermeneutic circles that come into being, would remain within a sphere of ambiguous ontology, in the phenomenological dream or inside the "transparency of being," if they were not conceived by us as supported by a strong ontological frame. But doing so we would run the danger of attributing to the person that substantiality which seems irreconcilable with the considerations made earlier and in the light of which the person is understood as a simple accomplishment of intentional acts. At this point the discourse comes into collision with the nature of traditional metaphysics and it raises the question of the very essence of philosophy.

We would like to propose a hypothesis concerning the substantiality of the person and the nature of metaphysics as an attempt to give consistency and meaning to the world we are and in which to live. The hypothesis of the person as a metaphysical reality, bearing in itself the image of a constitutive presence, encompasses, as we have seen, different stages of inquiry. If we conceive the person as a specific intentional sphere and simultaneously as a substance we run into a contradiction. Yet the contradiction is avoided if we admit the substantial nature of the person as a possibility.

The passage from possibility to the affirmation of reality goes beyond the domain of the discourse we have developed here. We do, however, maintain that such a passage is rooted in an original *life-world*, in which "rigorous science" gains a meaning.

NOTES

[1] Spinoza, *Cogitata Metaphysica*, II. C. 8: "Non fugit nos vocabulum quod theologi passim usurpant ad rem explicandam; verum, quamvis vocabulum non ignoramus, eius tamen significationem ignoramus, nec ullum clarum et distinctum conceptum illius formare possumus."

[2] Giovanni Gentile, *Sistema di logica come Theoria del Conoscere*, vol. 2, Florence, 1959, *Opere VI*, p. 238: "L'Io come noi l'intendiamo e come si deve intendere, e il Tutto, che si attua nell'attualita del pensiero; questo dramma divino, in cui nulla e pensabile che non vi partecipi con la totalita del suo essere. E' questa la vera emendazione dell'intelletto, a cui mirava Spinoza, che sempre concepí la sua filosofia come un'etica, ma non pote attuarla, invasato com'era dalla sua idea naturalistica del divino; la riforma del pensiero che, sdegnando ogni neghittosa contemplazione astratta, pensi, e pensi se stesso."

[3] *Il Formalismo nell'Etica e l'Etica Materiale dei Valori*, trad. it. ridotta di G. Alliney, Milan, 1944, p. 173.

CARDINAL KAROL WOJTYŁA

THE TRANSCENDENCE OF THE PERSON IN ACTION AND MAN'S SELF-TELEOLOGY

1. HISTORICAL INTRODUCTION

While referring to the rich program of the Conference of Arezzo-Siena*, devoted to the question of teleology in its multiple aspects and meanings, I wish to present the problem of man's self-teleology. In bringing up this problem, I refer to a few of my previous papers,[1] and above all I undertake one more attempt to develop conceptions contained in my study entitled *The Acting Person.*[2] That is why I have given this paper the title 'The Transcendence of the Person in Action and Man's Self-Teleology.' In the framework of the question thus presented I see the possibility of a new interpretation of a theme, traditional in philosophy, of man and of morality. I have in view an anthropology and an ethics, constructed on the lines of Aristotle's philosophy of being, in the framework of which τέλος was one of the four causes, i.e., sources of understanding reality, and one of the principles to explain it metaphysically.[3] On this basis "teleology," or the principle of final causality (*nota bene* τέλος means not only "end" or "aim," but also "limit"), was included in the interpretation of human action, the human πράξις and also in the interpretation of morality, the human ethos, ἦθος.

It seems impossible to deny the validity of the axiom: *Omnis agens* (and not only *omne agens*) *agit propter finem*; and that is why the teleological interpretation of human activity continues unceasingly in philosophy and science. A contemporary example, confirming this statement, is provided by the so-called "praxiology," or science of purposeful activity, or, as T. Kotarbinski has entitled his treatise, of *Good Work.*[4] From the point of view of the analysis of the human "praxis" there do not seem to exist reasons for doubting teleology as a principle of interpretation: for two centuries, however, it has been questioned in ethics as a sufficient principle of interpretation of the morality of the human ethos. The critique had been carried out by Kant,[5] and the direct basis for it was provided in the seventeenth and eighteenth centuries.[6]

Kant's criticism seems to have been in this case both valid and fertile.[7] He understood the defense of the essence of human morality, which utilitarianism

* This paper was submitted by the author for publication with the conference proceedings. It was not included in the program of the conference (Ed.).

Tymieniecka (ed.), Analecta Husserliana, Vol. IX, 203–212. *All Rights Reserved.*

tried to reduce to the level of pure utility. At the same time Kant defended the transcendence of the human person, which is not only realized within the bounds of human morality, but also seems to be visualized there and made evident in the most direct manner. However, it is not to Kant that we owe the key to cognitive visualizing of man's transcendence in the ethical dimension of his action and existence. The radical dualism of human cognition into two independent spheres, the sensuous "phenomena" and the intellectual "noumena" had proved to be an obstacle. If the criticism of Kant was fertile in regard to the utilitarian type of teleology in ethics, mentioned above – since it permitted perceiving anew and expressing in philosophical categories the transcendence of the human person – we owe this perception and expression to Kant only in an indirect fashion. Directly we owe it, in a much higher degree, to phenomenology.[8] We have in mind here not only phenomenology as the philosophy of "pure consciousness," but as a method which provides many contemporary thinkers with a new and deeper approach to the traditional, and yet never antiquated philosophical themes, especially in the fields of anthropology and ethics.

2. WE DISCOVER THE SELF-TELEOLOGY OF MAN THROUGH THE ANALYSIS OF SELF-DETERMINATION

Such an approach, and the trial connected with it of a renewed interpretation of such a basic fact of our experience as "I act – a man acts," allows us to both grasp and understand, as well as to visualize, the transcendence of man as the person in action. It seems that the traditional interpretation of the act, of the human act (*actus humanus*), did not reveal this transcendence in its proper meaning. The traditional interpretation of the *actus humanus* was strictly connected with the analysis of the *voluntarium* or act of will, understood, probably rightly, as the active basis and source (*potentia*) of man's tending to his aim. This surely corresponds to the character of human willing which always appears in the field of our consciousness as a striving "toward something." This striving arises spontaneously in the subject. To explain its spontaneous appearance, it is necessary to take into consideration, on the one hand, the potentiality of the subject, which in a broad sense we call "will"; on the other hand, we must accept the axiological nature of the end (aim), or in other words, of that to which this willing and striving spontaneously tend. Thus conceived, willing is nothing other than a spontaneous turning toward value.

It is not enough, however, to state the spontaneity of willing in relation to value, in order to explain the nature of the will, nor in order to interpret adequately the fact: "I act – a man acts." This fact itself may doubtless be explained by the dynamics of the will; what is more, human action (*agere humanum*) grows, as from a fertile ground, out of the spontaneous referring to value. But within the bounds of this action (*agere humanum*), of this human act, we encounter not only the spontaneity of willing alone. The spontaneous turning toward value in the process of the will is subjected, if not always, then in any case very often, to the specific "necessity" of choice, which finally gives to the human *voluntarium* the character of decision. In this form the *voluntarium* is contained in the experience of the deed, in the facts: "I act – a man acts."

When we speak of the "necessity" of choice, this necessity does not oppose liberty; quite to the contrary, it constitutes it. The liberty of the will appears in man as a "necessity" of choosing among values and deciding. Spontaneous willing, and the values which correspond to it intentionally, do not destroy that primordial (or fundamental) spontaneity of human willing and the turning toward values connected with it, but, on the contrary, draw on it and, so to speak, are accomplished "in it." The spontaneous willing, and the values which correspond to it intentionally, constitute the peculiar material of choice and decision proper to the will, which forms at the same time the specific dynamics of acts which we discover in the "I act – a man acts" type of fact.

We must add that this spontaneous willing, directed toward different values (very often in the rather passive form of sudden fancies, for example, "I feel like ...") appearing in consciousness, resembles peculiarly intentional acts. Arising in the subject from the various layers and centers of his potentiality, they are directed toward the objectivity, whose name is "value." This subject is differentiated as to its content, but most of all as to its quality; but in spite of this differentiation it may, and even must, be identified each time as a value. Willing by nature refers to value; it is never constituted in the subject and in consequence does not appear in consciousness except for this reference. There is nothing to prevent recognizing value, the object and reason of human willing, as an end (aim) while admitting its differentiation and its indispensable scale of values. In this fashion the axiological picture of the differentiated and hierarchized dynamics (and also potentiality) of man becomes a teleological picture. Of course, we do not touch upon the whole rich specificity of the cognition of values; we only accept the intentionality of willing, which consists in referring to values already cognized in some fashion.

However, we do not identify, and we cannot identify, this broadly conceived intentionality of willing with the will itself, with the *voluntarium*, conceived as the dynamic *constitutivum* of the act. Not the intentionality of willing, but the personal structure of self-determination is essential to the will. It is only the will that interprets the dynamics proper to acts, proper to what is contained in the experience of the "I act – a man acts" type of fact. It is not the willing itself as an intentional turning toward values that constitutes the dynamic essence of the act, but the determining of one's self, or in other words self-determination,[9] which engages the personal subject in a manner proper to himself. And so the true *voluntarium* is not contained in the very experience of willing something, but in the experience of determining one's self.

It is necessary to note that the term "self-determination" indicates both that only the personal subject, or the personal "I" is determining (and acting), and also, that this personal "I" is determining about himself as a subject. In this dynamic relation this "I" becomes an object for itself, an object of will considered as the faculty of the determining subject. The nucleus as it were of man's self-teleology is contained in this relation. We have stated above that τέλος means not only the "end" or "aim," but the "limit." The analysis of self-determination indicates that the *voluntarium*, as the interior dynamic structure of the person who constitutes the act, finds its proper "limit" not in the values toward which the human willing is tending intentionally, but in the subjective "I" himself, who through willing any kinds of values, through their definite choice, both determines himself and in some fashion wills and chooses his own self.

During this first approach the self-teleology of man is outlined as a closed cycle, constantly closing within the bounds of man as subject, who in the action becomes for himself not only an object, but also a limit and an end. This image may at first strike us as suggesting a certain solipsism, reminding us of a monad enclosed within itself. A more profound reflection, even upon what has already been said here, must lead to conclusions which radically modify this first spontaneous image. Self-teleology presupposes teleology; man does not become the limit of self-determination, the limit of his choices and of his willing, irrespective of all the values toward which these choices and willing tend. Man's self-teleology does not signify in the first place a closing up within himself, but a living contact with the whole reality and a dynamic exchange, characteristic of the structure of self-determination, with the world of values, hierarchized and differentiated within itself. The self-teleology of man indicates only that this contact and this life-giving exchange are accomplished on the level and by the standards of the personal "I"; in him it finds its starting point and its goal; to a certain extent it begins with him and is ultimately founded in him; it is from him that it takes its shape, and also shapes him.

3. THE TRANSCENDENCE OF THE PERSON IN ACTION GIVES THE SENSE TO HUMAN SELF-TELEOLOGY

In order to clarify the previous statement, and help us to discover the constitution of the very nucleus of man's self-teleology, which is fully expressive within experience itself, we must visualize more fully in what the transcendence of the person consists. Such a transcendence is contained in the facts of the "I act – a man acts" type. It may be revealed by penetrating deeply into the interior structure of these facts, and therefore must proceed on the basis of experience. It is self-evident that man, as the subject of action, by this action "transcends" himself, turning toward objects, or in other words, wanting different values and choosing them. However, as we have previously demonstrated, the intentional dynamics of the will alone, the dynamics of willing and choice, directed to values as an object and an aim, is not essential to the act. And that is why, when speaking of the transcendence of the person, it is he whom we have principally in mind. In relation to the acting subject the object of the action is transcendental, and so is the value, as the "end" or "aim" of willing or choice. However, the transcendence proper to the human act, connected with the personal structure of self-determination, is not yet exhausted by this, as it were, horizontal direction.

An exact analysis of the will as a personal structure of self-determination indicates that each of its decisions of full value, each choice or mature willing of a definite value presupposes an objective reference to truth. The simplest and most transparent confirmation of this fact is to be found in the analysis of the conscience, in the experience of human morality. It is the simplest, but not sole confirmation. Conscience pronounces its sentence upon the moral value, or in other words, upon the subjective and personal value of our acts, and therefore also upon our willing and our choices directed toward various objective values; it defines them as morally good or morally evil. The reference to truth, which in the case of the conscience is, above all, the truth of the good (or about the good), indicates a different dimension of the transcendence proper to a person than that which is expressed in the very transcending of the horizontal limit of the subject, when he tends to an objective value irrespective of the judgment of the conscience. This might be designated as the "vertical" transcendence, since by virtue of the reference to truth, by virtue of the conscience in which this reference is expressed and made concrete, the man as person achieves a peculiar domination over his action, his choosing and his willing. He takes his position as it were "above them".[10]

This dimension of transcendence, proper to the human person, constitutes itself through reference to truth, to good, to beauty in the transcendent and to a certain degree absolute meaning. This reference does not remain as an abstracted dimension of the human spirit, but enters into the real structures of the acting and existing of a person. It enters especially the personal structures of self-determination, and constitutes this structure in its own way especially through the judgment of conscience, in which it expresses the truth about the moral value of action, and also (although in a different way) through the judgments which express the truth of the logical value of thought, or the aesthetic value of production and creativity, etc.

For the sake of brevity we shall consider here only action itself, enhancing that transcendence of the person in action which proceeds from the moment of conscience. This moment is central and decisive for the constituting of the person's act in relation to truth. This moment is sufficient, as a real expression of the transcendence of the person in action, to obtain a more exact comprehension of the matter of the self-teleology of man. The expression τέλος itself signifies both "aim" and "limit." The self-teleology of man indicates therefore that man is also an end in himself. This conviction has its justification and basis in the analysis of human acts, and above all in the true dynamics of those acts, consisting in the *voluntarium* as the personal structure of self-determination. The precise analysis of this structure permits us to give to the self-teleology of man such a sense as results from the experience and interpretation of the transcendence of the person in action.

Man is in such a manner an aim for himself, as his acts, and the willing, choices, and decisions, contained in them find their limit in this man himself. This limit is found on the basis of the transcendent reference to truth (and with it also to good or to beauty, which we shall not analyze here).

In the reference to truth there is to a certain extent the "limit" of the personal structure of self-determination. When man, as the subject of this action, achieves this "limit," then to some degree he fulfills himself. The personalistic structure of that fulfillment corresponds in the field of experience to the self-teleology of man. It is not absolute fulfillment, although it has in it something of the absolute, since it is accomplished on the principle of the "absoluteness of good".[11]

It may be accomplished on this principle precisely because it arises in the act of the person from the foundation of the transcendent relation to truth, in the "vertical" profile of the transcendence of the subject. The greatness of conscience stems from its being righteous and telling the truth. If such a conscience is called "the voice of God," it is because it reveals in what way man not only "transcends" the horizontal limits of his subject, but also "outgrows himself," thus attaining simultaneously a profound conformity with himself. And precisely this conformity with oneself, which is realized on the basis of the transcendence of the person, of the transcendental relation to truth, enters as it were the definition of self-fulfillment, or in other words, the self-teleology of man.

Within the fulfilling of one's self, demanded as it were by the whole personalistic structure of self-determination, there is a "moment of the absolute" (this surely made Kant speak of the absolute imperative); and the fulfillment of self in the act is accomplished on the basis of the absoluteness of good. At the same time, the subject of that fulfillment keeps the full consciousness of his own nonabsoluteness, or in other words, of the contingency, of the limitations, of the relativity of his being. He does not take a stand, by virtue of the transcendence proper to him "beyond good and evil" – but only, as is expressly confirmed by experience, "outgrows himself" through his relation to truth (and therefore spirituality), in order to possess himself in the field of his willing, and thus also to fulfill himself. Man is constantly an assignment to himself; he is imposed upon himself as a task, and each time, in every action, willing, choice, and decision are imposed upon himself anew. On the basis of the whole human experience, and especially of the fundamental content of the fact of action ("I act – man acts"), we arrive at the conviction that above all the "self-teleology of the limit" which conditions all the "self-teleology of the aim" is proper to man.

4. SELF-TELEOLOGY IN THE DIMENSION OF THE "LIMIT" AND OF THE "AIM"

We have stated already in the previous stage of this analysis (in the context of the considerations on the personalistic structure of self-determination), that the self-teleology of man presupposes a whole teleology of willing and choices through which man is constantly included in an axiological (and subsequently teleological) structure of reality, within the bounds of which he exists and acts. It has also been said that the self-teleology of man does not signify a solipsistic closing up like an intransgressible monad, but quite to the contrary, an opening toward values, and potentially toward all. Self-teleology signifies only a personal mode (*modus*) of that opening. This mode must be many-sided, and so, e.g., different toward things and toward persons. As regards things it concerns, e.g., the whole enormous field of human work, production, technology, economics, civilization. In each of them it is necessary to presuppose and at the same time to postulate the self-teleology of man; the more these domains, while growing objectively, are prone to serve not only the fulfillment of man in himself, but also his alienation, the more necessary it is to presuppose self-teleology.

The relation to persons is in some measure entangled with the relation to things, but possesses above all its own dimension in the existence and action of every man. The self-teleology of the human person serves as a limit of reference, which essentially conditions the correct formation of the great community, both interpersonal and social. If a man spontaneously "transcends" himself toward another, toward others, toward the community (and in this "transcending" occurs the "outgrowing of one's self"), that is the proof that self-fulfillment, or in other words, self-teleology brings with itself an opening of the subject. Man fulfills himself "through others" and realizes his own self by living "for others." This emphasizes not only the transcending of one's self toward others, but most of all the outgrowing of one's self.[12] These matters have been analyzed many times and continue to be a matter of analysis, much required in an epoch such as ours, in which it is most necessary to balance the process of socialization by an effort of personalization.

In concluding these considerations on man's self-teleology, it is hard to resist one more thought, which is closely related to the double meaning of the Greek *τέλος*, signifying both "aim" and "limit." It is well known that in the long philosophical tradition of antiquity and the Middle Ages, anthropology and ethics, constructed according to the teleological conception, were closely connected with the category of happiness (*εὐδαιμονία*), as an aim toward which man ultimately tends through all he does and undertakes.[13] Christianity has built its theological synthesis on the same premise (see, e.g., *Summa theologica*, I–II). Kant questioned the principle of introducing eudemonism into ethics (while taking into consideration the version of eudemonism represented by utilitarianism). Owing to that, the self-teleology of man seems to have become to a certain extent more closely linked to the transcendence of the person in action.

While carrying out in these considerations the analysis of the self-teleology of man within the scope of the transcendence of the person in action, we may reach the conclusion expressed previously, namely, that the "self-teleology of the limit" is first of all existentially proper to the man who exists and acts in the world. It is this limit which is the truth of human acts. Conscience is the fundamental condition of the fulfillment of self. "The world" signifies for every man the necessity of a judgment, which arises from the very depth of his personal being. However, beneath this necessity and, so to speak, beneath this whole magnificent dimension of the "self-teleology of the limit," connected with the existential situation of man, existing and acting in the world, there is constantly being revealed as it were a second dimension, the "self-teleology of aim," which the ancient and medieval thinkers had connected with the concept of happiness, and the theologians endowed with a concrete shape, deriving it from Revelation, from the Word of God. Regardless of the manner in which we estimate the many aspects of Kant's critique, directed toward the teleological assumptions in ethics, we must not exclude the fact that owing to his critique we have been indirectly better prepared to understand also the human "self-teleology of aim," precisely as the final crowning of the person in action.

NOTES

1 I take the view that man's self-teleology is discovered through the analysis of self-determination.

2 K. Wojtyła, *The Acting Person* (ed. by A.-T. Tymieniecka), translated from the Polish (*Osoba i Czyn*, 1969) by A. Potocki, *Analecta Husserliana*, vol. X, D. Reidel, Dordrecht, 1979.

3 *Metaph.* i 3. 983 a 24–33; *Phys.* II 3. 195 a – cf parallel Metaph. V". 1013 a – 1014 a. (Indications according to the pagination in the work: *Aristotelis Opera*, ed. Academia Regia Borussica, vols 1–5, Berlin 1831–1870). Polish translation in: *Wybrane teksty z historii filozofii, filozofia starożytna Grecji i Rzymu*, (*Chosen Texts of the History of Philosophy: The Philosophy of Ancient Greece and Rome*), choice and preface by Jan Legowicz, 2nd ed., Warszawa 1970. PWN, p. 190–191.

4 T. Kotarbiński, *Traktat o dobrej robocie* (*A Treatise On Good Work*). Łódź 1955: 6th ed. Wrocław-Warszawa-Kraków–Gdańsk 1975, Zakład Narod. Im. Ossolińskich. *Sprawność 1 błąd* (*Efficiency and Error*), 5 ed., Warszawa 1970: *Teleologia a prakseologia* (*Teleology and Praxiology*), Prakseologia 1973, N° 3/4 (47–48); cf. also Henry Duméry, 'Philosophie de l'action et praxéologie,' in: *Fragmenty Filozoficzne, seria* II. *Księga pamiątkowa ku czci Profesora T. Kotarbińskiego w osiemdziesiąt rocznicę urodzin*, Warszawa 1967.

5 I. Kant, *Grundlegung zur Metaphysik der Sitten.*

6 Especially the theories of T. Hobbes and J. Bentham. See M. Ossowska, *Myśl moralna Oświecenia angielskiego* (*The Moral Thought of the English Enlightenment*), Warszawa 1966, PWN; see also Fr. T. Styczeń, *Zarys etyki* (*An Outline of Ethics*) cz. I, Metaetyka (Part I – Meta-Ethics), Lublin 1974 KUL, Zakład Małej Poligrafji.

7 M. Scheler, *Formalismus in der Ethik und die materiale Wertethik*, Bern 1954; K. Wojtyła. *O kierowniczej lub służebnej roli rozumu w etyce na tle pogladów Tomasza z Akwinu, Hume's i Kants* (*On the Directive or Menial Role of the Intellect in Ethics, on the Background of the views of Thomas Aquinas, Hume and Kant*), Roczniki Filozoficzne

KUL 6 (1958) z. 2, pp. 13–31; T. Styczeń *Problem możliwości etyki jako empirycznie uprawomocnionej i ogólnie ważnej teorii moralności. Studia metaetyczne.* (*The Problem of the Possibility of Ethics as an Empirically Legitimate and Generally Valid Theory of Morality. Meta-ethical Studies*) Lublin 1972, Tow. Nauk. KUL, ch. III; the same, *Zarys etyki*, (*An Outline of Ethics*), part 1 – Meta-ethics, Lublin 1974, Zakład Małej Poligrafii KUL, pp. 67–71.

8 Mention is due here especially to M. Scheler. Profiting by certain ideas of E. Husserl and F. Nietzsche, he created the well known "material ethics of value," being the most developed form of ethics presented by a phenomenologist. The views of Scheler, as is well known, were later developed by N. Hartmann; D. V. Hildebrand and others assimilate them creatively to Christian ethics and render them in the spirit of realism. *Cf.* K. Wojtyła, *Ocena mozliwości zbudowania etyki chrześcijańskiej przy założeniach systemu Maxa Schelera* (*Evaluation of the Possibility of Constructing Christian Ethics on the Basis of the Assumptions of Max Scheler's System*), Lublin 1959, Tow. Nauk. KUL; The same, '*O metafizycznej i fenomenologicznej podstawie normy moralnej, w oparciu o koncepcje św. Tomasza z Akwinu oraz Maxa Schelers*' (*On the Metaphysical and Phenomenological Basis of the Moral Norm, as Propounded in the Conceptions of Saint Thomas Aquinas and Max Scheler*), in *Roczniki Teologiczno-Kanoniczne* KUL 6, z. 1/2 (1959) pp. 99–124.

9 Cf. K. Wojtyła, 'The Personal Structure of Self-Determination,' translated by Andrzej Potocki, in: *Tommaso d'Aquino nel suo VII Centenario Congresso Internationale Roma-Napoli, 17–24 aprile 1974*, pp. 379–390.

10 Cf. *The Acting Person*, op. cit., part II: (*The Transcendence of the Person in the Action*), pp. 103–186, *Analecta Husserliana*, vol. X, D. Reidel, Dordrecht, 1979.

11 Cf. W. Tatarkiewicz, *o bezwzględności dobra* (*On the Absoluteness/Non-Relativity/ of Good*), 1919; the same, 'Four Types of Ethical Judgment,' *Proceedings of the VII Congress of Philosophy*, 1930.

12 In an analytical manner I have studied the problem of shaping the relation of the "I" to the "you" in my article: 'Participation or Alienation', *Analecta Husserliana*, vol. VI, D. Reidel, Dordrecht, 1977.

13 Arist. *Eth. Nicom.* I 6. 1097 a, 15–1097 b. 5. The problem of happiness may be approached from different points of view; cf. W. Tatarkiewicz, *O szcześciu*, (*On happiness*), 6th ed. Kraków 1962 PWN. Cf. also the same, *O doskonałości* (*On Perfection*), Warszawa 1976, PWN.

JEFFNER ALLEN

TELEOLOGY AND INTERSUBJECTIVITY

In the writings of Edmund Husserl the terms "teleology" and "intersubjectivity" are frequently used together. However, at no time does Husserl carry out an extensive analysis of the nature of the relationship between them. Accordingly, our task will be to undertake a clarification and amplification of several fundamental issues which, when considered together, will give us some insight into the relationship between these two themes.

Our clarification will first focus on teleology and mundane intersubjectivity, each of which is given to us as a fact (*Faktum*) of the *life-world*. In order to make thematic the basic issues that pertain to these two facts, our investigation will center around two problems, namely, how teleology is always already present in the *life-world*, and how the telic community of drives and instincts may serve to give rise to an understanding of the relation between the I and the other. By a shift in our point of view we will then focus on teleology and transcendental intersubjectivity as *eide* of transcendental consciousness. Here we will give a transcendental elucidation of the nature of, and the relation between, teleology and transcendental intersubjectivity. Our transcendental inquiry will emphasize the necessity of the relation between the *eide* of teleology and intersubjectivity, and will do so by examining, in particular, the issue of how the intersubjective community of egos may be understood as the "will to true being."

On the basis of our general clarification of issues we will then undertake an amplification of several clues that Husserl has given us concerning the relation between teleology and intersubjectivity, clues which focus on the problem and method of renewal. By elaborating on the phenomenon of renewal, and thereby delineating the basic elements involved in a phenomenology of "renewal," we may come to adopt and to affirm a Husserlian understanding of ourselves as telic and intersubjective beings. In so doing, our overriding claim will be that it is in the movement of renewal that we may awaken to and affirm the inseparability of teleology and intersubjectivity.

The framework for our undertaking will be that found in a text based on Husserl's notes of November 5, 1931.[1] This short sketch on the theme of teleology and intersubjectivity forms part of a series of working notes that Husserl wrote during the fall and winter of 1931 in order to explore the scope

Tymieniecka (ed.), Analecta Husserliana, Vol. IX, 213–219.

and limits of his philosophy.[2] The text is of special significance for us in that it articulates succinctly, although quite briefly, the major problems that are involved in working out the relation between teleology and intersubjectivity, a relation which, for the most part, is only hinted at in Husserl's major writings.

I. FACTS IN THE LIFE-WORLD: TELEOLOGY AND MUNDANE INTERSUBJECTIVITY

If we turn to the *life-world*, we find ourselves within a worldly, or mundane, teleology and intersubjectivity. Since this is the realm that first appears to us as living and experiencing beings, we may begin here and gradually build up those senses of the *life-world* that are especially relevant to an understanding of the relation between teleology and intersubjectivity. We may indeed take our starting point from the always already given facts of the *life-world*, facts which will be presupposed in the transcendental investigation that follows.[3]

Within this lived-context I find that I bear within myself a nucleus of fundamentally contingent, spatiotemporal being, of instincts, drives, and feelings. Here in the midst of my desires and fears, hopes and misery, I find that the constitution of the world appears to me as being already given in advance, that is, as being predelineated at the instinctual level. Moreover, if I follow a Husserlian train of thought I find that the world is pregiven, even at the level of my instincts, as being goal directed and, correlatively, that my consciousness is instinctively oriented toward a telos. Thus, I find that teleology is implicit in the *life-world*, prior to any attempts to reflectively thematize it.[4]

Yet we may ask, how can we determine the original legitimacy of the apperception which, according to Husserl, arises naturally within the instinctual level of the *life-world* and which, in consequence, universally and inevitably determines in advance all levels of meaning which are founded on it?

As long as our reflection arises from within the *life-world* we cannot step outside our factual being. Accordingly, in working to determine the legitimacy of our apperception we can penetrate deeply into the *life-world*, we can describe the most primordial levels of the *life-world*, but as long as we remain within the *life-world* we cannot jump over the shadow of what is always already present to us. I may, for example, notice that my personal unity arises only within the unity of my manifold strivings, all of which are goal directed. I may, in addition, note that overlapping the horizon of my own being are the horizons of others or, in other words, that my factual being extends to, and is bound up with, a horizon of really possible, but often wholly unknown, others. In so doing, my mundane horizon become a "play-space" (*Spielraum*) of possibilities for me.[5] For instance, within this play-space I may find that I, as an actual being in the *life-world*, can think every conceivable other, and can, in principle, be actual for it; that is, I may find that a totality of other subjects is included in my worldly horizon.

Yet, how can such a descriptive reflection help us to determine the original legitimacy of the apperception that teleology is always already pregiven in the *life-world*?

My exploration of my worldly horizon of possibilities is suggestive in that, whenever I look into the play-space of that horizon, I find that teleology, and also intersubjectivity, are ever-present in it. For instance, my drive for security, emotional fulfillment, love, etc., leads me toward the other. Often my drive is complemented by the reciprocal drive in the other. I find that my drive does not have its object, or goal, within itself; rather, the other to whom I am directed is the object of my drive. Thus, in my very drive there lies "the relation to the other as other"[6] and to the other's correlative drive. My drives that are related to the other may be uninhibited or reserved, etc., but whatever the case may be, they are at least implicitly directed to the other as an other. When, in consequence, our co-directed drives are fulfilled, "we never have two separate fulfillments of each drive in the one and the other primordiality, but rather one unity arises through the reciprocal fulfillment of the two primordialities".[7] Thus, by reflecting on such an inevitable interweaving of drives I come to understand that if I as a subject of drives and instincts and, correlatively, my world of drives and instincts, are to have any meaning, the other must always be implicitly there for me, as a factual ego within my actual world of experience.

Furthermore, my exploration of my worldly horizon of possibilities is helpful in that it prepares the way for a consideration of teleology and intersubjectivity from the viewpoint of transcendental subjectivity, a position from which we may more directly address the issue of the original legitimacy of the initial apperception that teleology is ever-present in the constitution of the world. I, the phenomenologizing subject in the world, constitute the fact whose *eidos* can be disclosed only by I, the transcendental subject.[8]

II. *EIDE* OF TRANSCENDENTAL CONSCIOUSNESS: TELEOLOGY AND TRANSCENDENTAL INTERSUBJECTIVITY

Let us now turn to what is last for us, although first in itself, namely, the transcendental structure of teleology and intersubjectivity. Here we enter into a realm which is unthinkable without teleology and intersubjectivity as facts of the *life-world*, but which may be thought independently of all such corresponding actuality. Our concern is no longer with a worldly, or mundane, ontology, but rather, with absolute, or transcendental ontology.[9]

An initial examination of the *eidos* of teleology shows that it is a form that encompasses all monads and is immanent in all aspects of transcendental intersubjectivity. As such, it is an all-embracing problem. It is, in fact, the "form of all forms,"[10] and thus is the ontological form within which transcendental intersubjectivity awakens and exists. Yet, how is this form to be understood?

We might first note that teleology is the form of all coming-to-be and passing away. Teleology involves a purposive coming-to-be which circles back upon itself, undercuts itself, and thereby transcends itself. For Husserl, the ideal at which such a telic process aims, but never reaches, is that of "infinite perfection" or "true self-preservation".[11] Moreover, teleology is also a universal form that points to an ideal unity, harmony, or true being, while at the same time giving rise to disunity, disharmony, and nonbeing.[12] If the impossible were ever to happen and the telic process were to achieve its ultimate goal, the *eidos* "teleology" would, of course, lose all of its meaning.

Nevertheless, since it is we who are the bearers of universal teleology, we must consider the idea of teleology with respect to its relation to intersubjectivity. As a propaedeutic to doing this, a preliminary transcendental reflection is first needed on the structure of intersubjectivity itself.

As I reflect, from the transcendental viewpoint, on the intersubjective community of drives and instincts within the *life-world*, I find that it is my being that serves as the apodictic ground for all being that has sense for me. Yet, if I come to see, within the play-space of possibilities in my horizon, the essential form of all other possible egos as being the same as the essential form of my own ego, then it will become evident to me that there is a community of others who are subjects similar to me and who are related to me. That is, within a system of drives I may come to see that every drive has its transcendent "goal" in the other subject. I may also see that, since the drives of all subjects overlap with one another, all subjects exist in and with one another and thereby participate collectively in a unitary, universal, and ever-developing community of drives.[13] Insofar as I find this to be the true state of affairs, I may claim that the *eidos* of intersubjectivity is always implicit in the *eidos* I.

If we now wish to determine how teleology is related to transcendental intersubjectivity, we may note that the intersubjective community, whose basic structure we have just presented, is always engaged in a meditative self-reflection (*Selbstbesinnung*), or investigation of its own sense. As transcendental intersubjectivity participates in an unending movement toward the ideal of ultimate self-understanding, its very being is defined by its goal-directed activity.[14] Teleology is the very form of the community's reflection, which progresses upward to ever higher levels, and which, ultimately, may bring the totality of monads to function as a genuine community; or, as Husserl writes, "The being of the monadic totality as flowing being coming to universal self-consciousness and already present in the self-consciousness is an infinite gradual process – universal teleology".[15]

However, we must also note that this telic process of self-understanding includes an element of willing. In the process of self-understanding, teleology and transcendental intersubjectivity are bound together in such a way that the self-reflection of the transcendental community of egos is not simply an abstract process; rather, there is an element of willing present in it. More specifically, for Husserl all willing within the intersubjective community is a form of resoluteness to attain an absolute, eternal, and infinite goal. Such a willing may be termed a "will to life" or "will to true being".[16] It is not a blind striving that aimlessly orders and reorders the world. Quite to the contrary, it takes place as the community of subjects awakens to its possible self-understanding and has as its goal the will to achieve a genuine humanity guided by the ideal of ultimate rationality, or radical self-understanding. Such willing is, like all telic movement, an aiming toward the ideal of coming to a contradictionless harmony with oneself.[17] Nevertheless, we must still ask: In what sense can the will to being, to harmony, to ultimate self-understanding, be said to awaken and to come to presence within transcendental intersubjectivity?

III. A PHENOMENOLOGY OF "RENEWAL": THE INSEPARABILITY OF TELEOLOGY AND INTERSUBJECTIVITY

If we are to determine how the "will to true being" is present in teleology and in intersubjectivity, we must amplify and expand those clues that Husserl has given us concerning the relation between teleology and intersubjectivity, insofar as such a relation pertains to the problem and method of renewal.

At times Husserl speaks of renewal in general terms. In this context he speaks of renewal as an eternal process that infinitely repeats itself. During this process existence renews its truth and willing awakens and is directed to a genuineness that is in accord with an idea of progress. New people also arise, people who understand the process of renewal and who will that this process recur eternally.[18]

Yet, when Husserl speaks of renewal in his article, "Renewal: Its Problem and Method,"[19] he speaks of a process that is needed in our own historical age. In this process, teleology and intersubjectivity are recognized as being bound together, and the reflective will to life and to true being is viewed as having a definitive task within the modern age.

More specifically, renewal appears as the universal call that rings out in our present age. The call is unanswered, for our feeble pessimism and "realism" without ideals have drained our culture of its vital energy. As we stand back and passively observe the decline of the West, we abandon the telic communal struggle for an ideal, for we find all ideals to be unattainable and chimerical. But, in the midst of such a state of affairs, something new must happen. It must take place in us and be carried out by us. This "something" is the actual process of renewal, the radical renewal of an entire human culture. In effect, for the movement of renewal to take place we must awaken to, adopt, and affirm, the relation between teleology and intersubjectivity. If we are to endeavor to elevate our present struggling community to the level of true humanity, we must understand and affirm the interseparability of telic goal directedness and intersubjectivity. In so doing, our ideal may appear as that of the life of reason, or of philosophy, the foundation of which "would itself be the necessary presupposition for an actual renewal, and indeed, would be a first beginning of its inception".[20] Accordingly, the movement of renewal is itself simply the activity of philosophizing. It arises within a community's process of self-reflection and "will to true being." In our present age it is primarily through renewal that we can understand and affirm the inseparability of teleology and intersubjectivity.

University of Florida at Gainesville

NOTES

1 Edmund Husserl, *Teleologie. Die Implikation des Eidos transzendentaler Intersubjektivität im Eidos transzendentales Ich. Faktum und Eidos, Zur Phänomenologie der Intersubjektivität, Texte aus dem Nachlass. Dritter Teil: 1929–1935* (ed. Iso Kern), *Husserliana* **XV**, Martinus Nijhoff, The Hague, 1973, pp. 378–386; hereafter referred to as *Teleologie.*

2 Iso Kern, 'Einleitung des Herausgebers,' in *Husserliana* **XV**, p. liix.

3 *Teleologie*, p. 385.

4 *Ibid.*
5 *Ibid.*, pp. 381, 382.
6 'Universal Teleology,' trans. Severin Schurger, *Telos* 4 (1969), 177. See also 'Universale Teleologie. Der intersubjektive, alle und jede Subjekte umspannende Trieb transzendental Gesehen. Sein der monadischen Totalität,' *Husserliana* **XV**, p. 594; hereafter referred to as 'Universale Teleologie'.
7 *Ibid.*
8 'Teleology,' p. 382.
9 *Ibid.*, pp. 380, 385.
10 *Ibid.*, p. 380.
11 *Ibid.*, p. 378.
12 *Ibid.*, pp. 378–380.
13 'Universal Teleology,' p. 176. See also, 'Universale Teleologie,' pp. 593–594.
14 *Teleologie*, p. 382.
15 'Universal Teleology,' p. 176. See also, 'Universale Teleologie,' p. 593.
16 *Teleologie*, pp. 378, 379.
17 *Ibid.*, p. 381.
18 *Ibid.*, p. 379.
19 'Erneuerung: Ihr Problem und ihre Methode,' *The Kaizo – la rekonstruyo*, III (1923), pp. 92–184. See also, 'Renewal: Its Problem and Method (transl. Jeffner Allen) in *Husserl*: *Shorter Works* (ed. Peter McCormick and Frederick Elliston) University of Notre Dame Press, Notre Dame, fortcoming.
20 *Ibid.*, p. 84.

M. R. BARRAL

TELEOLOGY AND INTERSUBJECTIVITY IN HUSSERL – REFLECTIONS

At this moment in history, the philosophy of Husserl is the center of much interest so that whatever one might say risks being repetitious. At any rate, to speak of phenomenology does not pretend to be original; if anything, it may be reflection in depth, or, in the spirit of Husserl, a tenacious return to beginnings in order to see better both the *life-world* and the investigating subject as consciousness engaged in the world. It is not then the case of exhausting a proposed theme, or of fully answering specific questions, but simply of reflecting on a few aspects of Husserlian philosophy with the hope of shedding light on some interesting points, engaging in daring hypothesis, or even, in the spirit of Husserl, doing phenomenology in a play of fantasy.

Therefore, more than presenting Husserl's philosophy, it will be useful to reflect on the possibilities which his thought opens to us, especially with the intuitions contained in the *Crisis of the European Sciences and Transcendental Phenomenology*, and, possibly, to discover its significance and importance for a philosophy of intersubjectivity and of man in the world.

Whether Husserl did radically change his thought to arrive at the teleological-historical reflections of the *Crisis* is an interesting speculation, which however, does not promise an easy answer to the question. However, this is not a problem. According to Husserl, every achievement is provisional – a milestone on the road. We seek nothing more.

Let us take our point of departure from Husserl himself, who says in the *Crisis*:

> The quite personal responsibility of our own true being as philosophers, our inner personal vocation, bears within itself at the same time the responsibility for the true being of mankind; the latter is, necessarily, being toward a *telos* and can only come to realization, if at all, through philosophy – through *us*, *if* we are philosophers in all seriousness. Is there in this existential "if" a way out? If not, what should we, who *believe*, do in order to *be able* to believe? We cannot seriously continue our previous philosophizing; it lets us hope only for philosophies, never for philosophy.[1]

Herein is outlined the serious duty to which Husserl feels obligated as philosopher; his personal responsibility toward humanity itself, whose being is toward a *telos* realizable only through philosophy itself – if philosophers

Tymieniecka (ed.), Analecta Husserliana, Vol. IX, 221–233.

will be able to do their work seriously, that is, if they will make of philosophy a rigorous science.

However, we are immediately confronted with a problem with which we have to deal. Teleology and *epoché* appear not only opposed to each other, but incompatible. And yet, here they must, in a sense, coexist – one could not say synthesize. Can teleology and *epoché*, as aspects of Husserlian phenomenology, truly coexist in his philosophizing? If, in the *Crisis*, the teleological aspect is assumed completely, does it not deny the ideal of philosophy as a science without prejudices, the *epoché*, the direct investigation of the "things themselves" as a unique source of the knowledge of man and of the world?

The problem does not present itself of an easy solution, because Husserl's thought is not without ambiguity. Reflecting on the existential "if" – "if they will be able to be philosophers in all seriousness" – does not one ask if it will be possible at least for him to maintain the purpose of a transcendental philosophy even considering the development of human history as an inevitable *telos*? However, the crucial point is this: Husserl feels unable to seriously continue to philosophize as before (he says, "we cannot"), because that has issued only in philosophies, not in philosophy, that is to say, universality. It is not at all clear whether Husserl speaks of himself and of his method, or if he refers only to modern philosophy "in crisis"; but then, why include himself in the criticism, as appears from that which follows. In fact, after having specified in what way one ought to reflect on historicity, and having traced a complete program of criticism regarding the ends and methods to be used to arrive at that which is fundamental and genuine in philosophy, he asserts that such investigation will bear fruits by apodictically conquering the will; that is, the will to proceed according to the *new* vision, in order to make of philosophy a true science, above all, in order to overcome its crisis.

Here Husserl modestly admits of not yet knowing exactly the way to follow, but proposes the methods used by him with strict controls for some decades. He plans then to work (with other philosophers – his contemporaries?) in order to discover the hidden teleological meaning of history. After the effort of reflection and research, it will be clear – he says – that philosophy, even if unconsciously, has always been interiorly oriented toward a new philosophic sense: precisely that which arises from the universal apodictic foundation of a philosophy which affirms itself in its execution.

At the same time, the problem of psychology will be made clear and it will be possible to understand the paradox of psychologism.[2] It is important to note the continued reference to "the new"; this already lends itself to two considerations:

(1) It seems clear that Husserl is intent on a path quite different from the one followed till then: reflection on the historicity of man and of the philosopher, considered not simply as a study of facts or events, but as the foundation of philosophy itself. He wants to give this new direction to his investigations because he has understood himself to be engaged in the philosophy of the past which has become the foundation of his search for truth.

This means that philosophizing cannot be the private work of a solitary consciousness intent on finding by itself the genesis of being and of knowing by privately questioning the world and the things in the world. It means also that neither world nor things can be found in isolation, but only and always correlated the one with the others, in relation with a multitude of consciousnesses which form an intricate network of communications and recalls. At any rate, to say *world* is already to signify all this exchange between consciousness and things; in fact, a whole network of relations, even at the levels of nature and of pre-theoretical knowledge. Is not this turning toward historical and existential reality already an acceptance of the world of experience in its dynamism? But, then, what happens to the search for pure essences?

(2) The second reflection regards the language itself used by Husserl in this passage of the *Crisis*; it reveals his methods of phenomenological investigation quite clearly with the usual terms: to question, to reveal, to control; to cut through the exteriority of historical facts in order to arrive at the root; to re-orient oneself; to discover new horizons; to formulate new problems, unheard of questions; to find oneself in unexplored fields; to make contacts never before imagined or even understood; to intuit that the very sense of philosophy must be essentially transformed.

All this seems to indicate the phenomenologist in search of pure essences, more than the philosopher immersed in the teleological dynamism of history and of human thought. Husserl also speaks of not wanting to take a teacher's attitude, but simply of wishing to share his insights, to *indicate and describe what he sees*, as one who has intimately lived the destiny of a philosophic existence.[3]

In this last affirmation, Husserl re-enters the realm of history and teleology, while, aiming at describing what he perceives, remains in his phenomenological method. And it is significant that precisely in the *Crisis* one cannot isolate passages which do not participate of both these aspects. This constitutes precisely the problem of the interpretation of the *Crisis*.

Therefore, we remain in ambiguity. On the one hand, he maintains that it is not possible to continue to philosophize as before; on the other hand, it is evident that he continues his reflections and investigations phenomenologically.

However, is it really possible to submit history, considered as a foundation or source of philosophy itself, to a pure phenomenological investigation? Would it not be tantamount to saying, "My being is rooted and immersed in historicity, but if I really want to know who I am, I must isolate myself from the flux of history, from the contingency of my existing in order to concentrate on my pure being, as it is in itself, without the temporality in which I find myself immersed from my genesis"? How can one be pure spectator of that in which one is actor?

Undoubtedly, there is validity in the objections raised by philosophers like Heidegger and Merleau-Ponty on the very possibility of the reduction. In fact, Merleau-Ponty says that "The most important lesson which reduction teaches us is the impossibility of a complete reduction".[4]

Instead, according to Husserl, phenomenological reduction can and ought to be done even here. If we want to understand our engagement in the world, we must suspend it (the world), take a certain distance from it, with a considerable mental effort (which actually hurts the sensitivity of consciousness). It is true, a distance is always needed to philosophize. However, even assuming that a complete reduction might succeed, if it is true that the philosopher is created by the philosophy which has preceded him, he will never be able to do on himself and on his world a radical reduction without altering some aspects at least of his being and of the being of the world in which he lives.

Yet, it is always a question of the original injunction, "back to the things themselves"; only that, in reference to the matter of history, the process is more complex. It is necessary to overcome not only natural prejudices, but also the philosophical-historical ones. So it will be necessary to see oneself, society, and the historical moment under consideration, not from the ordinary, natural point of view, but from the phenomenological one. And here Husserl's program is well defined: to have an exclusive and constant *theoretical* interest in the *life-world* as foundation of life in the world of men; particularly, to consider it in the manner in which it fulfills this general function of foundation.[5] This is the obligatory task of the new way adopted by Husserl: to keep the gaze riveted on the complex structure which is human life in its historical and profoundly teleological being, which one wants to understand in depth, but always through a penetrating and sharp phenomenological analysis.

It must then be said that Husserl proceeds towards the *new* horizon without departing from his method. It is indubitable that in the *Crisis* he expresses a new vision, brings his investigation to a field till then neglected, but his manner of investigation retains all the characteristics of phenomenological analysis. His intent is clearly expressed in the *Crisis*:

Our task is to make comprehensible the *teleology* in the historical becoming of philosophy, especially modern philosophy, and at the same time to achieve clarity about ourselves, who are the bearers of this teleology, who take part in carrying it out through our personal intentions.[6]

It is an ambitious program which immediately reveals the two aspects of his philosophizing:

(1) to render comprehensible the teleological process of the historicity of philosophy; that is, to make evident the laws which govern human becoming and the cultural forces which promote its development from the philosophic viewpoint – always, however, as rooted in the life-world; and,

(2) to obtain a profound insight into oneself as bearer of such teleology and as an active factor of the historic-philosophic process; a mysterious participation, one would say, because made of personal intentions which, of themselves, must be beyond any determination of nature's laws.

However, if for Husserl, this task must be accomplished always according to the phenomenological method, is it not licit to ask *how* he may be able to conduct an investigation requiring a radical reduction of a world, whose past was guided by a definite end and whose future is a process opening up on ever new paths, results of preceding developments, natural, personal, and cultural?

Here we must revert to the concept of the *life-world* – a basic idea in the project of the *Crisis*. For Husserl, life as lived, pre-objective and pre-theoretical, is the foundation and the basis of every human theoretical activity, and phenomenology seeks to discover its ontological being.

The structures of the *life-world* are difficult to describe: on the one hand, the world and things in the world, on the other, consciousness of things. The *Lebenswelt* as the foundation of all our activities, is always present to us who live it, it is always there before us. To live means to be alert to the world; it means also "to live with the certitude of the world".[7]

This investigation of the world of human life, keen phenomenological enterprise, brings into focus again the problem, already indicated, of the possibility of the historical reduction, which Husserl does not doubt. He proposes the reduction of historical processes, so that, by this means, it may be possible to re-constitute the hidden teleology of history and arrive at the deepest level of originary life, at the *life-world*.

Already Husserl had speculated on the possibility of finding a way of knowing "the concrete world as such" following the model of mathematics, which had proved useful for arriving at the knowledge of the things of the actual world.[8]

However, it is quite another thing to arrive at the world of lived experience.

Husserl contrasts it with the world of theory, understood as a development of intellectual Greek forms and with that of science which he conceived as that of scientific Europe itself. However, there are other structures of the world besides these: there is, in fact, a whole culture which, born in the *life-world*, also grows in the *life-world* itself, becoming in its turn the original world of the lived experience of the generations which follow. Then it is difficult to see this *life-world* as an originary world, simple, new, a-scientific.

One might then say that the *Lebenswelt* of any society is no longer a simple level of life, pre-theoretical and pre-scientific. Rather, it will be a world in which both scientific and cultural structure are already part of the lived experience, accepted without investigation and without theoretical reflection. Thus, the level of natural life, originary and always present, is already sophisticated and certainly not primitive.

Then two questions arise: in the first place, is it not possible that the *life-world* of the modern world, instead of being a world of raw life, so to speak, which scientific and cultural structures come to modify, is, instead, a sediment of culture and science which conditions the structures themselves which should be constructed upon it?

Secondly, it seems truly impossible to be able to arrive, by an historical reduction ever so radical, at the *life-world* as foundation and basis of life and of every structure as described by Husserl, because at this moment in history it does not exist any longer (exception being made for societies not yet touched by any scientific theory or sophisticated cultural development – if they still exist). On the other hand, how could one discover the *life-world* unless one already knew what it was and what are the structures built on it? A systematic "dismantling" of the strata of culture and science would be of no help as we have no experience of the original structures. One cannot relive a life not yet lived, a life not known, but merely presumed, or assumed as necessary foundation of one's experience – subjective or intersubjective, individual or communal.

It is possible to accept Husserl's phenomenological project as a theoretical task aiming at deepening the knowledge of the historical *telos* of life and the penetration of the *life-world*, even if it is foreseen that it will not be crowned with the anticipated measure of success. In fact, Husserl himself was aware of the difficulty:

The concrete *life-world* then is the grounding soil (*der gründende Boden*) of the "scientifically true" world and at the same time encompasses it in its own universal concreteness. How is this to be understood? How are we to do justice systematically – that is, with appropriate scientific discipline – to the all encompassing, so paradoxically demanding manner of being of the life-world? The paradoxical interrelationships of the "objectively true world" and the "life-world" make enigmatic the manner of being of both. Thus [the idea of] true world in any sense, and within it our own being, becomes an enigma in respect to the sense of this being. In our attempts to attain clarity we shall suddenly become aware, in the face of emerging paradoxes, that all our philosophizing up to now has been without a ground. How can we now truly become philosophers?[9]

It is, then, not surprising if the most radical historical reduction does not ensure the discovery of the fundamental basis of the *life-world*. There remains also the doubt whether the phenomenological analysis of the life-world be such as to justify the assertion that teleological-historical reflection is of itself the necessary introduction to phenomenology. Moreover, if the *life-world* is acknowledged as the indispensable basis for any theoretical activity whatsoever, ought we not ask *how* the phenomenological activity, which also arises from it, can nevertheless be independent of it – that is to say, that it be free to the point of appearing as bearer of the very *telos* of every *theory*, bearer, that is, of the interior strength of human reason? On this point, we may merely suppose: what if Husserl had conceived phenomenology as exempt from the laws applicable to all other theoretical activities, so that phenomenological reflection would be seen as activity directed to the *life-world*, but not deriving from it? But how? This is still enigmatic.

Indeed, Husserl leaves many questions unanswered, but this is not at all extraordinary. He himself admits that his investigation has brought him to such paradoxical situations as to almost make him doubt the possibility of philosophizing and of establishing a foundation for his philosophy. In the light of such comments from Husserl one may leave open the question of his success: whether he has or has not touched the depth of the *life-world*, whether he has truly discovered the essence of things – or perhaps abandoned the search. But the important thing is that his philosophy is incomplete, ever renewing itself; a perpetual begining anew – is rich in brilliant ideas and laden with insights for the founding of a theory of man in the world, the theory of intersubjectivity.

In a brief manuscript entitled *Universal Teleology*[10] Husserl seems to condense the results of his reflections on the Ego, the other, and the relation between himself and the other: intersubjectivity – at once indicating the universal teleology wich governs the world of monads.

The theory of intentionality is fundamentally a theory of communication. To communicate with another person is to be able to participate in the vision of the world of that person, it is to discover another world. It means to free oneself from solipsism and to form a community, be it even merely incipient, through which the world itself and the horizon of the world are unified, the various perspectives of the world are synthesized.

We will not stop to describe the phenomenological investigation by which Husserl establishes the essence of the intersubjective relation in the *Cartesian Meditations*, in *Ideas*, in the *Crisis*. However, it is important to specify how, for Husserl, intersubjectivity is directly tied to universal teleology. The subjectivity of the "I" is such only by its immersion in the world, by intersubjectivity, insofar as it is part of the world – the only existent world. This world of intersubjectivity is operative already at the originary level of reality: "As primal ego, I constitute my horizon of transcendental others as co-subjects within the transcendental intersubjectivity which constitutes the world".[11]

All that which the Ego perceives and feels is subjective, individual, totally his own; but this only because in each monad is present the originary life which attains in the intersubjective world, by which every subject becomes conscious of his own individuation and transcendence.

The teleological theme appears in the "being of the monadic totality as fluent being which arrives at 'auto-consciousness,' which is already in auto-consciousness, a gradual process to infinity – a universal teleology".[12] Since the ego is "auto-conscious" and individuated, he can reflect on the pre-conscious and on the pre-individual. However, there is always a return to that life which has preceded both consciousness and time – the *life-world*. There is an originary time and there is a temporalization of the world and of the subject in the world. The universal and originary *telos*, as power in pre-individual life, is actualized vis-à-vis the world (in the subjects whose inter-relation constitutes intersubjectivity) and is the origin of meaning – sense – for man who constitutes the world, and of history itself.

Both the teleological dimension and intersubjectivity are founded in time, contingent and therefore finite, which, however, is always related to the infinite from which it arises. Universal teleology is revealed in the gradual process by which the totality of all beings (monads) tends, intentionally, to constitute itself as a rational society. Rationality is a terminal point; man at first reveals himself as intersubjective impulse.

In particular, Husserl points out the impulse of the one towards the other, the sexual impulse which, as with the hunger impulse (and as with any other impulse) is determined by the object which satisfies it. The determination may be indefinite, as in the case of a generalized urge, or be directed to an object already known and desired. However, this intersubjectivity is possible because man and woman are already one in the other, they already live a "reciprocal union".[13]

Husserl, in describing phenomenoloically the "within" of procreation, notes that the impulse may be present in various modalities: expression, renunciation, or aversion. Originally, the impulse is free and spontaneous; the intentionality is constituted by the one through the other in a reciprocal correlation. However, procreation is profoundly immersed in mystery: the problem which Husserl poses to himself is whether – and how – the intentionality of the impulse towards the other (both sexual and social) ought to be sought at a more profound and more general level, in the constitution of the world – a reference to the *life-world*.

In fact, intersubjectivity as human or as sexual relation is steeped in originality, as vast as the world of monads, as if buried in the primordial being, bearing in itself the whole past of the monads which yet contain the totality of living things.

One could say that here Husserl indicates the ultimate, primordial level of man and of the world. However, it is always a divining more than a *knowing*, because one cannot recognize that of which one does not have experience. It is always the same problem, but in a new garb: here, since reciprocal feelings of human beings arrive at the mystery of the person, one is more keenly aware of the profound roots from which intersubjectivity arises. Psychology then reveals (although already as theoretical structure) that there is a whole world of the "pre-conscious" from which human consciousness emerges, all a weaving of myths, symbols, intentionalities which form the underground of the human essence, opening up in the intersubjective relations of love, sociality, sexuality – basic expressions of the human *telos*.

Husserl seriously considers the possibility of a universal intentionality of the impulse, such that it might constitute every presence as a temporal union, so that the original impulse may pass from one presence to the other until the *telos* is fulfilled. This would bring about, by interrelations, the joining of all the monads (already originally involved with one another) into one single intentionality. Such concentration of intentionality would lead to a permanent centralization of primordiality through the I-pole, to arrive then at the attainment of a universal teleology.

Husserl places the "auto-conscious" I as center of the activities which arise from a world of "beings," bearing within themselves their own proper horizon, in the activity of a communal life. The I-center is, however, conscious of being part of the intentionality deriving from the universal teleology which is the foundation and impulse of life from a past of which the genesis is unknown.

Thus, human life in the world is conceived with its roots profoundly imbedded in "pre-time" and bursting forth from a vital source which extends, so to speak, in the underground of existence. Thus Husserl says: "Before the world, is posited the constitution of the world, is posited my auto-temporalization in the pre-time, and the temporalization in the intersubjective pre-time".[14]

The intersubjective relations which are instruments of concrete life arise from the pre-temporal level and bring to completion the *telos* of the impulse: sociality, love, sexuality. The act of procreation involves man and woman in intersubjective relations which are both original and unique. Two lives are interpenetrated, following an impulse, wholly individual and personal, yet participating of an entire world – the human monads are in a continuum of communion with the other forms of life in the world.

The intersubjective act of procreation, an encountering and compenetrating of two animated lives, does not take away anything from the one or from the other; neither is there in the act of love a loss of the one to the other (if there were, it would be a perversion of love and of the impulse, which, of itself, is a positive act). On the other hand, the impulse, arising from the depth of the life-world, participates of the intentionality of the world itself. It gives and it receives; it is all a process of reciprocal actions between man and the world.

Procreation and maternity are two aspects of life which are not limited to the intersubjective act that motivates them: "Teleology circumscribes all the monads, and that which occurs in the maternal sphere, does not occur only there, but is reflected in everything".[15] There is an unlimited participation of the world in the mystery of the genesis of a human life. Maternity is not an individual thing concerning the intersubjectivity of two persons in a given social-sexual act, but a cosmic event, as if every new life were arising from infinite and profound waters, fertilized by a breath of universal life.

The significance of these intersubjective and intervital relations between man and the world, among the various levels of human and animal life, is very important for specifying the function of man in the world and in society. This means that there is a continuity of relations among the various modes and levels of life in the world, an influx and an interpenetration, the whole governed by a *telos* which defines itself totally at the human level. The *telos* is the affirmation of reason, the development of *sense-meaning* – in the world. This projects itself as the "auto-consciousness" of the subject in human society.

In this human society, intersubjectivity opens to an infinity of aspects, first among them that of the reciprocal acknowledgment as subjects, which, therefore, prevents their ever being objectified. Secondly, having a common origin becomes the basis of solidarity; one feels himself part of all the others, and egoistic solipsism gives way to the intersubjective impulse, which implies on the one hand, individual affirmation and, on the other, rejection of identification one with another. Thirdly, the value of liberty is affirmed with the acknowledgment not only of subjectivity, but of the personal element in every individual, which demands reciprocal respect and relations of justice; that is to say, rejection of both slavery and the abuse of others. Above all, the value of life is affirmed in the intersubjective relation of love, be it as *Eros* or as *caritas*; here the roots of the impulse are deepest and the most inscrutable, but also the most *fecund*.

The reference to the *life-world* with its original richness seems to make one forget the temporality and the contingency of life – so evident at the level of theoretical reflection. And yet, it is still this temporality which influences intersubjective relations. The presence of irreversible time and the anticipation of the end of individual life have a bearing on reciprocal relations of love, of solidarity, and even of simple communication. Man as contingent being feels the anguish of his precarious position and suffers for it. The impulse of the one towards the other can also be defined as a reciprocal defense – or an effort to get free – from sorrow, from the fatality which threatens life.

The value of the perception of the *telos* in human life consists in knowing how to overcome anguish, directing subjective intentionality at a future which projects itself as promise of security and as desirable end. However, does not intersubjectivity open the door also to negative relations? No doubt it does. For every answer of the subject to the existential call of the other, there is also the possibility of a negation of the value of life. While there is a giving of self in positive intersubjectivity, there is always a way open to egoistic solipsism.

Since no life is solitary, but each one is woven, body and soul, with other lives, the positive and the negative, good and evil, touch each other constantly. The true end of history consists in *realizing* intersubjectivity, that is, in affirming life in its positive values – freedom, love, justice, truth – and in negating or overcoming its negative aspects such as the objectification of man, egotism, oppression of the other, refusal of the truth. This is the teleology of history as universal intentionality. If and how this intentionality is already evident at the pre-theoretical level is difficult to ascertain.

However, if one considers the *life-world* as life, lived at the level of any modern society, one must say that this teleological intentionality, aiming at the type of intersubjective living described above, is always an actual possibility which tends to transform negative aspects in positive values in the society itself.

It is to be noted that this description of intersubjective relations is already a description of interpersonal relations, and nothing more would be needed for the formulation of a philosophy of man in the world than to explicitate its content and the implicit ramifications. Husserl's reflection reveals clearly the human intersubjective structures and the opening to social and communitary life. It should not be difficult, with further phenomenological reflections, to gather and synthesize Husserl's insights into an explicit theory of interpersonal relations in the world.

There remains, however, the problem of a solid foundation for this theory; it seems that notwithstanding the phenomenological effort, the last basic level of human reality is still elusive; but, perhaps it is a foundation that cannot be reached as one cannot analyze the ultimate level of being. If there is mystery in life and in the world, it is comprehensible that there also be mystery – and for a stronger reason – in the realm of human consciousness and in all that which touches it.

Husserl himself, who travelled a long and difficult path from pure essences to a teleology of history – whether knowingly or not is still in question – admits to being still in the dark about many things, and declares himself to be unsatisfied. But then, he says:

> What autonomous thinker has ever been satisfied with this, his "knowledge"? In that obscure "knowledge" . . . the historical is concealed . . . the spiritual inheritance of him who philosophizes. . . . In the same way, obviously, he understands the others, in whose company, in critical friendship and enmity, he philosophizes.
>
> For what autonomous thinker in his philosophizing life, has "philosophy" ever ceased to be an enigma?[16]

NOTES

[1] E. Husserl, *The Crisis of European Sciences and Transcendental Phenomenology*, tr. with an introduction by D. Carr, Northwestern University Press, Evanston 1970, p. 17.

[2] *Crisis*, pp. 17–18.

[3] *Crisis*, p. 18.

[4] *Phenomenology of Perception*, trans. Colin Smith, Humanities Press, New York 1962, p. XIV.

[5] *Crisis*, p. 155.

[6] *Crisis*, p. 70.

[7] *Crisis*, p. 142.

[8] *Crisis*, p. 33.

[9] *Crisis*, pp. 131–32.

[10] Husserl, E. III 5. *Teleologia Universale*, 1933, trascr.: Biemel, 1952, in Paci, *Tempo e Verita*, pp. 256–269.
[11] *Crisis*, p. 184.
[12] *Teleologia* (Paci), p. 257.
[13] *Teleologia* (Paci), p. 261.
[14] *Teleologia* (Paci), p. 267.
[15] *Teleologia* (Paci), p. 267.
[16] *Crisis*, p. 394.

FILIPPO LIVERZIANI

TELEOLOGY AND INTER-SUBJECTIVITY IN RELIGIOUS KNOWLEDGE

There is a certain scientistic mentality which, while elevating scientific knowledge to the highest possible level, undervalues pre- and extrascientific existential knowledge and, in particular, exhibits little regard for any kind of metaphysical or religious knowledge. At most this viewpoint concedes that such knowledge has a merely psychological, subjective and private value. However, my investigation of Husserl's phenomenology has led me to the view that this knowledge may well offer us a means of introducing a right sense of proportion into the claims of scientism. It will give us a more objective and considered view of science within the limits of its specific function and aid in the reappraisal of pre- and extrascientific knowledge having as its object the life-world; we shall then be in a position to give a noetical-objective reappraisal of metaphysical and religious knowledge. Space does not allow a historical overview of the way the scientistic mentality has come about; nor will I be able to explicate Husserl's phenomenological analysis of the way science has come into being and its connotations. My intention here is confined to raising the following problem: how can a phenomenology based on the Husserlian method help to provide a philosophical justification of an authentic cognitive status for religious experience (and in a wider context, for metaphysical and mystical experience, and for interior and spiritual experience in general), to meet the claims of a certain scientistic mentality which denies any objectivity to such experience and limits it to the subjective or the human. It does not appear that Husserl actually tackled this problem as such; but he certainly did deal with the wider and more general issue, namely, whether one can attribute a real and serious cognitive status to pre- and extrascientific knowledge. It would appear, moreover, that a correct approach to this wider problem, in the manner of Husserl, might be able to pave the way toward a definition of the more particular problem of the cognitive and objective status of religious knowledge. This is a consequence that we ourselves can draw from an interpretation of Husserl's method which, despite its theoretical and personal nature, will sufficiently reflect the spirit of Husserl's phenomenology. Moreover, Husserl's phenomenology sets out to be a method, rather than an acquisition of results; a method to be applied to ever-new data which lie beyond the results of Husserlian research, and beyond the issues that Husserl

Tymieniecka (ed.), Analecta Husserliana, Vol. IX, 235–247.

concerned himself with, hence, beyond his own personal objectives and interests, and beyond his own cultural horizons.

For Husserl, in contrast to the objectivity of the "thing" which is investigated and determined by the physical sciences, the "thing" (*Ding*) which is perceived by the senses is something that is "purely subjective" (*als etwas "bloss Subjectives"*).[1] But this does not mean that this subjectivity and apparency should be understood as if the things perceived were pure and simple "experiences" (*Erlebnisse*)[2] or mere "signs" (*Zeichen*),[3] and as if there existed some "totally unknown reality" (*eine völlig unbekannte Realität*) lying on the opposite side of objectivity – a kind of hidden cause which could be characterized only via the detour of mathematical concepts.[4] The "physical thing" is in no way extraneous to what is perceived by the senses: on the contrary, it announces itself to some extent by revealing itself through its appearance (*Das physikalische Ding aber ist kein dem sinnlich-leibhaft Erscheinenden Fremdes, sondern sich in ihm, und zwar a priori (aus unaufheblichen Wesensgründen) nur in ihm originär Bekundendes*).[5] In other words, Husserl feels that, from the phenomenological point of view, there does not exist a hiatus between scientific knowledge and pre- and extrascientific knowledge: more precisely, there does not exist that hiatus which modern science uses to distinguish between scientific knowledge and pre- and extrascientific knowledge, somewhat in the manner of the Greek distinction between rational and sense knowledge. This modern distinction leads to an exaggerated evaluation of scientific knowledge and an undervaluation of pre- and extrascientific knowledge. If we admit that scientific knowledge is the grasping of something essential (and it is a debatable point whether it really is essential, and to what degree) we do not thereby mean that pre- and extrascientific knowledge is merely an appearance of reality. Husserl maintains that even in this case, a being, a real being (*ein wirklich Seiendes*), is grasped; and if it is grasped imperfectly, subsequent knowledge will in some way be able to identify it in a more rational manner.[6]

Something real is grasped in any form of pre- and extrascientific knowledge. It is grasped in an original fashion (*originär*) even though this may not be adequate (*adäquate*).[7] One can have an immanent and "adequate" intuition only of the "being-as-experience" (*Sein als Erlebnis*). On the other hand, the thing (*Ding*) can be perceived only as far as it "adumbrates" itself (*sich "abschattet"*):[8] a thing gradually presents itself only under one or another aspect, and announces itself (*"bekundet" sich*) only through modes of appearance (*Erscheinungsweisen*).[9] In each "thing" one can distinguish a nucleus (*Kern*) which actually coincides with what is really presented (*wirklich Dargestelltem*); and it is surrounded by a whole horizon of other elements which are given together with the nucleus, although they cannot be considered as authentic data in the strict sense of the term; then there is a rather vague area of indeterminacy (*Unbestimmtheit*).[10] Out of this indeterminacy, new perceptions, different perceptions, continually take shape, and pass into each other all the time until they are fused into the unity of a perception; and it is within this unity that the "thing," despite its duration and despite its permanence, continues to present itself to us under ever-new aspects. In this way, the indeterminacies receive an increasingly precise determination, even though one cannot prevent the occasional occurrence of the opposite process.[11] The experiences need not necessarily be always concordant: they may also be partly concordant and partly discordant; and, lastly, they may also give rise to absolutely irreconcilable discordancies, such as to induce us to doubt, to modify our judgments.[12] Apart from these incidents, one can establish, as a general principle, that through its manifold apparitions (*Erscheinungen*), through what Husserl calls his perspectives or adumbrations (*Abschattungen*), one and the same thing appears with growing perfection:[13] one can thus speak of an experience that bears witness to the thing in an ever-richer manner,[14] and therefore one can speak of a teleology in the true sense of the word: of the "teleology of the development of an increasingly perfect truth" (*Teleologie der Entwicklung der immer vollkommenerer Wahrheit*).[15] Here it is not the "later" as such that judges the "former." One would rather say that the subsequent evidence springs from the critique of the earlier, and it is in this sense that it corrects it. It is in this progression of our knowledge of the thing that Husserl sees the possibility of an "unlimited development" (of a "*Grenzenlosigkeit im Fortgange*").[16]

During this process of gaining knowledge of things, I, the subject, immediately become aware of the fact that other subjects are simultaneously acting and knowing.[17] Everything that exists, always exists by virtue of an act of awareness; in this sense, therefore, subjectivity constitutes the world. But if all of this is true, it is equally true that, given a plurality of subjects, subjectivity can perform this constitutive function only through intersubjectivity.[18] In this paper I have no intention of attempting to deal with Husserl's approach to the "constitution" of things and the world, but simply want to consider this world of things, this *life-world* (*Lebenswelt*), as it is revealed to us and the manner in which this revelation can be described in phenomenological terms. This world in which we subjects find ourselves to be living, and which constitutes the unitary and permanent object and the background and horizon of every cognitive act we perform, this world is revealed to us as something that we all possess in common: in this sense, says Husserl, "the ontological form of the world is the world for everyone" (*Die ontologische Weltform ist die der Welt für alle*)[19] or, putting this into other words, a "world for all real and possible subjects" (*Welt für alle wirklichen und möglichen Subjecte*).[20] Although to each one of us the world appears in a different manner, one may yet affirm that it is one and the same world that we all know and that gradually takes shape for us, becoming continually realized through one intersubjective community.[21] To live a personal life means to live as oneself and as ourselves within a common horizon.[22] The existence of others as co-subjects is something that we learn and experience through "empathy" (*Einfühlung*) or "empathic attribution" (*Eindeutung*),[23] which is an intuitive act, although not an original one.[24] This being so, it is absolutely impossible to reduce it to analogical reasoning,[25] since it is, by its very nature, a form of experience.[26] We can have an original experience of somebody else's body – through "external perception" (*äussere Wahrnehmung*) – but not of other people's spirit, consciousness, or interior life. In the same way, Husserl maintains that there will always remain a radical difference between "internal perception" or "self-perception" (*innere oder Selbstwahrnehmung*) – an equally original form of perception that we might have of ourselves and our own states of consciousness – and the knowledge or awareness that we can have of other people's states of consciousness through empathy; but this does not mean that we cannot have any kind of certain knowledge of others or of their psychic life as something that is "there, in person" and inseparable from their physical or bodily presence (*Das Andere und sein Seelenleben ist zwar bewusst als "selbst da" und in eins mit seinem Leibe da, aber nicht wie dieser bewusst als originär gegeben*).[27] While not concerned here with an "original experience," we are nevertheless concerned with an experience. And it is within this intersubjective experience that one can have a continuous mutual realization (and it is here that it becomes possible *sich wechselseitig zu korrigieren*). And it is also here that, following our different experiences or discordancies, we can come to a communicative agreement (*in der kommunikativen Einstimmigkeit*) and thus arrive, step by step, all together as a community, at truth (*und so in der Gemeinschaft zu einer Wahrheit zu kommen*).[28] The subjectivities, in their various interactions of agreements and disagreements, constitute a single psychic whole (*so gibt es nur einen einzigen seelischen Zusammenhang, einen allheitlichen Zusammenhang aller Seelen*);[29] they are not in a relationship of exteriority but are rather mutually contained within each other (*alle nicht äusserlich sondern innerlich*).[30] Each subject, by virtue of empathy, experiences awareness of the others precisely because they perceive the same world.[31] The same reality of communication, the same language, all this both presupposes and bears witness to intersubjectivity. A community of experiences, a community of thought and action, becomes established among the "co-subjects".[32] It is a "*communis opinio* of science"[33] and a corresponding "community of scientists";[34] but one attributes a "common consciousness" also to the "society of philosophers,"[35] and even works of art have an intersubjective existence (*Kunstwerke haben ihr intersubjektives Dasein*).[36] One can say that "the people who create culture" are all acting within the totality constituted by "acting personal humanity" (*die leistende personale Menschheit*).[37] Husserl particularly stresses the "new interior community" (*neue und innige Gemeinschaft*)[38] that comes into being among philosophers, where each thinker appropriates and absorbs the contributions made by those who have preceded him in time, reinterprets their works in his own manner, gaining fresh and original insights every time he studies them anew.[39] Thus the history of philosophy as a whole, notwithstanding its involutional and regressive phases, shows a constant trend toward an ever-increasing degree of "self-understanding" (*Selbstverständnis*) in the rational and philosophical sense, and therefore shows a "unitary teleological structure" (*eine einheitliche teleologische Struktur*).[40]

These remarks about the manner in which Husserl conceives knowledge of the "thing" permit us to conclude that his vision undoubtedly leaves room for pre- and extrascientific knowledge with a certain value of objectivity, although the objectivity in question is obviously very different from that of the sciences. Knowledge of the "thing" gives rise to both a certain objectivity and a certain intersubjectivity. In Husserl's works, in fact, we find that this word "thing" (*Ding* or, more rarely, *Sache*) oscillates between two meanings: one of these, the narrower one, designates a material reality, while the wider one is used to describe any kind of reality that can be perceived or grasped by any type of intuition whatsoever (and therefore also a value or a phenomenon of one's own or of other people's consciousness). In this second sense, which is the one most widely used by Husserl, the term "thing" includes the "experiences" (*Erlebnisse*), while in an another and narrower sense – although not necessarily limited to that of a material reality – "being-as-thing" (*Sein als Ding*) is placed in contraposition to "being-as-experience" (*Sein als Erlebnis*)[41] and excludes it. Husserl's concept of "thing" may be of interest for our particular line of thought by virtue of this third and intermediate meaning (which does not include the pure *Erlebnisse* as such) and which at the same time excludes also a mere identification with material reality.[42] When we take the term "thing" to have this intermediate meaning, one of Husserl's passages assumes particular importance, namely, when he writes that everything one can say

> in connection with the natural objectivities as "mere things" must also apply to the axiological and practical objectivities (aesthetic objects, cultural formations, etc.) that are derived from them. Just as, in the last resort, it must also apply to all the transcendencies that constitute themselves in relation to consciousness (. . . Alles, was wir uns hinsichtlich der Naturobjektivitäten als "blosser Sachen" klargemacht haben, gelten muss für alle in ihnen fundierten axiologischen und praktischen Objektivitäten, ästhetische Gegenstände, Kulturgebilde, usw. Und ebenso schliesslich für alle sich bewusstseinmässig kostituierenden Transzendenzen überhaupt).[43]

As is brought out particularly clearly in *Ideen II*, not every reality, not every "substance" or "thing," is necessarily material, extensive, corporeal; in every man, and more generally also in every animal, there is a psychic reality that is qualitatively distinguishable from the corresponding bodily reality, even though it is necessarily and indissolubly associated with this latter. There is a way of considering things, animals, and even men themselves, that permits us to grasp the material, the physical aspects, and this is the manner characteristic of the scientist. But there is also another and different way of considering things, and, in particular, men: a way that permits us to see in them the persons, the spiritual phenomena, as well as everything that forms part of the world of the spirit and of culture.[44] All this is grasped, perceived, in a personal relationship and in an experience of the existential type. It is in this type of active and participating knowledge, by immersing oneself in the interior life of other subjects, that one grasps their motivations.[45]

The road by which one accedes to this kind of knowledge – as we have already mentioned – is empathy. This road, as we have also seen, is not an original one but rather, in a certain sense, an indirect one: everything that relates to the psyche and the spirituality of other subjects we learn solely and exclusively through the experiences that we may have of their bodies. As far as Husserl is concerned, "the body is the means of all perception, it is the very organ of perception, and necessarily participates in every type of perception" (. . . *Ist der Leib das Mittel aller Wahrnehmung, er ist das Wahrnehmungsorgan, er ist bei aller Wahrnehmung notwendig dabei*).[46] Every cultural form is necessarily bound up with some form of bodiliness in which it finds its peculiar physical expression: thus, in the bodiliness that is the essential attribute of every human being, in the notes emitted by a violin, in a manuscript or a printed document, just as in any tangible work of the human spirit, there is always, on the one hand, a founding "physicalness," a basic "somaticity" (*fundierende Leiblichkeit*), and, on the other hand, the meaning that "animates" it (*der beseelende "Sinn"*).[47]

This twofold affirmation made by Husserl is of importance for the purposes of the theme I propose to develop in this paper: he is in fact saying that there exists a spiritual reality, and that it can be known, experimentally in a certain way, through a type of experience that is qualitatively different from the one that enables us to learn the material realities. In getting to know the spiritual realities we once again have a progression, a teleology, and at the same time an intersubjectivity. But it is important to note on what this intersubjectivity of spiritual knowledge is based: we can say quite decidedly that, as far as Husserl is concerned, the intersubjectivity of the knowledge that we can realize of the spiritual realities is invariably founded on the fact that each of these is necessarily connected with a material reality, is linked to it in an indissoluble manner, and that only this corporeal aspect is originally grasped in an intersubjective manner. The spiritual aspect is grasped by means of empathy, but always on the basis of this sensorial knowledge that is the sole original knowledge. *Originally*, therefore, knowledge becomes intersubjectivity only insofar as it is knowledge of corporeal, bodily realities. Husserl regards nature as an intersubjective reality potentially open to all human subjects, who nevertheless can enter it only through their own bodies.[48]

This specific statement made by Husserl cannot but represent a grave limitation for anyone who wants to see whether it is possible to widen the Husserlian phenomenology in order to apply it to experience of the sacred: considering it in its originality, after all, the object of religious experience cannot but be spiritual. The divine communicates itself to all the realities, even the physical ones, and in this participation can also express itself through a physical reality; but if we are to base ourselves on the universality of the religious testimonies, the divine *in se* cannot be anything other than a spiritual reality: the divine is spirit, it is spirituality itself in all its original purity. All the testimony of religious men, of mystics, of saints, shows that they have always conceived experience of the sacred as a "contact knowledge," as an immediate experience – albeit an imperfect one, and even abysmally imperfect and inadequate – of a purely spiritual reality and, what is more, grasped directly in its spirituality. Here we come up against a serious limitation in Husserl's work: true, he does conceive a pure experience of *Erlebnisse* immanent in consciousness; but he does not even vaguely succeed in conceiving any possible experience of spiritual phenomena that present themselves as transcendent, unless and until they communicate themselves through the mediation of experiences of bodily realities. For Husserl, therefore, it is wholly inconceivable that there can be any experience of a pure spirituality, of a sacred reality, of an absolute that can reveal itself as a "totally other," as a *ganz Andere.*

One must hold that this difficulty is not an essential one, necessarily connected with the phenomenological search as such; in actual fact it is connected only with Husserl's personality, with his mentality, with the streams of culture and thought absorbed by him, and also with the particular spiritual sensitivity that took shape within him as a result of all these factors. On the whole, therefore, we can say that Husserl closed himself to experience of the sacred, and that it was essentially this closure that prevented him from developing his phenomenology to include a phenomenology of the sacred. But this does not mean that an appropriate development of certain forms of spiritual sensitivity cannot induce a given subject to open himself to such an experience and to the corresponding phenomenology. Such a subject would thus discover – in fact, directly and personally, experience – Husserl's error in assuming that spiritual knowledge can be obtained solely and exclusively through the mediation of corresponding sensorial experiences of physical realities.

Having stated this important and essential reservation with the greatest possible clarity, we can now return to the problem that we set ourselves from the beginning, which is simply this: in view of what has been said here about Husserl's manner of conceiving pre- and extrascientific knowledge, what relevance can all this have in connection with the recognition that religious knowledge has a certain objectivity and intersubjectivity? Let me say right away that it certainly is relevant. It is true that Husserl did not explicitly come to grips with the problem we have set ourselves, but it is equally true that religious knowledge, by virtue of certain of its specific connotations, can well form part of this type of pre- and extrascientific knowledge that Husserl characterizes in his work. Religious knowledge has certain very peculiar characteristics, and from these I shall here abstract with a view to bringing out only certain features that it has in common with every pre- and extrascientific knowledge, with what Husserl calls knowledge of the "thing." It must be very clear by now that this word "thing" (*Ding*) – which is possibly not the most satisfactory among all the various terms that one could use in this wider acception – designates a *quid* that can be neatly distinguished from phenomenologically "reduced" experiences, and includes every kind of reality, be it material or spiritual; it therefore also includes the sacred, the very thing of which, albeit in widely differing forms, religious knowledge provides some experience.

In certain respects religion is a practice, but in other respects it is also knowledge. (In any case, even a practice, if it is to be truly such, i.e., a way of acting and not just a blind and instinctive way of moving, presupposes a certain awareness, and therefore it is knowledge.) This has to be clearly recognized, because Husserl, when all is said and done, seems to place a somewhat excessive emphasis on the practical (or, as he says, mythico-practical) character of religious experience, thus placing it in contraposition to the theoretical attitude of *θαυμάζειν*; as far as he is concerned, philosophy – in sharp contrast with myth – draws its origin from the *thaumazein*; it is almost as if the religious attitude and even the very formation of myths were underlaid by an analogous feeling of astonishment, of marvel – by an analogous *thaumazein*; or if the factor *contemplation* had to be considered excluded from religious life, whereas in actual practice it appears to be closely connected with it. Undoubtedly, religion has an appreciable pragmatic aspect; but the same thing, in different respects, may also be said of the sciences, which not only develop in close relationship with the corresponding technologies but, even when they are considered as pure sciences, as forms of pure research, tend to manipulate reality rather more than they mirror it (and this aspect of science as a manipulating practice is being brought increasingly to the fore by the development of the critique of science). And does not philosophy itself make similar use of concepts to manipulate reality in one way or another in order to give it a meaning, a significance? We know that phenomenology tends first and foremost to be receptive to the pure revelation of things as such, doing this by means of a rigorous suspension of every "natural attitude," and also of every manipulating conceptualization: in this sense, therefore, phenomenology tends to assume the connotations of the most theoretical of philosophies. Even with allowance for the obvious differences, one may well ask oneself whether a very similar attitude does not characterize the religious who endeavors to make himself as receptive as possible to divine revelation by eliminating the temptation to manipulate it for his own purposes, or even according to his own mental habits, and thus to frame it in the cultural schemes to which he is accustomed and in which he finds himself at ease. It would seem that this receptivity of the religious vis-à-vis divine revelation, this *epoché* to which he subjects every pragmatic motif, this suspension of all judgment in order to place himself in a state of pure listening, and therefore also in a pure speculative and theoretical attitude, is very similar to the substance, to the primary and essential motivation of the attitude that the phenomenologist assumes vis-à-vis the pure manifestation or revelation of things.

If we now pass on to consider what one may call the phenomenological knowledge of the "thing," we may note that the various passages in which Husserl has endeavored to characterize this knowledge, and which have been briefly sketched in the earlier part of this paper, can be used with equal validity to characterize religious knowledge, at least in those essential aspects that it has in common with every existential knowledge, with every pre- or extrascientific knowledge. Basing oneself on what one can extract – and translate into ordinary language – from the attestations common to all the religious, even religious knowledge presents itself as a concept of a reality that in some way transcends the subject. This sacred reality once again expresses itself in "appearances," "adumbrates" itself in *Abschattungen*, "announces" itself in them. And here, too, one can say that every appearance, in some way or another, expresses or announces a being, or in some way participates in it (to the extent to which a phenomenon is authentically religious, it is always a manifestation of the sacred, and this no matter how strange or aberrant certain modalities of the phenomenon may be). There is thus some kind of objectivity (not a scientific one, obviously) in every religious phenomenon. However, it is an object that to some extent always escapes us: it is a reality that transcends the subject and, consequently, the knowledge that the subject can have of it will, in a certain sense, always be inadequate, and this even though, at one and the same time, one may well say that the "modes of appearance" of that object grasp the very object itself and not just a mere image of it, a mere symbol. In every religious intuition there is the revelation of a nucleus, and this is a halo of indeterminacy that tends to become more and more determinate in a succession of "adumbrations," successive approximations that gradually reveal that nucleus better and become more adequate to express it. In other words, in this case, too, we have at least the theoretical possibility of an "unlimited development" of religious knowledge, in the sense of an unlimited deepening of this experience, which is a mytical experience and at the same time an ethical one; an experience of prayer and action, of encounter with the Eternal and temporal commitment; an experience of sanctity to the extent to which this can be actuated in this earthly human condition; an experience of beatitude and of a beatifical vision; a vision that can be realized in a different condition according to the various revelations and theologies. Even in the field of religious knowledge, therefore, one can speak – in the Husserlian sense – of a teleology of the development of an ever more perfect truth, and here, too, one can speak of an intersubjectivity. We are always concerned with a reality that in its way is objective, transcends the subject, and presents itself in different modes of appearance in each one of us. Everybody has his own experience, but in a certain way these experiences are always analogous and can therefore be communicated by means of language, so that there can be not only an exchange of experiences, but also mutual stimulation and mutual correction. A common horizon of both spiritual experience and faith may thus be created. Moreover, in this field one can again speak of empathy, because each one of us can have a certain insight into other people's religious experiences. The remarks about a kind of communion of spirits that Husserl makes when he affirms that the various subjectivities jointly lead toward a knowledge of a world in which all constitute, as it were, a single psychic unit – by virtue of which the souls are not in a relationship of exteriority, but rather one within the other – has particular relevance in the religious field: one need only think of the evangelical image of the vine and its shoots and of the "communion of the saints," a theological concept, an article of faith and a dogma that the Husserlian analysis – quite obviously – does not grasp in all its richness and profundity, although it can in some way introduce us to it and, as it were, lead up to it from a phenomenological point of view.

If the Husserlian phenomenology of pre- and extra-scientific knowledge is correct, and if religious knowledge (or metaphysical, or mystical, or spiritual and interior knowledge, or any other name by which one wants to call it, emphasizing one or another of its many aspects) can be included within the ambit of this knowledge, if both these premises are valid, then it follows almost as in a syllogism that religious knowledge has all the qualities required to be considered as knowledge in the true sense of the word, with a certain objectivity and intersubjectivity of its own, even though neither is of the scientific type. Pre- and extra-scientific knowledge – with its "appearances" or "modes of appearance" in which a "transcendent essence" is not just signified or symbolized, but rather "announces" itself – manifests itself directly and in the first person; and it represents the starting point from which scientific knowledge comes to constitute itself the kind of knowledge that is concerned exclusively with highlighting the "primary qualities", i.e., those phenomena that lend themselves to objective verification, measurement, and calculation. Certainly, scientific knowledge has an objectivity and an intersubjectivity of its own and also its own peculiar type of development, a teleology of its own, always within the ambit and the limits of its peculiar function. And, certainly, scientific knowledge is also a knowledge by full right, even though its object is not the original revelation of the phenomena, but rather the manipulation of these phenomena for very specific ends. Indeed, Husserl himself devotes a rather acute phenomenological analysis to the rise of modern science, an analysis that shows that science is not an exclusive or even privileged type of knowledge, but simply something derived from pre- and extrascientific knowledge by means of a progressive manipulation of its data with a view to peculiar and distinct ends of its own. It therefore follows that it is wrong to say that scientific knowledge not only grasps the original manner in which phenomena reveal themselves, but also grasps them in such a manner as to find within itself a criterion for judging other forms of knowledge and establishing these latter as either less valid than itself or not valid at all. It follows that religious knowledge, insofar as it can take its place by the side of the other forms of existential pre- and extrascientific knowledge, and insofar as it can stand with them as something that is equally original, has full rights of citizenship among the various types of knowledge, and can consider itself knowledge by full right, possessing all the characteristics of objectivity and intersubjectivity that are attributed to knowledge and identify it as such. Further, like every other form of knowledge, it has a teleology of its own: the trend of an unlimited development toward the limiting idea of a truth that all the forms of knowledge pursue by working together, and that none of them can claim to seek on its own.

Pontifical Gregorian University, Rome

NOTES

[1] *Ideen I,* § 52, p. 122. Citations of the various Husserlian texts refer to the appropriate volume of *Husserliana; gesammelte Werke*, Nijhoff, The Hague, 1950–
[2] *Ideen I,* § 52, p. 122.
[3] *Ibid.; Ideen II,* § 56, p. 234.
[4] *Ideen I,* § 52, p. 123.
[5] *Ibid.*, p. 126.
[6] *Krisis*, Beilage X, p. 422.
[7] *Ideen I,* § 149, p. 365.
[8] *Ibid.*, § 42, pp. 96–97.
[9] *Ibid.*, § 44, p. 100; § 61, p. 146.
[10] *Ibid.*, § 44, p. 100.
[11] *Ibid.*; see also *ibid.*, § 41, and *Ideen III,* § 20, p. 104.
[12] *Ideen II,* § 15, pp. 44–45.
[13] *Ideen I,* § 42, pp. 97–98.
[14] *Ideen II,* § 15, p. 44.
[15] *Krisis*, Beilage XXV, p. 491.
[16] *Ideen I,* § 149, p. 366.
[17] *Cartesianische Meditationen,* § § 55–56.
[18] *Krisis,* § 50.

[19] *Ibid.*, Beilage XX, p. 469.
[20] *Ibid.*, § 71, p. 259.
[21] *Ibid.*, Beilage I, p. 349.
[22] *Ibid.*, Abhandlungen, Die Krisis des europaïschen Menschentums und die Philosophie, p. 314.
[23] *Cartesianische Meditationen*, § § 42–62; *Ideen II*, § § 43–47; *Ideen III*, Beilage I.
[24] *Ideen I*, § 1.
[25] *Cartesianische Meditationen.*, § 50.
[26] *Ibid.*, § § 52–54.
[27] *Ideen I*, § 1, p. 11.
[28] *Krisis*, Beilage XVIII, p. 464; see also Beilage VIII.
[29] *Ibid.*, § 71, p. 258.
[30] *Ibid.*
[31] *Ibid.*, Abhandlungen, Naturwissenschaftliche und geisteswissenschaftliche Einstellung, pp. 307–308; Beilage III, pp. 369–370.
[32] *Cartesianische Meditationen*, § § 55–58.
[33] *Krisis*, Beilage XXVI, pp. 492–493.
[34] *Ibid.*, Beilage XXVII, p. 507.
[35] *Ibid.*, Beilage XXV, p. 495.
[36] *Ibid.*, Beilage XXVII, p. 506.
[37] *Ibid.*, Beilage III, p. 381.
[38] *Ibid.*, Abhandlungen, Die Krisis . . . , p. 334.
[39] *Ibid.*, Beilage XXVIII.
[40] *Ibid.*, § 73, pp. 273, 276; Abhandlungen, Die Krisis . . . , 347–348; Beilage XIII, pp. 435, 442–443.
[41] *Ideen I*, § 42, p. 95.
[42] As, for example, in *Ideen I*, § 44, p. 101; § 52, pp. 129–130; and generally in the paragraph as a whole.
[43] *Ibid.*, § 52, pp. 129–130.
[44] *Ideen III*, Beilage I, § 3.
[45] *Ideen II*, § 51, pp. 190–191, 198–200.
[46] *Ibid.*, § 18, p. 56; see also *ibid.*, § 46, p. 167, as well as *Ideen III*, § § 1–3 and Beilage I, § 2.
[47] *Ideen II*, § 56, pp. 220, 240; Beilage IX, pp. 320–321; *Ideen III*, Beilage I, § 4, pp. 117–118.
[48] *Ideen II*, § 18, p. 86.

FERDINANDO L. MARCOLUNGO

THE PHENOMENOLOGICAL HORIZON AND THE METAPHYSICS OF THE PERSON ACCORDING TO GIUSEPPE ZAMBONI

Giuseppe Zamboni's work (1875–1950)[1] can be placed within the context of the conflict of positivism and neoidealism which occurred between the two world wars in Italian philosophy. Professor at the Catholic University in Milan between 1921 and 1932, Zamboni represents a significant example of a deep and keen thinker – and one who has remained unknown. During his life he was compared with Husserl and called the "Italian Husserl," not because of any exchange or contact they had (actually they had none) but because of the close resemblance on some essential points at which the two philosophers had arrived independently.[2] It is not my intention here to provide a comparison of the two men's work. I will instead limit myself to some essential points of *Gnoseologia pura* or *Filosofia dell'esperienza immediata, elementare, integrale* in the light of the theme of the present volume: "The Teleologies of the Husserlian Phenomenology."

The phenomenological attitude finds its expression in the radical bracketing (the Husserlian ἐποχή) that it is necessary to make not only with respect to ordinary and spontaneous thought but also with respect to scientific convictions and philosophical theories. In the analysis of experience it is necessary to reach what is *present and manifest*, what does not presuppose anything else, but finds its justification in the originality of its givenness to pure consciousness. This is also Zamboni's idea. Zamboni writes:

> All that is immediately present to the consciousness, there is indeed, is existent, is real, is not a deception of illusion. And as the contents presented by senses are real, so the contents presented to the consciousness of the ego are real and not vain illusions.[3]

Referring to the criteriology of Mercier, whom he had known at Louvain before 1921, Zamboni asserts that the reference to what is present and manifest is the last criterion of truth: it is unnecessary to fix any criterion of evidence beyond the immediate givenness, understood in the integral experience: "What is immediately present and manifest, does exist, at least in the measure and in the form in which it manifests itself".[4]

Consciousness is according to him originally open to what one offers to it in experience: if we refer our propositions to the givenness in its purity, we have reached that primordial ground from which every philosophy ought to

Tymieniecka (ed.), Analecta Husserliana, Vol. IX, 249–254. *All Rights Reserved.*

start. Diversity among philosophies, often incompatible, is due, according to Zamboni – as well as to Husserl[5] – to insufficiency of that gnoseological analysis that should precede every system: experience must be received in all its components, in the integrality of its original context: this is the only way to escape unilateral absolutizations and to conform as much as possible our thought to lived experience, to *vecu*, to the life-world, to use Husserl's expression in *Crisis*.[6]

Pure gnoseology therefore makes an accurate inquiry into what is "present and manifest" to consciousness. The phenomenological horizon is assumed in its whole richness, without any presuppositions or preclusions. The concept of "phenomenon" assumes a radical and original meaning: "phenomenon" is the "reality that presents itself and manifests itself": "This is the principal meaning of the word 'phenomenon': not an empty appearance, not an empty aspect, but a content that presents a true objective reality to a subject in cognitive function".[7] By using the term "reality" we do not intend to undertake our research by presupposing at the start a metaphysical realism: "real" is all what *appears*, in that it appears, i.e., in that it is present and manifest within the limits and in the ways in which it manifests itself. Zamboni writes: "In order to avoid confusions, I will call *being, in the proper sense*, 'substance'; in an *analogical sense*, the 'created substance' and 'God'; and I will call *'reality'* or *'real in general,'* being understood in the gnoseological sense".[8]

The metaphysical expansion of the Zambonian research is expected to take its starting point from the phenomenological ground envisaged in its fullness without any limitation. In opposition to the positivistic and psychological reductionism, the intentionality of knowledge appears then with clearness, in that consciousness is just "the sphere of all that which is present and manifest to the subject, immediately".[9] The Husserlian ego is specified in Zamboni as *self-conscious ego*: "The root of every intellectual work consists in self-consciousness; it is in the consciousness where being and thought coincide: thought is fundamentally the self-manifestation of being".[10] What is present and manifest is such in virtue of the *intellectual light* of self-consciousness that intimately illuminates the given and objectifies it, i.e., makes it an object of attention, and therefore of knowledge:

> 'The perception of the presence of a sensitive datum,' 'the distinct perception of the presence,' and 'the mental objectivation' are synonymous expressions. Here we evidently touch the limit between sense and intelligence, between brute sensation and intellectualized sensation; and we can form a clear concept of the difference between the human psyche and the pure animal psyche.[11]

The given datum, as that object of attention, becomes "this," and in "this" one distinguishes individuality and structure, existence and essence. Zamboni utilizes the classical doctrine of abstraction and underlines its relevance within the phenomenological plan: abstraction is not "to extract a part from the whole," but to catch essence and structure of the given, apart from the only individuality *hic et nunc*. Essence is called precisely *tality*,[12] in the sense that it expresses all that makes a thing to be one thing and not another: as regards other possible data of experience, the process of universalization does not add anything to the essence of "this" datum – the object of attention – but it expresses the possibility of further references. The validity and the necessity of our cognitions are founded on the possibility of reaching the essence of what is given to us in the phenomenological perspective.

The ego remains, in self-consciousness, however, at the center of the whole field of cognition. In this light Zamboni interprets Kant's transcendental apperception:

All these first operations of intelligence because of their possibility suppose such an ambient as is not presented but by a consciousness, at the same time one and multifarious, and completely penetrated with the intellectual light of self-consciousness; and it is exactly that which permits not only the synthesis, but also the analysis of the complex of consciousness (subject and object) and of the datum (essence and existence). In my opinion, this is the meaning of the transcendental apperception, self-conscious unity, a light, not only synthetical, but analytical.[13]

Thought, however, does not create what is present and manifest to it; that is, consciousness is intentionally open to receive being: "Thought is the manifestation of being; therefore, being is fundamental with respect to thought that reveals it".[14] It is up to the researcher to investigate the limits and the conditions of what is present and manifest, without indulging in empty speculations.

Among the original contents of consciousness the analysis of integral experience indicates not only the data of sensibility, but also the states and the acts of ego. The fact that – on the ontological level – these latter belong to a knowing subject, does not affect their cognitive objectivity. Up to this level being is understood by Zamboni as a simple *to be* (*esserci*), i.e., as simply "to be present and manifest to the consciousness"; as a reality in the gnoseological sense. But, in addition, there appears on the surface of the self-consciousness an element of the energetic order. This element is also originally present and manifest. It is real and actual and it comes to appear in its evidence when acts and states of ego, purely remembered, are compared with the same states and acts being actually performed.

Only by this *existential energy*, that appears on the surface of self-consciousness, is the passage from the phenomenological level to the metaphysical depth possible. At this deeper level being qualifies itself no more as a simple *to be* (*esserci*), but as an *act of being, actus essendi*, an expression taken from Thomas Aquinas. Writes Zamboni:

> Being belongs to the order of action, act or energy, and not to the order of quality and essence; therefore, in order to keep something in mind, it is not sufficient to wait until an image presents itself by means of association, but it is necessary to rebuild a complete psychic process, in which being and existence may enter as energy in actualization of this something.[15]

Telos – which in Husserl is reflection which leads to the constitution of transcendental *ego* – expresses itself in Zamboni's thought in the affirmation of the focal and fundamental role of the act of being with respect to the whole phenomenological field of *to be*. Zamboni reaches, however, a realistic conclusion founded on his metaphysic of being, in opposition to the idealistic Husserlian position, although the concept of being is not given by the experience of the external senses but by the experience of the *ego*: *ontological experience*. Hence we acquire the knowledge of the others and of the things in our mental development. From the experience of the senses, elaborated by intelligence, we gain the first principles of the reason, mathematics, and logic; but it is from the experience of the *ego*, elaborated by the formal functions of the intelligence, that we obtain the cognition of ontology, or the metaphysics of the substantial being.[16]

Experience acquires sense and ontological dimension only if referred to the self-conscious ego, concrete and individual. It is not a question, however, of discovering *a priori* principles, but rather of deepening experience in its concreteness. Self-consciousness does not consist in applying universal concepts of individual and substance and being to grasp concrete experience, but, on the contrary, it is the experimental nucleus from which these concepts derive. Then they may be applied to what sensibility presents us with, that is, to the exterior aspects of external things, e.g., of substances, plants, animals, and men. "The subject *knows* these realities as substances" – writes Zamboni – "through consciousness that he has of himself as an individual real substance".[17]

The human person is not characterized in a psychological or personalistic sense, but in a metaphysical one. The authentically metaphysical instance that supports the affirmation of the ego, as self-conscious substance, finds its expression in the act of being, as the principle and basis of the experience, that is, of the being as appears at the phenomenological level. On this point

Zamboni is again close to Husserl. Indeed, knowledge of the *human person*, in its structure and its developments, includes the whole experience in its largest meaning as well as the knowledge of the cognitive faculties and means of evaluation and judgments. Thereby the metaphysical extension of knowledge relies upon the results of gnoseology.[18]

In the Zambonian program, metaphysical investigation of experience leads to the affirmation of the *human person*: the focal point and the *telos* of this inquiry reside in the metaphysical pregnancy of the act of being, contained in the self-consciousness of the ego.

The idea of a '*telos*' which finds expression in the pure gnoseology partakes of the *radicality* of the inquiry as well as of its *metaphysical pregnancy*: it is the human person, in the whole richness of its components, as the center toward which the gnoseological perspective tends and from which the spheres of ethics and of the research into the Absolute depart.

The act of being accomplishes, in the explication of this *telos*, an essential function. The human person is the self-conscious subject in which the act of being and its foundation are combined, and all developments of our knowledges tend to explicate them.

University of Padua

NOTES

[1] Among the numerous works by Giuseppe Zamboni, we indicate the most significant volumes in chronological order: *La gnoseologia dell'atto come fondamento della filosofia dell'essere. Saggio di interpretazione sistematica delle dottrine gnoseologiche di S. Tommaso d'Aquino*, Vita e Pensiero, Milan, 1923; *Introduzione al corso di gnoseologia pura*, Vita e Pensiero, Milan, 1924; *L'origine delle idee*, Soc. Ed. Veronese, Verona, 1924; *Sistema di gnoseologia e di morale. Basi teoretiche per esegesi e critica dei classici della filosofia moderna*, Studium, Rome, 1930; *Studii esegetici, critici, comparativi sulla "Critica della Ragione pura"*, La Tipografica Veronese, Verona, 1932; *La gnoseologia di S. Tommaso d'Aquino*, La Tipografica Veronese, Verona, 1934; *Metafisica e gnoseologia. Risposta a Mons. Francesco Olgiati*, La Tipografica Veronese, Verona, 1935; *Il realismo critico della gnoseologia pura*, La Tipografica Veronese, Verona, 1936; *Realismo – Metafisica – Personalità*, La Tipografica Veronese, Verona, 1937; *La Persona Umana soggetto autocosciente nell'esperienza integrale. Termine della gnoseologia – Base della metafisica*, La Tipografica Veronese, Verona, 1940; *La "filosofia dell'esperienza immediata, elementare, integrale" per la completa autoconsapevolezza dello spirito umano*, La Tipografica Veronese, Verona, 1944; *Itinerario filosofico dalla propria coscienza all'esistenza di Dio*, La Tipografica Veronese, Verona, 1948; *La dottrina della coscienza immediata (struttura funzionale della psiche umana) è la scienza positiva fondamentale*, La Tipografica Veronese, Verona, 1951. For a bibliography see Tiziano Adamoli, *Bibliografia*

zamboniana, in *Verona a Mons. Giuseppe Zamboni*, Fiorini, Verona, 1969, pp. 179–240. For critical studies see in particular *Studi sul pensiero di Giuseppe Zamboni*, Marzorati, Milan, 1957; L. Zerbini, *Giuseppe Zamboni e critici. Terminologia diversa e dottrina convergente su problemi gnoseologici*, Pont. Ateneo Antoniano, Florence, 1963; Demetrio Da Crema, *La questione del mondo esterno nella filosofia di Giuseppe Zamboni*, Milan, 1965; Giovanni Giulietti, *La filosofia del profondo in Husserl e in Zamboni* (*Uno studio comparativo*), Canova, Treviso, 1965; *Verona a Mons. Giuseppe Zamboni*; F. L. Marcolungo, *Scienza e filosofia in Giuseppe Zamboni*, Antenore, Padua, 1975.

[2] For a comparison of Husserl and Zamboni see Paolo Valori, *Il punto di partenza della filosofia in G. Zamboni ed E. Husserl*, in *Studi sul pensiero di G. Zamboni*, pp. 667–74; Novato Picard, *Gnoseologia pura e fenomenologia trascendentale*, pp. 675–88; Giulietti, *La filosofia del profondo in Husserl e in Zamboni*; for an account of meetings between the two philosophers see Zamboni's discussion of conversations which occurred at Freiburg i.B. after the publication of Husserl's *Meditations Cartesiennes* (Zamboni, *Interesse egocentrico – Interesse teoretico – Interesse storico*, Verona, 1947, p. 8).

[3] Zamboni, *Il realismo critico*, p. 65.

[4] *La dottrina della coscienza immediata*, p. 54; surprising is the analogy between Zamboni's passage mentioned and Husserl's "principle of all principles" (*Ideen*, I, Section 24). For the comparison with Mercier, see Ignazio Bonetti, *Dalla criteriologia alla gnoseologia*, in *Studi sul pensiero di G. Zamboni*, pp. 611–21.

[5] See Valori, *Il punto di partenza*, p. 670.

[6] "The theoretical research into the bases of the cognitive processes must be conceived as a progressive discovery, that will be completed only when one can scientifically realize the total psychical life of the human nature; i.e., when the gnoseological thought system will equalize the mental lived organism (*vecu*, really tested)" (Zamboni, *Sistema di gnoseologia*, p. 6). For a further critical deepening of this theme, see Picard, *Gnoseologia pura e fenomenologia trascendentale*, pp. 686–87.

[7] *Metafisica e gnoseologia*, p. 75.

[8] *Sistema di gnoseologia*, p. 75.

[9] *Il realismo critico*, p. 65.

[10] *Introduzione al corso di gnoseologia pura*, p. 60.

[11] *L'origine delle idee*, p. 48.

[12] See *La Persona Umana*, p. 179.

[13] *Studii esegetici*, p. 248.

[14] *L'origine delle idee*, p. 54.

[15] *Ibid.*, p. 62.

[16] *Realismo – Metafisica – Personalitá*, p. 93.

[17] *Itinerario filosofico*, p. 93.

[18] *La Persona Umana*, p. 11.

EUGENIO BORGNA

THE MELANCHOLIC CONSCIOUSNESS OF GUILT AS A FAILURE OF INTERSUBJECTIVITY

In recent psychopathological investigations some fundamental anthropological structures have been exposed which can best be appreciated and described in their intentional and teleological signification through the use of phenomenological methods.

In the present study we propose to vindicate the clinical phenomenon of melancholia, freeing it from any ideological and scientific-naturalistic reification, and to recover it in its *sense*-horizon and apart from considerations of the causal or genetic-formal *raiṣon d'être* of melancholia as a psychotic experience. We turn, to this effect, to Husserl's phenomenology and to the phenomenological investigations of such scholars as Binswanger,[1] von Gebsattel,[2] Minkowski,[3] Ruffin,[4] Kurt Schneider,[5] Schulte,[6] Straus,[7] and Tellenbach.[8]

Melancholia is regarded by clinical psychiatry as the emblematic evidence of the metamorphosis (ultimately a metaphysical one in Nicolai Hartmann's understanding)[9] of the endogen. It is marked, in its empirical *raison d'être*, by a "vital" or "psychic sadness," as Scheler would say;[10] by a "vital inhibition," as Straus would see it; and particularly by an intentional experience of guilt which is the constitutive thematic region (that is, the essence) of the most radical melancholic metamorphosis. The deep signification and the phenomenal articulation of melancholia can be grasped in its essence only when we transgress the validity of all objective sciences,[11] and the thematic area of the life-world emerges.

Our study aims at a description of the life-world of a melancholic person and of the epxerience of guilt in particular – as being formed at the level of the life-world; it focuses upon the distortion of intersubjectivity.[12] Obviously, the first *epoché* to be performed in our investigation is the reduction to the life-world: guilt as well as anguish and hope as well as despair are parts of it.[13]

In the course of our inquiry we intend to describe and analyze the melancholic guilt phenomenon, first, in its sense content (*Sinngehalt*), and, second, in its sense-genesis (*Sinngenese*) – the latter sense rooted in Husserl's concept of the life-world.

The phenomenology of guilt (moral guilt and existential guilt, in fact) has been masterly outlined by Ricoeur,[14] who brought to light the deep reasons

Tymieniecka (ed.), Analecta Husserliana, Vol. IX, 255–265.

and the framework of reference of an unfathomable sense of this phenomenon. Guilt is "la conscience d'être accablé par un poids qui écrase" and it is "la morsure d'un remords qui ronge du dedans." Guilt also is "une manière de se tenir devant une sorte de tribunal invisible qui mesure l'offense, prononce la condamnation et inflige la punition." The description of the significance implied in the experience of guilt as a consciousness and as the value experience is carried on by Ricoeur with unremitting semantic depth. "Le sentiment de culpabilité est la conscience d'être inculpé et incriminé par le tribunal intérieur"; that is, finally, to say that guilt "est devenue ce malheur sans retour décrit par Kafka: la condamnation est devenue damnation." But just at the point at which Ricoeur's analysis reverses to its depths, the *sense* of 'guilt' indicating its infinite life-world expansion and its emblematic referential dependence, we read that "la coulpe, *culpa* in latin, est l'auto-observation, l'auto-accusation et l'auto-condamnation par une conscience dedoublée."

The phenomenological analysis of melancholic guilt compels us to face guilt seen as essentially comprising this boundless (dehistorified) condemnation and this infinite (eternal) damnation that changes in melancholia the life-world into a desert, destitute of sense. It appears as intended by consciousness (the experience of consciousness). But the teleological (and temporal) horizon of the person seems distorted; its intentionality and its relationship with alterity appear shattered.

Melancholic guilt, then, is understood as coupled with a burning relentless condemnation and damnation (under the jurisdiction of an "unshakable logic of the law," in the sense of one of Kafka's words); it is seen as a deformation of a shattered intentional and teleological horizon, of the *conscience dèdoublèe*. The latter can neither restore the distorted temporal sequence of experience nor live it and transform it by means of an orientation toward an end, a *telos.*[15]

So described the phenomenon of melancholic guilt emerges with dazzling transparency from the words and the testimony of one who falls into Kierkegaard's "deadly disease" (melancholia, that is), as it occurred to one of our patients, Maria Teresa. As a shadow that, with an upsetting, breathtaking deceleration extends and reabsorbs into itself every mundane sphere, so the guilt experience has incinerated the intentional horizon[16] of Maria Teresa, who cannot experience anything but an implacable (and inexhaustible) *cogitatum* of a frozen, and parched, *cogito*, Here is Maria Teresa's emblematic self-testimony:

The truth is that there has really been the guilt. The consummated guilt. A very painful truth. Let them trample on me. I have this huge guilt gnawing into my soul. Mine is a guilt that cannot be erased. Any guilt can be erased, but not this one. Do not waste your time with your treatment: I only deserve prison and punishment. Men neither forgive nor punish. I do not pray any more because I know I am no longer forgiven. I have no more faith nor hope. One cannot live without hope as one cannot live without peace, but me, I have lost my peace and my hope forever. I want to attend myself the trial being tried against me. I did not say the truth. Nor are those whom I slandered guilty. I maintain this, although it might appear that I have stuck to a truth that is no truth. But I will go and defend myself in court. I made unjust charges. It is senseless for me to confess: my sin is unforgivable.

There is no other value horizon (no other value horizon is possible) for Maria Teresa but this one: guilt as a disfigured image of "a truth that is no truth." Guilt is no longer part of an intentional horizon that might constitute with reference to a teleological meaningfulness (as of redemption, or of ransom in forgiveness). In the collapse of intentionality as teleological orientation, which the psychotic experience inexorably drags along with itself, the experience of melancholic guilt does not refer to a goal transcending it. It thus is transformed into an experience that is unrelated to, and cut off from, any (possible) metamorphosis in time. Melancholic guilt no longer has a history, nor does it evidence a history: it is burned and plunged into the abyss of solitude's horizon of radical impossibility. There is no forgiveness, and there is no longer any hope; guilt impending as a dehistorified event and a "trial" incalculable in its pending relentless presence. Confession has become entirely destitute of any of its cathartic functions: Maria Teresa's religious experience burns in the fire of a hopeless desolation, as her own words admirably and tragically bring to evidence.

The phenomenology of melancholic guilt, and of the experience of guilt consciousness, finally meets its emblematic counterimplication of sense in the conception of the description of the "common surrounding world" outlined by Husserl.[17]

In general terms, the surrounding world is not the world 'in itself' but a world 'for myself'; it is precisely a world surrrounding *his own* subject-self, experienced by him, present to his consciousness. A world placed within a particular intrinsic content of sense of its own through the intentional experiences [*Erlebnisse*] of the subject himself. As such, it is always somehow becoming, in a constant self-generation through sense evolutions and ever-new sense formations involving inherent positions and eliminations.

The melancholic consciousness of guilt allows no other sense-giving (of any *Sinngebung*) except for the content of guilt, and of *this* particular guilt. There are no longer any sense evolutions or ever-new sense formations; there is no longer, indeed, a "becoming," but a radical destitution of "becoming." Intentional experiences make "a" surrounding world (the one of damnation in guilt), separated from any other sense-giving than of guilt, as the last unlimited and petrified horizon of transcendence. Admittedly, in this thematic district one may wonder whether – in the intentional guilt experience taking place in melancholia – one is dealing with "fulfilled acts" or with "unfulfilled acts," which are, according to Husserl, acts failing to be accomplished or "fallen out of fulfillment" ("motions of acts"). Now, under psychopathological conditions of the *Wahnstimmung* in the *Vor-Feld* of a delusional psychotic experience, the distortion of the intentional system with its interrelations, that is, of the complete sense-horizon, draws along the laceration of the experiences. The self *does not* live in them as the "acting subject," in relinquishing any fulfillment of acts and producing "unaccomplished acts." Nevertheless, when the *Wahnstimmung* dies out like a burned-up comet and changes itself into a delusional guilt consciousness, such collapse in act-accomplishment does not happen. Guilt appears then as an experience of unshattered consciousness, that is, a consciousness not destitute of some meaningfulness.

In the preceding discussion we have tried to delineate the melancholic guilt phenomenology in its *sense content*. Let us now try to analyze its *sense genesis* in the *life-world* horizon. We propose to analyze and describe the experience of melancholic guilt in its deeper signification as a fundamental structure of the life-world, and in its emblematic relational and teleological *raison d'être* as a failure in intersubjective communication, in meeting the other self.

The phenomenological foundation of the melancholic consciousness of guilt considered in its sense genesis as a radical failure of intersubjectivity (of intermonadic subjective life) will be investigated also within the thematic context of the melancholic life-world. Obviously this is *one* possible viewpoint from which the problem of guilt as a failure of intersubjectivity can be scanned. The life-world is, after all, dialectically and inexorably implicated in the region of the ego. As Szilasi pointed out:[18]

> Auf Grund der immanenten Transzendenz hat jedes Ich seine eigene primordinale Welt (primordinale Transzendenz), auch der pathologische Mensch (zum Beispiel der Melancholiker), indem er die Weltfülle der Aktivität seines reinen Ego entsprechend für die ihm eigene Lebenswelt reduziert. Es wäre für dieKonstitutions-Phänomenologie eine lehrreiche Aufgabe, diese Reduktionsschritte zu verfolgen.

We will not probe into the thesis proposed by Szilasi, but instead shall confine ourselves to the "first" Husserlian reduction. We will orient our phenomenological analysis starting from the melancholic life-world by understanding the experience of guilt consciousness as a failure of intersubjectivity. Nevertheless, even when approaching the issue within the horizon of meaning of the life-world, we cannot lose from sight the teleological signification of the problem of melancholic guilt. In point of fact, according to Waldenfels, "Welt, Wir und Ich fügen sich zusammen zu einem einzigen Geschehen in der mundanen und sozialen Teleologie".[19]

The life-world is given in originary experience which is beyond any scientific and philosophical category. As Husserl put it, "in the life we live together, we have in common an already given world, the world which is there and that is valid for us, the world which we are a part of, even in our living together, the world for us all, the world already given in this sense of being." The *life-world*, "for us who live awake in it, is always already here, is always already for us, is always the "ground" of any praxis, either theoretical or extratheoretical." In the *life-world* "there are also men who, by their human acting, by their doing and suffering, and by their social bonds, live within the common horizon of the world and who know they are in this world." There follows the well-founded assertion: "Von der Lebenswelt zu sprechen, wie Husserl es tut, heisst, sich der primären Realität unserer unmittelbaren Erfahrung zuwenden, der Welt der Bedeutungen, wie sie sich handelnden Menschen direkt darbietet" (Natanson).

Let us analyze the *life-world* of the guilt pathology in relation to the thematic horizon of the intentional conception of Husserl's life-world, with its ramified interrelation system, to see *how* it deflects itself in a melancholic experience submerged in, and committed to, the intentional experience of guilt.

According to Paci, "*Life-world* structures must be grasped from within, that is to say, they must be resolved into subjective structures in which the subjective transcendental life is a life that experiences the world." In a melancholic *life-world*, the experience of guilt consciousness appears as a sense formation that excludes any other possible variant of the intentional system. There is in fact no other psychopathological experience that could so radically transform (and reabsorb) the intentional structures of the *life-world*, drawing along with it the radical metamorphosis of the structures of temporality, of mundane diaphaneity, and of intermonadic subjective life.

Guilt as content of conscious acts (as *sense*) permeates the complete spread of consciousness, which becomes petrified into an inert dehistorified immobility. This flooding and infinite stagnation of guilt in the intentional horizon of consciousness is possible only in the wake of a radical metamorphosis of temporality. Temporality indeed is truncated in its phase structure. In the temporality of the melancholic *life-world* there is, in fact, no phase of protensions; experience is reabsorbed instead into the abysmal dimension of inexorable retentions, hence there is no more becoming. When the project-oriented protension vanishes, with it vanishes historicity; there remains only a stultified, guilt-devoured patient.

In the melancholic *life-world*, the mundane structures of meaning are also deformed in a radical metamorphosis. They are emptied of "familiarity," and of "light," and consciousness is changed from a positing to a merely allusive mode. The world is from its natural existence transformed into an enigmatic and allusive landscape. Lightness of experience changes into heaviness, heaviness of glancing and suffering, motion and speech; light changes into dazzling darkness. The melancholic *life-world* crystallizes in the terrifying experience of a desert without *any* hope, terrifying with its icy silence of human and mundane realities, and submerged by this guilt which in itself has no boundaries nor aim or reason.

The world to which melancholic consciousness refers is a world resulting from this inexplicable sense-giving; consequently it is terrifying by its surrounding sense of the unknown. "Familiarity is extraneity at the same time. As universal familiarity, the world is, in a sense, the possibility for extraneity to become unknown; the possibility for the unknown to change continuously into known".[20] In the melancholic *life-world* there is no longer any familiarity, but precisely the unspeakable scanning of infinite extraneity. And yet, "Die Erkenntnis des Vertrauten und des Fremden [writes Natanson] ist eine transzendentale Voraussetzung für die Möglichkeit von Erfahrung."

The radical impoverishment of experience, which occurs in any melancholic consciousness, here finds some possible phenomenologico-transcendental explanation. However, within the strictly phenomenological framework, melancholia could not be conceived as experience since the distorted guilt experience absorbs and nullifies all the virtualities of consciousness, any possibility of another type of experience. In fact, a mode of an infinite distance peculiar to melancholy distorts also the spatial expansion of the life-world and moves any other form of *life-world* away.

Finally, we intend to show how the melancholic life-world is related to the intersubjective foundation and teleological horizon of constitutive consciousness.

Melancholia as a psychopathological and clinical experience is characterized by a condition of radical autistic solitude – particularly investigated by Kranz.[21] This condition does not allow for existence of any realistic mundane and interhuman inclination, plunging it into a region of inaccessible Utopia. In fact, the experience of melancholic guilt shows the clinical signification of the autistic condition in its expanded and definitive form radicalizing it. If we try to grasp the profound *raison d'être* of the melancholic autism and ultimately its peculiar *sense* formation, we have to relinquish a mundane (natural) attitude once again and recover the phenomenon in its distorted form of being-lost-in-the-world over against the normal intentional experience. In fact, within the intentional horizon of the real world vindicated in its intentional and transcendental signification, the autistic condition is in melancholia nothing else but the clinical metaphor of an unfathomable (and uncommunicable) metareality. Only the phenomenological *epoché* allows us to grasp it in terms of the thematic region of melancholic life-world as the emblematic manifestation of the failure to constitute intersubjectivity. In fact, the failure to constitute an intersubjective life is nothing but the failure to constitute the other-than-oneself.

The melancholic life-world devoured in its existential richness by the guilt experience is a *life-world* in which intentionality, as we have seen, is distorted and intersubjectivity is destitute of any teleological horizon. And yet, we can say with Paci that "Teleology has its place in existential and historical life. Man's existence is teleological existence. To be human means to be *telos*-oriented (*Menschsein* is *Teleologischesein*)."[22] In the melancholic *life-world*, on the contrary, we are obviously unable to find a sense-horizon, nor can we constitute the other in an intermonadic community[23] founded on a *telos* that would project them into the future.

In the distorted intentional scheme of a melancholic consciousness, intersubjective life is *not* constituted. The encounter with the other human being is restricted to the inert enigmatic alterity that does not carry on an intentional relation between the life-world of those who are in melancholia and those who exist within a world common to all. Human history, the history of all men, is not the history of melancholic consciousness, which is, in itself, radically dehistorified. Where the constitution of an intersubjective world fails to come about, it fails also the *raison d'être* of the empathy: the natural links with the *life-world* of the other person are broken. The intentional sense-giving links which we project toward the other who is encapsulated in his melancholic *life-world*, encounter rationality radically different from our own; hence the silence of communication.

On the one hand, as Merleau-Ponty puts it: "La communication ou la comprehension des gestes s'obtient par la reciprocité de mes intentions et des gestes d'autrui, de mes gestes et des intentions lisibles dans la conduite d'autrui".[24] In the melancholic metamorphosis of meanings, this is not possible. The other is constituted by melancholic consciousness *not* as an alter ego, but as a shadow walled up in his own solitude and silence; hence the impossibility of any communicative (and dialectic) reasonance.

On the other hand, the other constitutes melancholic consciousness in its lack of freedom and in its stiffened intentional relation without a horizon of transcendence. The radical destruction of intersubjective life and communication that take place in the melancholic experience of guilt, obviously find here their ultimate constitutive reason.

Furthermore, in melancholia, in which consciousness is overflowed by guilt, any intercourse (any association of individuals in which "there is constituted in original manner *societas*" and in "I come forth out of my solitude only as a person among persons")[25] which otherwise focuses upon a teleological orientation, that is, always a projection into the future, either, as we have already mentioned, fails to take place, or develops with an intense resistance. The dialogue experience turns into its opposite, that is, the monadic solitude. Just in radical opposition to the dialogue, which "Dans l'expérience du dialogue [asserts Merleau-Ponty], il se constitue entre autrui et moi un terrain commun, ma pensée et la sienne ne font qu'un seul tissu, mes propos et ceux de l'interlocuteur sont appelés par l'état de la discussion, ils s'insèrent dans une opération commune dont aucun de nous n'est le créateur." The failure to constitute an intersubjective life ultimately means that *this* dialogue (communication at large) is not possible.

As for the unsuccessful communication, it would have to develop – and to become articulated – within the life-world. As Natanson emphasizes, "Von der Lebenswelt zu sagen, dass sie vor allem eine intersubjective Welt sei, heisst nicht nur behaupten, dass Kommunikation darin stattfindet, sondern dass Kommunikation sie erst ermöglicht." But in the radical transformation of the communication structure, the *life-world*, which is founded on communication, collapses; communication is truncated. Here the clinical definition of the radical psychopathological state of the melancholic autistic condition, which sees it as "the loss of the world" (*Welt-losigkeit*), finds its constitutive reason. In "the loss of the world" is summarized the desperate melancholic denial of the world and of any other-than-oneself.

The exfoliation of the nature of melancholia as related to the experience of guilt has allowed us also to see that the melancholy metamorphosis of the *life-world* has itself a deep significance for the understanding of intentionality. The intentional analysis of melancholia consciousness (as submerged by the guilt experience and as besieged by autistic solitude) is, in fact, the conclusive evidence for Husserl's thesis that teleology and intersubjectivity are founded in the temporal nature of consciousness. In the dialectic circularity of our reflections, various modal aspects of this thesis have reemerged. We have obviously refrained from investigating the possible reasons for the clinical onset of the experience of guilt; and from asking *why* one can suddenly be swept away by this experience that transforms the *sense* of a life, and of the intentional system of consciousness. We have restricted our scope to a description of the tragical state of melancholic guilt. The experience of guilt is the phenomenon that, in its emblematic eidetic evidence, establishes the melancholic life-world. Nevertheless, guilt in itself, and the failure to accomplish intersubjective life as well as the disappearance of the teleological perspective of existence, which become apparent in guilt, are not possible and understandable without relating them to the metamorphosis of temporality, especially to the disappearance of the phase of the future.

The last problem now before us is to ask whether such failure of direct communication in melancholia that distorts the possibility of the intersubjective *life-world* could be incorporated into some type of communication for the sake of an hermeneutical and a psychotherapeutical approach.

In fact, leaving aside the mundane dimension and attempting to transcend our life-world in the perspective of "participation" and "disconcealment" of the world of the one who suffers in the anguish of melancholia, it is still possible to grasp the sense, or the nonsense, of an existence transformed by guilt into autism and still longing (despite everything) for the testimony of the *other*. The emblematic expression of the *sense* of a witness with its essential significance as well as with its *limitations* lies in the following words uttered by Maria Teresa: "I am in a prison: only with you, who ask after my sadness and my suffering, I think I live. But out of here, there is only prison." Concerning the perspective of the sense formation, in this tension between the looked-for and repeatedly lost dialogue, we see the lack of freedom. One is *out* of prison, and one is *in* prison: one steps in and out, in a spiral of a dilemmatic tension. When the patient is incapable of communicating directly the act of *listening* to the patient enlarges into communication and dialogue. Even when the metaphysics of the unknowable and the darkness of hope petrify the process of realizing, Maria Teresa's human existence and fate in a prison has nothing metaphoric in itself. In an attitude tending to restore some sense to Maria Teresa's lost experiential intentionality, the rational, meaningful intervention can be only that of seeking, despite everything, to introduce into it a precarious and naturally unstable element of an intersubjective life. Maybe we could restore the intentional links to carry the reciprocity and to animate the truncated intentions which still remain as a far-off indefinite teleological perspective of the other. To recover the elements of such an intentional horizon and restore the basis of a natural sense formation is the last available *raison d'être* of the encounter with a person who is in the state of the melancholic metamorphosis of guilt. This is only possible if any specific objectivating *life-world* attitudes are abandoned. And yet the encounter of "patient" and "psychiatrist" takes place in the thematic region of everyday life, and it is just "ein funktioneller Moment in der infraszientifischen Realität der Lebenswelt".[12] The *life-world* (to repeat a remark by Husserl) "is always 'the ground' of any praxis, either theoretical or extra-theoretical."

Ospedale Psichiatrico di Novara

NOTES

[1] L. Binswanger: *Melancholie und Manie, Phänomenologische Studien*, Neske, Pfullingen, 1960.
[2] V. E. von Gebsattel: *Prolegomena einer medizinischen Anthropologie*. Springer, Berlin, 1954.
[3] E. Minkowski: *Le temps vécu. Etudes phénoménologiques et psychopathologiques*. Delachaux et Nestlé, Neuchâtel, 1968.
[4] H. Ruffin: 'Melancholie,' *Dtsch. med. Wschr.* **82**, 1084, 1957.
[5] K. Schneider: 'Die Aufdeckung des Daseins durch die cyclothyme Depression,' *Nervenarzt* **21**, 193, 1950.
[6] W. Schulte: 'Nichttraurigseinkönnen im Kern melancholischen Erlebens,' *Nervenarzt* **32**, 314, 1961.
[7] E. Straus: *Psychologie der menschlichen Welt*. Springer, Berlin, 1960.
[8] H. Tellenbach: *Melancholie. Zur Problemgeschichte, Typologie, Pathogenese and Klinik*. Springer, Berlin, 1961.
[9] K. Schneider: *Psychiatrie heute. Eine Rede mit einer Anmerkung*. Thieme, Stuttgart, 1960.
[10] M. Scheler: *Der Formalismus in der Ethik and die materiale Wertethik*. Francke, Bern, 1966.
[11] E. Husserl: *La crisi delle scienze europee e la fenomenologia transcendentale*. Traduzione di E. Filippini. Il Saggiatore, Milan, 1961.
[12] M. Natanson: 'Philosophische Grundfragen der Psychiatrie I: Philosophie und Psychiatrie', In: *Psychiatrie der Gegenwart*, herausgegeben von H. W. Gruhle, R. Jung, W. Mayer-Gross und M. Müller. Bd. 1/2, *Grundlagen und Methoden der klinischen Psychiatrie*, S. 903. Springer, Berlin, 1963.
[13] G. Brand: *Die Lebenswelt. Eine Philosophie des konkreten Apriori*. W. de Gruyter, Berlin, 1971.
[14] P. Ricoeur: *Le conflit des interprètations. Essais d'herméneutique*. du Seuil, Paris, 1969.

[15] E. Paci: *Funzione delle scienze e significato dell'uomo.* Il Saggiatore, Milan, 1963.
[16] E. Husserl: *Ricerche logichè. I e II.* A curä di G. Piana. Il Saggiatore, Milan, 1968.
[17] E. Husserl: *Idee per una fenomenologia pura e per una filosofia fenomenologica.* A cura di E. Filippini. Einaudi, Turin, 1965.
[18] W. Szilasi: *Einführung in die Phänomenologie Edmund Husserls.* Niemeyer, Tübingen, 1959.
[19] B. Waldenfels: *Das Zwischenreich des Dialogs. Sozialphilosophische Untersuchungen in Anschluss an Edmund Husserl.* Nijhoff, The Hague, 1971.
[20] G. Brand: *Mondo, io e tempo nei manoscritti inediti di Husserl.* Traduzione di E. Filippini, Bompiani, Milan, 1960.
[21] H. Kranz: 'Der Begriff des Autismus und die endogenen Psychosen,' In: *Psychopathologie Heute*, herausgegeben von H. Kranz, S. 61. Thieme, Stuttgart, 1962.
[22] E. Paci: *Tempo e verità nella fenomenologia di Husserl.* Laterza, Bari, 1961.
[23] E. Husserl: *Meditazioni cartesiane e i discorsi parigini.* Traduzione di F. Costa. Bompiani, Milan, 1960.
[24] M. Merleau-Ponty: *Phénoménologie de la perception.* Gallimard, Paris, 1945.
[25] E. Paci: 'A Phenomenology of Eros,' In: *Facets of Eros*, edited by F. J. Smith and E. Eng, p. 1. Nijhoff, The Hague, 1972.

C. FINITENESS AND THE "FORM OF ALL FORMS"

Section I: Telos of History

BIANCA MARIA CUOMO D'IPPOLITO

THE THEORY OF THE OBJECT AND THE TELEOLOGY OF HISTORY IN EDMUND HUSSERL

Reading phenomenology on the basis of the eidetic insight understood as a return to originary forms of experience means not only linking its epistemology to the traditional concept of "theory" (Horkheimer), but also understanding the teleology of history as the repetition and conservation of an idealized past by means of the privilege of originality (Foucault).

Philosophical reflection about science in modern times has combined "science" and "system" into a unity manifesting a precise program that has been pursued ever since Descartes. The "system" is a logically necessary coordination of true propositions. Before a body of thought can propose itself as a science, its propositions must all be true and, further, internally coordinated on the basis of a unifying principle that constitutes the foundation of the truth of each proposition and orders it in relation to all the others. Husserl, in the work of his mature stage (*Crisis*), was to see the Cartesian *ratio* as the origin of a science which, setting itself up as technical reason with a view to dominating things, leads even itself back into the fetters of the instrument-object relationship of domination. However, even at the very beginnings of phenomenology, he used the theory of the object to delineate a different vision of scientific knowledge from that of the nineteenth-century philosophy.

In the present study we intend to investigate Husserl's conception of scientific knowledge as founded upon his theory of the object on the one side, and teleology of history, on the other side. Our argument falls into three parts: (1) that there exists a substantial correspondence between the manner in which one conceives "object," "science," and "system" and one's manner of conceiving history; (2) that Husserl, introducing an innovation in the manner of understanding the role of "theory," opens a new vision of history; and (3) that Husserl raises the historical problem of this correspondence and its coverage neither at the level of a "history of knowledge," nor at the level of the dialectics between structure and superstructure, but rather in the sense of a "critique of culture" (*Kulturkritik*) that opposes the *ratio* with practical reason. Practical reason is here the phenomenological-historical subject that even epistemology requires as the condition that makes it possible. In the theory of the object we thus rediscover the key to the relationship between theory and history.

Tymieniecka (ed.), Analecta Husserliana, Vol. IX, 271–274.

Right from the very beginning, Husserl places himself in a line of thought "that knows itself to be" *scientific*; his first operation consists of recognizing that the ambit in which his thoughts are moving is circumscribed by a horizon of *objectivities* that are the product of deliberate formalizations in accordance with established rules. *Operational rules, computational operations*, and *numerical quantities* configure a "system" in which logic and mathematics constitute a single solidary whole. This is an area of clarity and no true problems of an epistemological nature arise within it. The question regarding the origin of numbers opens an epistemological problem only when the numerical object, the operations and the rules, i.e. the "logical concatenations," are situated in that "frontier zone" that binds them to the "psychological" concatenations. The error that leads Husserl to consider that the origins of the numerical objects are constituted by these "psychological concatenations," which are really products or objects of scientific operations of another type, leads one to identify the true problem of origin as the calling into doubt of the entire logical relationship or "system" and thus opens the road to philosophy. Philosophy comes into being in a zone of nonclarity, within the shadow zone that both separates and unites the objective clarities of the scientific "systems." This zone has to be clarified not so much as an area of an extrinsic causality among objectivities of different levels, but rather as an area of historical mediations in which the systems of knowledge arise and have a meaning. "Science," "system," and "object" are linked by a relationship that tends to turn itself upside down: science is a system of objects that is the object of a nonsystem: it is a system, but of operations that produce objects in accordance with given rules, and it is an object produced by nonscientific operations. The development that causes research on the origin of numbers to become a theory of the object along the road traveled by Husserl also marks the first beginnings of the epistemology of history.

The problem of the logical object, which Husserl raises in *Philosophy of Arithmetic* and *Logical Investigations*, very quickly presents itself as the problem of the whole of logic as object, that is to say, as a "system" with a structure of its own, with its own peculiar rules and modes of connection. The epistemological rupture is brought about via the thematization of the twofold meaning of the term "object": the object as a set of elements with its own rules and modes of articulation, and the object as "object of," i.e. as the term of an epistemic relationship. During the course of the analysis one becomes aware of the need for the logical object to appear in relation to the object of experience or perception, and this within a concept of "objectivity" compatible with both the indicated meanings. In this operation the very *ideality* of the logical being will be rediscovered in the object of experience, as also the multiform possibilities of this latter in the object as logical being.

The nucleus of the logical object is its ideality. Husserlian ideality is a new relationship between the object and space-time. The object is in time without being time, it remains the same as it changes, it is the discontinuity that reasserts itself in the very tissue of the continuous. The idea is not that which can be *seen* in the sense of intuitive immediacy or of the being of the lived within life (*Erlebnis im Leben*). Rather, what can "be seen" is that which is idea or, putting this into other words, the possibility of every object is in the *discontinuity* by virtue of which time ceases to be "life," "continuum," and gathers within itself the object as something that is not identical with life. The presence of the object is the de-realization of life. "Originary" time is not *real duration*, but *intentionality*. The concept of intentionality introduces logic into the ambit of phenomenology, because the object as logical structure has necessarily revealed itself as an "object of" constituted in intentionality. Space, on the other hand, is not the dominion of the exteriority of matter, the abstract product of quantification. "Lived" space, which opens the road to existential analysis in psychiatry for Minkowski and Binswanger, is the entire ambit of one's own body. It is one's own body as sensitivity and movement, the living subject of perception.

The analyses of *Ideas II* show how the epistemic relationship between an "object" as logical structure and a subject as a system of *a priori* forms (Kant) breaks up into two directions. On the one hand the passage of the *human body* from quantifiable spatial object to intentional subject radically transforms one of the poles of the epistemic relationship: the corporeal "situation" does not imply only the *finiteness* of a Kant read by Heidegger, but also impotence, pain, and death understood as objects purged of and from knowledge, an irreducible manner of human presence in the world, the originary need for reading the world as a world of passion (as Sartre quite rightly pointed out). On the other hand, Husserl's analysis performs – in the true sense of the word – a *destructurization* of the object in a twofold sense: the object as an empirically given something of the world of natural attitudes becomes dissolved in view of a constellation of possibles; the object as essence is brought back to the interplay of the spatial and temporal "horizons" that have their point of origin in one's own body and become sedimented there. The body is therefore the memory and the language of essence itself, the story of its origin and oblivion. The most radical of the imaginary operations performed by Husserl, i.e., the hypothesis of a destruction of the world (*Weltvernichtung*) conceived in the sense of a nonconcordance of the perceptual glimpses, shows that the essence as "a nucleus of identity that remains" is not something originary in the absolute sense, but rather the result of an operation that connects, that postulates relationships and correspondences, that preannounces and prefigures. The work of a *ratio* that is protended toward overcoming the multifarious, and with it the nullity of the world, the fear of death, the essence is bound up with a significance of identity and projects it as *telos*. This operation contains in itself an ideology of conservation because, through the correspondence and interchangeability of origin and *telos*, it aims at maintaining the dominion of a past that has been grasped once and for all as "value," "essence," "significance." The development of the theory of the object and the hypothesis of the destruction of the world, on the other hand, supply a very different indication. While the *ratio* that constructs the essence "rationalizes" in a Freudian sense, the "heroic" reason – which is protended toward the possible right up to the human possibility of nothingness – rediscovers mediation between knowledge and history in the precariousness of the bodily gesture. To the extent to which the body is the ambit of interhuman relationships, and therefore a social object, it bears the very stamp of society as the unconscious dominion of something that *a priori* objectivates and imposes upon history the false teleology of a repetition of the past. Striving toward the essence, however, the thought that dares to visualize a variation will encounter the possible, the unconscious and the play of the imagination between the fixity of the "fact" and the identity of the "essence."

The affirmation of reason victorious – over the *ratio* that covers the possible, the nothingness, and the dominion by rationalizing and reifying – is a *telos* that does not rely on the guarantee of an immanent entelechy, being in the last resort the task of philosophy as a reflection that connects the epistemology to the ambit of those historical mediations to whose silence it owes the pretended eternity of its forms.

University of Naples

†JOSE LUIS SANTALO

THE DESTRUCTION OF TIME BY HISTORY

I. TEMPORALITY

Saint Augustine was the first person to remark that when people did not ask him what time is, he knew, but when asked for an explanation, he did not know. In this context it is to be noted that the time with which we are concerned is not strictly chronological time in the sense of "before" and "after" – as in the classic definition given by Aristotle: *numerus motus secundum prius et posterius*. We shall see that this observation does not apply to existential time. Referring to the thesis expounded by Heidegger in *Being and Time* that the original unity of the structure of "concern" resides in temporality, we find that time, to be precise, is that which makes possible the unity of existence, "facticity," and the "fall." The elements of "care" are not simply accumulated but are revealed to us as a multiplicity of modes of being of the "being-there."

It is in this sense that we can say that the world reveals temporality. Thus, temporality is revealed by making itself present to us. But at the moment it shows itself to us also in the modes of past, and simultaneously, originating the concern, it reveals the becoming to us. We recall that dread and concern lie in the future, never in the present.

As Heidegger says, "the characteristic of 'time' that is accessible to common understanding, consists, among other things, precisely in that in it, as a pure sequence of nows without beginning or end, the static nature of the original temporality becomes leveled." And this, basically, is because the present has its roots in the past and in the future. Therefore, it may well be said that time is a becoming that is being transformed into something that has been.

In this way time acquires a new dimension, more important even than that which is offered to us by Kant's *Critique*. It is no longer merely an internal and *a priori* condition of our representations and lived-through experiences but now presents to us those same lived-through experiences in a pressing manner, obliging us to give them an adequate response. All these live-through experiences, irrespective of their nature, that come before us have to be judged with the whole vision of the cosmos which we possess. That is, undoubtedly,

Tymieniecka (ed.), Analecta Husserliana, Vol. IX, 275–280.

the decisive sense of time: that it obliges us to adopt a vision of the cosmos and an attitude of judgment with respect to life.

To the extent that it has a meaning time is, first and above all, a vital dimension that is not counted by revolutions of the earth around the sun, that is, by days or years. This vital dimension is ours and for that reason it is different in each man who is present in the world.

To consider time as a vital dimension is to consider it as a means of contradiction and of life in which multiplicity of existence is established, which is the same as saying that time has a psychological dimension. This is, at least in a certain sense, a direct consequence of Kant's position in accordance with which time is a necessary representation that serves as a basis for all intuitions.

The vital dimension is not found either in the past or in the future, in that which was and is not, or in that which is not but will be; only in the present, which is what we are living through, will be given fundamentally that dimension that characterizes time.

As a vital dimension time offers us three categories the intensity of which will very in accordance with what may be called internal "tempo" and in accordance with the rhythm of life. Such categories are "at the same time," "contemporaneity," and "the moment."

The category of "same time" is what makes it possible for man to be "several things" at the same time, and consequently determines that "he is nothing" (Kierkegaard). The category of "contemporaneity" is that which makes it possible for man, of any time and of any generation, to be contemporaneous which can pass from the temporal to eternity. Therefore, we said above that there is an internal "tempo" that is different for the man situated in "at the same" from that of the man situated in "contemporaneity"; in this case the rhythm of the life is also different. The more rapid is this rhythm, the more man will feel eager to be "many things" at once; the slower is this rhythm, the more he will be disposed to situate itself in "contemporaneity."

Hence also the other sense of time: to help us to understand in what way objective "Truth" is subjectivized and is realized in each and every one of us and in what way also, on interiorizing it, it makes us free. And since we live in time, the fundamental requirement will be that of knowing and understanding the time in which we live, because it is at that time and in no other, that we must internalize this truth in order to live and to die for it.

At this point we find ourselves with the third category of which Kierkegaard spoke: the category of the "instant." Man, wrote Kierkegaard, is a synthesis of time and eternity, and both are found in what Kierkegaard has called the "instant" or the "moment." This conjunction does not occur precisely in the Greek interpretation, in which it was related to the category of the invisible, because for the Greeks two concepts were lacking which are fundamental in this case, those of "temporality" and of "spirit." We are concerned here with the Christian dimension, according to Saint Paul, who tells us that the world will perish at the opening and closing of an eye, in an instant in which time and eternity will meet, to conclude time and to commence eternity.

Therefore, the "instant" is the dialectic category of decision, because we confront in it eternity and we risk, so to speak, our future. To live in the "now" is not the not being that counts only as an intersection between the past and the future. He who lives in the "now" is in danger of being situated in finitude, because he does not care about either the past or the future.

On the other hand, the "instant" helps us to understand the "present" abolished in succession, because we are continually exercising the decision that assures for us the possession of eternity. The "instant" helps us also to realize our destiny insofar as it reveals to us our image and likeness with respect to God who at every moment exercises His decision with regard to us and with regard to the world. In a wink the world will perish, but in a wink we can gain or lose that eternity, depending on whether or not we manage to understand that from the nonbeing of time, we pass to the nonbeing of eternity.

Hence comes the decisive role of time in our life and the reason of its importance in a philosophy of existence: because we live, we move, and we act in time. And every day we have to gain it by means of our decision and help our fellowmen also to gain it, even though the decision is a strictly and wholly personal act.

II. HISTORICITY

It was Heidegger who stated the problem of historicity in a direct and immediate relation with that of temporality. Here it is advisable to distinguish between "history," "historicity," and "historiography," which concepts very often are involved and confused one with another. It is accordingly necessary that we consider each one of them separately.

What Heidegger calls the "continuum of life" between the birth and the death of a sequence of lived-through experience "in time" is a sequence in which only the lived-through experience "before the eye" is "properly speaking real." However, the mobility of existence is not the movement of something "before the eye" but is defined by the prolongation of the "being there." This specific mobility of the prolonged prolonging itself is what we call the "historic gestation" of the "being there."

It is essential for this historic gestation to open itself and interpret itself. From this form of being of the entity which exists historically stems the existential possibility of an express opening and apprehending. Later will come historiography insofar as it proceeds ontologically from the historicity of the "being there." In summary, the "being there" only exists and can exist historically because of its being temporal in the essence of its being.

Thus, the origin of history – of what is historic – lies primarily in the "being there" and secondarily in what it confronts within the world: not only the useful "at hand," in the widest sense, but also the nature that forms part of the surrounding world, as the "ground to history." For this reason Heidegger could write that with the existence of the historic "being-in-the-world" is already included in each case in the history of the world the "at hand" and the "before the eyes."

Thus between historicity and truth there arises a close correlation insofar as the former has its origin in the "being there," which is the first truth that we find as discoverers. On the other hand, in the idea of historiography as a science it is implicit that historiography has taken for its particular task the disclosing of the historic entity. It is only by starting from this "open state," which is the truth of the historic existence, that it is necessary to expound the possibility and the structure of historiographic truth. But this does not always occur. Historiography and truth do not always go hand in hand. And the fact is, that on conceiving it as a process, the basic idea is overlooked and instead of the being being "open" it is still more closed; it hides behind a cascade of words. Consequently we shall see how history destroys time.

One of the fundamental concepts in the temporal dimension of man, and above all, in his constitution as a historic entity is that of *traditio*. This word, with its etymological sense of "deliver" or "transfer," is included and interpreted by Heidegger when he considers the subject of historicity and explicitly recognizes its decisive meaning in the specific case concerned.

For Heidegger, the state of resolution in which the "being there" withdraws toward itself, opens the factual possibilities of existence proper in the case, starting from the "heredity" that it takes on itself with respect to abandonment. To regress to this state of abandonment implies a "tradition" of possibilities transmitted. Still more, in the "individual destiny" the "being there" becomes a tradition of itself in a possibility inherited but, nevertheless, chosen insofar as there is a well connected finitude of existence. And with the "individual destiny" is the "collective destiny," an expression by which Heidegger designates the historic gestation of the community.

What then would be traditional? Simply knowing how to assume the position of abandonment that arises in each individual destiny, but being conscious at the same time that we are for our time, and to cause there to dwell together in each one of us death, doubt, moral conscience, liberty, and finitude in its original form. Therefore, in this "tradition" that becomes the individual destiny the "collective destiny" is simultaneously established; by this, then, we are to understand the "historic gestation" of the "being there" in the "being with" others.

All that has been said aids us decisively in understanding the ultimate sense of history and of the historic. It show us that it is a sequence of lived-through experiences until death and that the root of our present life is both in the "been," from which we received a range of possibilities that have been transmitted to us and that we freely choose, and in the future, insofar as abandoned, we are directed forward to the extent to which, in a state of "open" and "resolved" exercising our freedom and our power of decision, we realize these possibilities.

This sense is the only one in which we can say that history, not so much perhaps historiography, is the master of life, because it teaches us to live for our time, but without forgetting the almost infinite range of possibilities that we have received from those who preceded us in their "existence in the world."

The concept of "repetition" was utilized by Kierkegaard to express a determined situation, but also (in his works published under a pseudonym) to explain dialectically his rupture with Regina Olsen. Heidegger also used it in *Being and Time* as a manifestation of tradition.

However, between the two writers there is, in this respect, a certain difference. While Kierkegaard refers to the problem of free choice as the beginning of the life of the spirit, for Heidegger it is a question of the express tradition, that is to say, of the retrogression to the possibilities of "being there" and "been there." A detailed exegesis would be necessary to understand up to what point there is a coincidence between the Danish writer and the German, if not in the manner of presenting the problem, at least in the consequences which it is possible to reach in one case or the other.

What is certain, nevertheless, is that if we accept the position of both Heidegger and Kierkegaard, we could state that the more decisive and transcendent the acts or the attitudes of man, the less they are repeated. And let the same be said of the great events of an historic nature: the birth of Christ and the discovery of America are not repeatable. Therefore, the phrase "history repeats itself" has to be taken *cum mica salis*, because neither the circumstances of time and of place nor the men are the same or identical.

A careful analysis of all that is expounded throughout the preceding pages leads to a conclusion that may seem paradoxical, but which, if it is analyzed in detail, necessarily imposes itself on us. I refer to history as the destroyer of time, and this in various senses.

In the first place, history destroys time by means of the tradition in which the historic "being there" goes back in search of an interpretation of this range of possibilities that it is receiving continuously. In this sense the historical is "what has been," but it is also the "what is becoming" insofar as man is an abandoned being with respect to what is to come. But, in both cases, it is a question of something "not alive" leaving here a question which does not bear on the problem of memory or of imagination. In this sense, the historical destroys time to the extent that it erases the frontier of the past, of the present, and of the future. With regard to history it is no longer possible to say in an absolute manner that time is a becoming that is transformed into something that has been.

Furthermore, historiography, history as a science, destroys time in turn on constructing a system: the ancient age, the modern age, the contemporary age. Systems constitute by their nature something dead that serves to learn but not to live. My own lived-through experiences from the moment that they are transferred to paper continue being lived-through experiences while I write them, but once expressed they have ceased to be so, since they are not lived by the person who reads about them. They are "live-through experiences" insofar as they are being communicated by being read about but they are not insofar as the readers have not lived that which is communicated as written on the paper.

Moreover, while I relate it, history is a pantheon of dates, of names, and of events; but these do not in any way constitute lived-through experiences. I cannot "live" the Stone Age, or the landing of the Pilgrim Fathers. And if I live them it is by means of a work of the imagination and of fantasy, not as a direct lived-through experience.

Let us leave aside the problem of history as empty talk – as Kierkegaard wrote in his *Journals*. Empty talk contributes, as Heidegger wrote, to the lack of authenticity of human existence. In this sense, also, history as a narrative would become something "dead" even though it did not go so far as to destroy time, as in the preceding hypotheses.

Madrid

PAOLA RICCI SINDONI

TELEOLOGY AND PHILOSOPHICAL HISTORIOGRAPHY: HUSSERL AND JASPERS

In 1910, Jaspers read Husserl's essay 'Philosophy as a Rigorous Science' and was overwhelmed, as he later recalls,[1] by a profound sense of intellectual rebellion. He considered unjustified Husserl's pretence to elevate to philosophy that which was merely method, and refused, as inacceptable, the need to base philosophy in the apodictic and universal which is characteristic of the scientific process.

Jaspers met Husserl three years later and was praised by him for the clarity of the phenomenological analysis which he accomplished in the fields of medicine and psychopathology.[2] When asked by Jaspers about the real meaning of phenomenology, Husserl replied, "But you make excellent phenomenology in your writing. You do not need to know what it is, since you make it. Think only to go on with it!"[3] This seemed to be the start of a fruitful contact, but other biographical data shows no evidence of it. Husserl never mentions Jaspers in his writings, only indirectly and controversially, while Jaspers seems to forget the incisive cultural presence of Husserl's philosophy. Only in *Nachwort* (1956), which precedes the third edition of *Philosophy*, does Jaspers again express his critical opinion:

> The reading of the essay 'Philosophy as a Rigorous Science' was an illumination for I felt I understood that this was the point in which, by right of rigorous science, all that could be called philosophy, ceased. During the period that Husserl was a professor of philosophy, I felt that philosophy had been betrayed in the most naive and radical way.[4]

When Husserl had attained a method capable of clarifying unobserved presuppositions of psychical phenomena by considering things in their hidden reality, he went beyond the schematisms, conventions, and prejudices of psychologism; in this respect Jaspers followed him;[5] a point which is confirmed by some of his articles about psychopathology and by his imposing work *General Psychopathology.*[6]

Jaspers failed to penetrate the essential dynamics of phenomenological method[7] since such analyses established clear philosophical presuppositions, providing the space where the transcendental subject's constitutive activity was allowed to emerge. According to Jaspers, this particular project that

Tymieniecka (ed.), Analecta Husserliana, Vol. IX, 281–299.

pretended to found each scientific research, mystified the role of philosophy itself, conceived as a means of existential realization.

In this way, there was a continuous dialectical dispute between philosophy and science and the attempt to recuperate philosophy, without objectivistic pretensions, in its authenticity as it had developed in the course of *philosophia perennis*.

Husserl, too, attaches great importance to the authentic dignity of philosophy,[8] but for him this is philosophy raised to the epistemological solution of the absolute, fundamental principle of both logic and science: "Its meaning, the meaning of philosophy, implies a radicalization of the foundation, the reduction to an absolute absence of presuppositions, a fundamental method through which a philosopher, who is just beginning, is guaranteed an unconditional field which must be understood in full clearness as the absolute presupposition of all the obvious presuppositions in the common sense."[9] Husserl considered it necessary to recover the primary idea of philosophy which is the basis of European philosophical and scientific culture. He argued that "philosophy is, in my opinion, a universal and rigorous science in the most radical meaning".[10]

Aware of actuating a new theoretical beginning that separates itself radically from positivistic scientism and historicist relativism, Husserl follows with interest the birth of "philosophy of existence", which came forth from a development of phenomenology.[11] However, he soon noticed the different, even opposite, philosophic prospects which were emerging and, in February, 1937, in order to be safeguarded against any possible misunderstandings, he wrote to Karl Lövith as follows:

> Maybe you realize that Scheler, Heidegger, as well as all former followers, have misunderstood the true and profound meaning of phenomenology, and its only possible transcendental sense, with all that it implies. . . . Maybe you are able to understand that following an inner necessity, not pressed by obstinance, I have gone my way alone for so many years . . . and for which reason I have thought that *the obscure mysticism of the fashionable existential philosophy* and of historicist relativism, with its professed superiority, are the weak failure of a mankind, without strength, which has shirked the enormous task that the collapsing of modern age in its whole has given, and continues to give us.[12]

The dispersion, formalism and superficiality of contemporary philosophy, of "this unhappy age" seeming to succumb to the "sceptical flood" undermines "the basic beliefs and a sure faith in true philosophy".[13] It is time, therefore, to fight, with a new, radical choice, the false enthusiasm for the scientific progress of mankind and the obscure anguish of those assuming "grandiose, existential attitudes"[14] on behalf of a philosophy whose task ends in the egocentric fight of man for his own hypothetical existence.[15] We are still confronted by a world view (*Weltanschauung*),[16] Husserl continues, a sort of "personal, religious faith"[17] purporting to provide a *wise* outlook on life, and for this reason able to define itself as radically different from science. But our age recalls us to overcome, responsibly, individualism and to fight "under the direction of certain principles" for a philosophic science "having no need of wisdom, but of theoretical gifts" and holding to the "collective formation of the work performed by generations of scientists".[18]

According to Husserl, it is indispensable for us to become a part of history, "not to lose ourselves within the bonds of development where great philosophies have emerged, but for philosophies themselves and their spiritual content to inspire us". Consequently, it becomes essential that we be rooted in the authentic tradition and remembering also that "we do not become philosophers through philosophies" nor by becoming "dazed by the greatness of a Name".[19]

The urge to research comes from things and problems, without any kind of prejudices and with the consciousness of a boundless field of work sustained by philosophic intuition, by "*the direct phenomenological prehension of essences*" which "without using indirect symbolism or using mathematics, without display of syllogism and proofs, still acquires a great deal of rigorous knowledge that is decisive for *any* further philosophy".[20]

Thereby, the distance between Husserl and Jaspers seems deepened, leaving no room for mutual, constructive communication and for a historic comparison that would result as intrinsic and essential as possible. These are the two philosophic choices: one devoted to research on the scientific nature of philosophy, the other given to distinguishing it deeply from any objectivistic contamination. Both of them, when considered in a historical perspective, justify not only the lack of a fertile collaboration, after a first and fruitful meeting,[21] but the almost entire extraneousness of the two philosophers.

However, it is interesting to study them in parallel, to discover some unexpected convergences, seeming, at first sight, entirely arbitrary. Both Husserl and Jaspers are convinced, each one in his own way, to become part of the most authentic, philosophical tradition; each thinking to belong to that *philosophia perennis* which, born with Plato, has impressed all the European culture in an original and indelible way. And what can be said about the mutual effort to go beyond psychologism and scientific objectivism, and about the growing attention to read philosophically the history of philosophy as the ethical-rational way of mankind? Regarding this question, there is a trend to determine the common grounds of philosophical research, so that it becomes an essential, ethical solution of the crisis of a whole civilization.

I. PHENOMENOLOGY AS HISTORY'S TELEOLOGICAL HORIZON

Through the continually renewed exercise of the transcendental reduction, each result is *ever again* regained, each stage opens up new horizons: the only rigorous, evident, apodictically certain thing remaining is the intentional consciousness, as constitutive centre of the transcendental Ego. Husserl has charged to it the task of acting as guide in a long and difficult road: that is, the trend towards a radical transformation of philosophy, resuming both Descartes and Kant's effort; the return to the first beginning, going beyond scientific knowledge in order to ground it critically; the recovery of the most profound meaning of both history and learning.

Crisis marks a basic stage of this itinerary: phenomenology seems to be able not only to bring about a reform of the analysis of consciousness and with it an epistemological re-establishment of every science, but to examine critically the profound reasons for the collapse of European culture. In search of a "radical self-intelligence", philosophy thus widens its bounds to the whole historical-social reality.[22]

According to Husserl, the crisis of Western civilization is caused by the crisis of science that is unable to give itself a basis and to provide modern man with a theoretical and practical guidance. Firmly persuaded that truth lies in what can be expressed by a system of true propositions, science has provided man with techniques that, by reducing reason to a merely instrumental meaning, has made barren even the sense of learning.[23]

The history of this progressive mystification is marked by Husserl as an entirely extrinsical "naive learning" which can be and must be overcome by carrying out a phenomenological reduction through which the nature of learning changes and returns transparent in its constitutive form.

It is here that a living being immediately recognizes himself as a transcendental subject, discovers to be enriched with "constitutive sense" and becomes aware of being identical to the subjective ground of life.[24]

To carry out the reduction in a phenomenological way means, therefore, a "return to the self" within the sphere of a true learning, considered as "constitution", as "Being" that is, the living Being of the absolute subject.[25] History, vouching for this continual recovery of the "source" from the utmost alienation, is the genetic-dynamic process[26] that expounds the various and progressing, historical stages of learning authentication. The first degree of historicity is entirely naive and natural, and is sustained by popular traditions.[27] It is replaced by a more critical and conscious outlook. With the birth of theoretical motivations, philosophy originates as well, and, in its historical realization, constitutes the subsequent degree towards "the self-responsibility of learning". But only at a further degree of its maturing, history, with the help of "phenomenological learning", reaches a new philosophic consciousness and becomes the constitutive form of the spirit, so that it is able to put into effect "a greater philosophic self-comprehension of both humanity and its historical existence, making possible a human existence full of sense.[28]

In Husserl's opinion, the respective degrees of historicity, that is, the mythical, philosophical and phenomenological, turn into the particular history of the single *Dasein*. It often, but discontinuously, passes from a naive stage of life to one sustained by conceptual reflection; so that in the end it arrives at a primary dimension, that achieves meaningful operations and that is constitutive of transcendental subjectivity.[29]

The radical link to life in its whole is realized only through the phenomenological reflection, that, after reduction, is able to postulate a new type of transcendental science "that invests subjectivity through which the world is pre-given".[30] The focusing of the historical horizon of "life-world" becomes the new task of transcendental phenomenology, claiming an explanation of the pre-given world, in the conviction that the consciousness of a "Being" in the world necessarily implies the world-consciousness as historical horizon.[31] Since "the pre-given world is the fluent-constant horizon that embraces all our ends, transitory or permanent, and also embraces preliminarily the intentional consciousness of the horizon",[32] it is deduced that the horizon of the *life-world* is nothing but the horizon of universal history. The consciousness of such a horizon implies the horizon of history and so "the constitutive operations of transcendental subjectivity must imply the conditions of the possibility of experiencing, in its historicity, the life-world as historical world".[33] Once the 'a priori' foundation of the *life-world* has been investigated, as "a general and invariable structure", it is possible to suture the historical break between the world of the *epistemè*, peculiar to philosophy and science, and the obvious world of the *doxa*, peculiar to the *life-world*. Consequently, it claims to found a "new science", as absolute foundation of the whole objectivity.[34]

Considering the transcendental constitution, the science of the *life-world* recovers and, as it were, legitimates every aspect of human life in its peculiar activity as transcendental Ego. Such absolute subject, omnitemporal and pre-individual, becomes particular by entering time, that is, constituting itself as social being in history.[35]

After becoming science of the *life-world*, phenomenology turns towards people who no longer believe in the exercise *epoché* and are satisfied with the results obtained while a progressive loss of sense occurs. If its aim is to "bring back oblivion to presence and its task is the task of making present",[36] phenomenology can explain its pretension to absoluteness, since the transcendental foundation of science implies in its depths both life and 'telos', the teleological sense of history, the meaning of humanity as *Entelechy*. In the past, it recovers a "return to the original foundation"; in the present and the future, it recovers an infinite, ideal principle, a recovery of the teleological essence peculiar to the Ego, an intersubjective absolute as "expression of humanity in its infinite road towards the self-creation of its authentic being".[37]

The history of humanity, as the history of the existent man, mirrors the continuous and dramatic struggle for the search for an intentional unity, in the *immer wieder* recovery of the original foundation, to the "return to mothers" that the phenomenologist aspires through his tiring and often enigmatic work to bring to light in a clear and comprehensible way.[38] This theoretical trend that, in Husserl's opinion constitutes "the hidden wish of the whole contemporary philosophy",[39] in historiography, is realized in the comprehension "of teleology that is innate in the historical becoming of philosophy".[40] This hermeneutic act directly involves the philosophical existence of the interpreter[41] so that it becomes possible to understand past thinkers better than they themselves did.[42]

A past philosopher is understood only because he can enter into the personal "cogitation" of the interpreter[43] who must be a philosopher, able to throw light behind the so-called "historical events", and to investigate the hidden and always present unity of history, in the intentional interiority that is constitutive of past thinkers.

After being clarified as fundamental research on reason's universal teleology in history, phenomenology goes beyond the old historiographical conceptions that extrinsically examine the history of philosophy and reduce the philosopher's living to mere historical events.[44] Moreover, phenomenology carries out an authentic link to tradition by throwing light on

> a sole unitary and systematic tract with no historical-critical purpose, since the essential intent of this systematical research is to rise to the level of transcendental phenomenology and to prove that it is the only transcendental philosophy which can be founded, starting from the original sources of knowledge, and, therefore, the only possible philosophy.[45]

Descartes and Kant's philosophies can be taken as examples:[46] they are valid inasmuch as they have clarified the need for a radical change in philosophy. However, they are limited since Descartes was not able to grasp, in its depths, the transcendental Being of the *Ego cogito*, whereas Kant's prospect did not acknowledge the historical horizon of the *life-world.*[47]

Husserl's historiography frees itself in this way from historical-critical preoccupations, which, with verbal definitions clarify the mere historical crust of events, but kill from within the 'telos' peculiar to philosophizing.[48] Husserl defines his own *superhistorical* methodology as follows:

through the philosophical person and the intersubjective and supertemporal community of philosophers, who are spiritually united and mutually motivated, *runs a necessary teleology, oriented towards the pure configuration of the task, and towards its fulfilment*, but in such a way that the teleological task peculiar to the process of development necessarily takes the form of a manifestation and a perfect explanation of the true idea of task and method.[49]

The overcoming of chronological time opens the new dimension of the "living present" as essential moment of the meeting between the interpreter's philosophic existence and the philosopher's "generative community".[50] Consequently, the new hermeneutic tasks are clarified:

in this way, since there is a philosophical historiography, the history of philosophy is, in turn, a supertemporal process through which one learns from history ... Scientific history transforms the history itself of philosophy and creates the supertemporal present peculiar to all philosophers and all philosophies, and of all times which are scientifically disclosed and that can regain evidence and validity through scientific history.[51]

By clarifying the true philosophical method, historical research leads to the revelation of philosophy itself, not of "a" philosophy, Husserl specifies, but of "the" philosophy,[52] that is impressed by the seal of eternity. Moreover, it is able to lead to fulfilment

a humanity conceiving itself as a rational humanity, realizing to the rational in the *will-to-be-rational*, understanding that this means the infiniteness of life and of the efforts towards reason, that reason indicates that towards which man, as man, is inclined to in his intimate ..., *that being man implies a being-teleological as well as a must-be and that this teleology prevails over every action and every egologic plan.* Humanity understands that reason, through self-comprehension, recognizes in everything the apodictic 'telos' and that this knowledge of the extreme self-comprehension can take no other form but that of self-comprehension according to the 'a priori' principles, of a self-comprehension in the form of philosophy.[53]

II. PHILOSOPHY AS AN ENCOMPASSING HORIZON OF HISTORY

Husserl's ideal of the absolute purity of transcendental consciousness, as principle and constitutive foundation of world-being, is radically rejected by Jaspers. He claims a more complex and problematic question about Being, which exhausts itself neither within the consciousness of the subject nor in the universality of the scientific object.

First of all, the research moves to verify the possibilities of a thorough investigation of Being within the horizon of the world (*Weltorientierung*), but the limit, the failure of this attempt, reveals the inexhaustible dimension of "existence" which, though in an enigmatic and inscrutable way, becomes a privileged centre of philosophical analysis (*Existenzerhellung*).[54]

The validity of rational learning, refused as being unable to grasp the existential dynamics of Being and confined within the limits of objective science, is somehow recovered on behalf of a research that, far from exhausting itself in an empty monologue, investigates the means for an "existential self-ascertainment" and claims a wider space for reason, made barren and reduced by science to a mere instrumental function. All this explains the radical difference between the two processes: what science sees as a negative limit, a paradox and an enigma becomes, in the philosophic research, a positive incentive to the clarification of the space of existence, that intellect is unable to grasp and fix objectively.[55]

The transcending movement that permits the empty "being there" (*Dasein*), to free itself from its objective dimension and to establish itself as center of freedom in existence, can be understood only if it is able to involve existentially, and if it is oriented in the infinite idea of reason.[56] The two poles of reason and existence enable us to grasp the clear and unitary source of Being, the "encompassing" (*Umgreifende*), that appears as horizon of all the horizons of Being and as an encompassing bond grounding all the prospects within which Being is realized.[57]

"Existence" represents a privileged centre, "as life and foundation of the encompassing,[58] and its freedom and historicity is clarified by philosophy become living style; philosophizing, essential space for the reading of the "cipher-script" of the historical manifestations of transcendence, *Metaphysic*. The "rational alogical" (*Vernuftige Alogik*) describes the proceeding of existence, that always aims at transcending the mortifying, static dimensions, the wall of "boundary-situations", and all that forms the historical basis of our existential finiteness, in order to reach in that space of "absolute silence" where transcendence manifests itself.[59]

This progression towards existential authentication can be found in history as gradual manifestations of the manifold expressions of truth embodied in time and only the essential character of historicity peculiar to philosophy guarantees the sole possible attaching between the absolute truth, that is never a definitive possession, and the historical stages where truth has manifested itself.[60]

Against the omnipresent, objectivistic temptation, and against any pretension to absoluteness, the existential style of philosophy becomes the incessant appeal to a unity and to a universality essentially different from that suggested by the instrumental reason of science, and to a unity that can be simply shown, enlightened, deciphered by the encompassing space of reason. "Philosophizing is a being between origin and end, a being in the middle (*Zwischensein*), therefore, something intermediary (*Mittlere*);[61] thus, it is an essential instrument of infinite tension between the original manifestation and the final fulfilment of truth.

The reflection on the "rational alogic", considered as evidence of reason becoming aware of itself in its motion towards the One, intimately merges with the research of history presented as the only great phenomenon of the revelation of Being. Thus Jaspers has avoided both the evanescence and the irrationalism to which he seemed to lead inevitably with his *Philosophie* and reproposes a radical re-foundation of philosophy by means of tradition and its connection with communicative logic.[62]

Jaspers insistently calls our attention to be loyal to the "origin" (*Ursprung*); which means to receive "that original inclination as pre-objective foundation of both Being and thought".[63] It also means the essential inclination to the manifestations of the One, the comprehension, and assimilation of tradition, the heritage of truth able to enlighten the true, radical beginning[64] and finally, the continual appeal to the moral responsibility towards our historical present.[65]

Jaspers has a peculiar interest in the hermeneutic problem:[66] seeking a self-justification, philosophy becomes historiographical methodology, considered as the existential comprehension of the single philosophers,[67] and becomes also, the exhaustive instrument for the ambitious plan of founding a world-history (*Weltgeschichte*).[68]

By refusing the chronological and cultural aspects, as regulative principles of history, Jaspers focuses his attention on the "great men" of philosophy leaving them, as it were, "suspended", outside time, in a new, superhistorical dimension that permits them an existential relation with the interpreter.[69]

To approach a past philosopher, therefore, is the same as to re-understand the "great man", reliving the stages of his theoretical beginning that have led him to go beyond the spirit of his time and to make him eternal-contemporary to us.[70]

This kind of existential appropriation (*Aneignung*), renders possible to us the meeting with the "being himself" of the other existence, to make us of the "becoming with the One" in the distinction, to perceive that "I am only through the other and yet I am always myself", and to feel that "withdrawal" makes impossible any exclusive pretension to truth. That meeting, becoming existential verification, enables us to attain the clear contact with transcendence.[71] To divine the uniqueness of the great man is to convince ourselves of the constitutive character of truth, which, by historicizing itself in time and shattering itself, as it were into manifold manifestations, is defined in its being "eternal One".[72]

A historical-theoretical reconstruction is necessary: Firstly, try to recompose their intentions, to think with the philosopher within his idea, to make present his totality through the intellectual operations offered by him".[73] Then it is possible to establish the true relation, animated by "polemics" and "fight", and by the movement of question, listening, and answer that is not only reconfirmation and rational judgment, but metaphysical and existential judgment. Only by taking as a model this style of "essential communication", existence is educated, as it were, to *philosophizing* and, rooted in history, it is aware of participating and constructing the "*philosophia perennis*". It is the theoretical proceeding covered by humanity through the stages of thought which has arrived, from intellectual clarification, at the original intuition of cipher-world.

As an example, Plato's intuitions have proved that "philosophical life is the way through which I go beyond the characteristic of intellectual thinking to understand what leads me to myself in my inner acting, throwing light on the obscurity I am for myself as soon as I meet myself".[74]

Similarly, Laotse's Tao has a meaning why "it turns to an origin inside us, that remains concealed by intellect and the purpose of will" and "directs itself towards the encompassing foundation existing inside and outside us".[75]

The total view, not only towards the history of Western philosophy but also in the direction of Indian and Chinese philosophy, reproposes the task of discovering the sole historical source of universal philosophy that demonstrates, in Jaspers' opinion, the essential unity of the spiritual development of mankind.[76] The objectivistic temptation reappears and pretends to interpret unity as a whole, that can be universally known, and to turn the historical course into a pre-constituted, philosophical idea. Against this hypothesis

Jaspers reproposes a historical task that is research, a working hypothesis sustained by the faith in reason" as faith in man and in his concrete possibilities of self-improvement".[77]

It is interesting now to point out, still in a hypothetical and problematic way, some unexpected convergencies between Husserl's teleological project and Jaspers' historiographical proposal. These convergencies appear even more unexpected if we think of their different, if not opposite, philosophical priority choices, stirred by theoretical trends, apparently irreducible. But a more careful consideration of their speculative development shows how, in a mutual desire to refound radically philosophy, Husserl progressively gains a new kind of science, more compound and problematic than it was initially.[78] This is proved by the recovery of the complex historical dimension of the *life-world*.

For his part, Jaspers, who had always refused any objectivistic contamination, on behalf of a true existential verification, is compelled to submit to rationality, even if in a particular sense, and to recover it as the support of existence and of its historical course.

Husserl and Jaspers seem to meet within the historiographical perspective: We need only think of the mutual conception of the "appropriation" of philosophical history, as true linking to the tradition to which they feel to belong, and to the *philosophia perennis* in continuous authentication of itself and soaring in a supertemporal space constituted by the living community of the philosophers, who are animated by essential communication. Can the hidden and still present teleology of the historical becoming of philosophy be compared to the progressive authentication of Jaspers' reason embodied in tradition and in continuous return to its source?

Putting aside these critical problems that need further investigation, it seems useful to focus on the mutual, ethical prospect that, both in Husserl and Jaspers' opinion, is able to cope with the crisis of European civilization.

III. CONCLUSION

The inevitable limits of Husserl and Jaspers' "historiographical reflections" seem to be exhausted in the a-historical recalling of philosophers, through a sort of theoretical re-interpretation in view of the teleological-rational ideal present in the historical becoming of philosophy. Such limits do not impugn the clearness and the rigour of the analysis on the philosophical and historical motives that have determined the profound germ of the crisis of the whole European civilization.

The contemporary world is deeply divided: man has lost his own essential identity, that is, his own position towards Being-totality, since an authentic relation with world and Being that is the world's foundation failed him. Thus, he has forgotten the essential point of reference for that self-comprehension that "means not only the crisis of the theoretical interpretation of the world, but, in the same measure, absence of a practical orientation within this world".[79] Neither science, reduced to a mere technological instrument, nor philosophy, that has progressively neglected its task of "self-responsibility of mankind"[80] have been able to overcome it.

In Husserl's opinion, therefore, the priority-function of phenomenology is to find out the profound reasons of such crisis, not through analyses generally extended to the whole humanity, but through the exclusive examination of *our* history.[81] "Only in this way will it be possible to decide whether European humanity has within itself an absolute idea or if it is a mere anthropological and empirical kind as 'China' and 'India'."[82] The subsequent historical analyses have identified the European spirit as the privileged bearer of the teleological value of Greek philosophy; which value the phenomenologist tries to bring to light through a new historical-theoretical recomprehension.

> Then this task has no longer only a factual necessity, but becomes a task truly *assigned* to us, as we are philosophers. We are exactly what we are, as we are officers of modern, philosophical humanity, heirs and bearers of that orientation of will that *goes through* it; and we are all this on the basis of a primary foundation that is at the same time a re-foundation and a transformation of the primary Greek philosophy. It constitutes a *teleological beginning*, the true birth of the European spirit in general.[83]

Only the Renaissance has been able to provide Western man with an absolute, radical autonomy, so that, after freeing himself from the bonds of myth and medieval traditions, he constructed a new dimension of sciences that became "dependent branches of a sole philosophy".[84] The Renaissance gave back an encompassing sense to reason, both in the theoretical and practical fields, by consigning to the scientific learning the function and dignity of a leader, the absolute clearness of the "sense" of existence and of the world, the profound contact with the model of ancient humanity.[85]

Later the faith in the ideal of philosophy and method became uncertain; reason became particular and was, as it were, fractioned into a series of metaphysical ethics and scientific problems, so that the always deeper causes of the failure of philosophy appeared, while philosophy was unable to unify harmoniously the manifold expressions of human spirit. Positivism is an example of it. After "beheading" philosophy, it concluded by giving a reductive and mystifying meaning of reason to particular sciences. "Physicalistic objectivism" is, therefore, the most responsible for the radical crisis of European humanity, that is "existentially unsuccessful" and

whose enthusiasm about progress, whose ideal of a cognitive dominion over nature to be carried out through an endless progress, and whose ideal of a technical dominion to promote without end, pursues indeed a tranquility (*Sekurität*), and, in some way, a security (*Assekuranz*), against destiny. Correctly considered, this guarantee, that is, the ideal of apodictic art, means a vile attempt to elude the responsibility and the verifications that human existence requires.[86]

Crisis, matured in the critical, political atmosphere lying on Europe, represents the unmerciful denunciation of the excessive power of mystifying reason, certainly not on behalf of a "destruction of reason", but rather

a refinding of primary 'telos'. Through this, a varied, though unique, philosophy took place, in the sole development of Western humanity, with the double aspect of a disinterested theory and of a practice actively capable of abnegation and of altruism.[87]

In front of the contemporary, disappointing situation, that passively witnesses the alternation of philosophies lacking in historical incisiveness,[88] the phenomenologist does not give up his faith "in the possibility of philosophy as a task", and in the certainty that "we *know* to be *called* to this task as we want to be philosophers seriously".[89] These same live in the enigmatic, problematic situation of the present[90] and investigate "in the self-responsibility of learning the ethical grounds for a conservation and the development of the social community sustained by cooperation and love." Therefore, animated by a profound and true ethos, philosophy embodies "*the constant function of fully-becoming-man and of realizing-a-full-humanity*".[91] In his latter activity, Husserl has been exposed to the legitimate critical questions about the possibility of turning his phenomenological science into a life-mysticism or, better, into a *world view.*[92] His profound humanistic charge leads us again to the center of our analysis, that is, to the research of that inner convergence of ethical motives, alive and present also in Jaspers' existentialism.

The cause of the profound crisis of western civilization intimately merge, in Jaspers' opinion too, with the development of its theoretical vision. It is the philosopher's duty, not only to investigate, though critically, the precariousness and the fall of "the spiritual situation of our time",[93] but to give valid, philosophical alternatives, capable of a profound historical efficaciousness.

The critical analysis seems to become here more radical, in comparison with Husserl's. He, at least, is able to keep the faith in European civilization, as historical bearer of rational teleology.[94]

Western, philosophical culture, Jaspers stresses, since Plato, has mystified and disrupted forever the direct relation between Being (*Sein*), and Being-There (*Dasein*), giving the absolute pre-eminence to the "Being-There" that, instead of reading the cipher-script of Being, has devised it, "mastered" it and for this motive, has definitively lost it.[95] Western history thus became history of the progressive loss of sense, and the way leading to nihilism that for the first time is radically announced by Nietzsche.[96] Another negative character peculiar to West is constituted, for Jaspers also, by the aberrant innovations introduced by science and technique to which man, and the whole civilization, have committed their existential destiny, with the result to be slave to their creation.[97]

The fundamental equivocation translates into the rejection of Transcendence, and with it of freedom and existence, and into the total entrusting to imminent reason. This reason (*Vernunft*), reduced to mere intellect (*Verstand*), is capable only to bend Being, by engulfing it into the web of a strict logic, that "conceals that *presence* to which Being-There is always referred".[98] It is impossible to go beyond Western, nihilistic logic if a "revolution of the way of thinking"[99] is not realized, together with a completely new beginning of philosophy. This philosophy, going beyond the knowing (*Erkennen*), peculiar to Being's-There objectivity, assigns a privileged role to thinking (*Denken*), "that discovers Being-There as message of Being, and thinking through (*quer*)". "There-Being is realized as ascertainment of Being".[100]

To overcome the spectre of Nihilism and achieve a real and universal communication, Jaspers leads his research, left uncompleted, on the possibilities of world-philosophy,[101] by way of an analysis on the Chinese and Indian cultural world, whose connection with the West could prove the presence of a sole source full of sense.[102]

There's nothing left to do for the philosopher conscious of these duties but to make a passionate appeal to those who care very much about the salvation of philosophy and the entire civilization,[103] and to make an appeal for the "philosophical faith" as inexhaustible confidence in "reason's" presence.[104]

In this prospect, Jaspers brings to maturity his ethical-political analyses on the crisis of German people[105] and on the coming of the Atom Bomb.[106] This last investigation on man's destiny facing nuclear weapons represents the work that better than any other mirrors Jaspers' ethical purpose and seems to summarize the philosophical conviction on the radical faithfulness to reason's ideal (*Vernunftglaube*), that is peculiar to Husserl also. By analysing the progressive realization of atomic armaments, in the service of the major powers' politic-economic ends, "the possibility that, in the future, people will face this choice: totalitarianism or the Atom Bomb"[107] cannot be hidden. In this prospect there is nothing else to do but to strengthen the appeals to a never-ending communication, and to the theoretical trend always seeking new possibilities of political actions and renewed ethical dimensions.[108]

Having confidence in "reason", as man's original disposition, transcending the necessitated occurrence of history through choices and decisions, it will be possible to realize a "rational" community being the kingdom of free men.[109] In front of the objection on the impossibility of realizing the utopian character of reason, Jaspers, like Husserl, reproposes the essentially ethical aspect of philosophy:

We must go our way today, immediately, within reality, and this with uncertainty which is so difficult to endure, but inevitable because it belongs to human destiny. *But we can choose it in the certainty of reason that must-be*, even if it fails. Reason cannot wait, but must immediately look for its realization everywhere, even in a very small sphere.[110]

University of Arezzo, Siena

NOTES

1 K. Jaspers, 'Mein Weg zur Philosophie', in *Rechenschaft und Ausblick; Reden und Aufsätze*, München 1951, p. 328.
2 H. Saner, *K. Jaspers*, Hamburg 1970, pp. 32–33.
3 K. Jaspers, 'Mein Weg zur Philosophie', p. 328.
4 K. Jaspers, *Philosophie*, vol. I, 3, Aufl/; Berlin-Göttingen-Heidelberg 1956, p. XVII. Hans Saner, Jaspers' assistant in Basilea, tells that Jaspers no longer read Husserl, so that he kept this negative and certainly unobjective opinion.
5 K. Jaspers, 'Mein Weg zur'. . . . , p. 327.
6 K. Jaspers, *Allgemeine Psychopathologie*, Italian translation edited by R. Priori, Roma, 1964, p. VIII; p. 58.
7 M. Heidegger, 'Anmerkungen zu K. Jaspers' "Psychologie der Weltanschauungen" (1919–1920)' in *K. Jaspers in der Diskussion*, hrsg. von H. Saner, München 1973, p. 94; L. Binswanger, 'K. Jaspers und die Psychiatrie', in *K. Jaspers in der Diskussion*, p. 26.
8 A. Lichtigfeld; 'Jaspers' Kantian Step from Objective to Non Objective Thinking', in *Akten des 4. Internationalen Kant-Kongresses* Mainz 6–10 April 1974, Teil II, p. 770 where the Author draws Jaspers' investigation on the "grounding of philosophy" to Husserl's research on the "Science of Beginning".
9 E. Husserl, 'Postilla alle "idee" ', in *Idee per una fenomenologia pura e per una filosofia trascendentale*, Italian translation edited by E. Filippini, Torino 1965, p. 935.
10 *Ibid*., p. 915.
11 Cf. P. Chiodi, *Esistenzialismo e Fenomenologia*, Milano 1963, pp. 52–57; E. Garin-E. Paci-P. Prini, *Bilancio della Fenomenologia e dell'Esistenzialismo*, Padova 1960.

[12] K. Löwith, 'Eine Erinnerung an Husserl', in *Edmund Husserl 1859–1959*, Haag 1959, pp. 49–50. (The underlining is the commentator's).
[13] E. Husserl, *Cartesianische Meditationen und Pariser Vorträge*; Italian translation edited by F. Costa, Milano 1960, p. 49.
[14] E. Husserl, *Die Krisis der europäischen Wissenschaften und die transzendentale Phänomenologie. Eine Einleitung in die phänomenologische Philosophie*, Italian translation edited by E. Filippini, Milano 1972, p. 447. (*Krisis*)
[15] *Ibid.*, p. 536.
[16] It must be remembered that since 1910 – date of the first edition of *Philosophie als strenge Wissenschaft*, (translated into Italian by F. Costa, Torino 1958, pp. 59–78) – Husserl turns his criticism towards the philosophies of the *Weltanschauung*.
[17] *Krisis*, Appendix XXVIII, p. 536.
[18] Cf. *Philosophie als strenge Wissenschaft*, p. 81.
[19] *Ibid.*, p. 83.
[20] *Ibid.*, p. 84.
[21] Cf. K. Jaspers, 'Mein Weg zur Philosophie', p. 328.
[22] L. Landgrebe, 'Die Bedeutung der Phänomenologie Husserls für die Selbstbesinnung der Gegenwart', in *Husserl und das Denken der Neuzeit*, Haag 1959, pp. 223–225.
[23] *Ibid.*, p. 225.
[24] E. Fink, 'Welt und Geschichte', in *Husserl und das Denken der Neuzeit*, p. 166 following.
[25] *Ibid.*, p. 164.
[26] H. Hohl, *Lebenswelt und Geschichte. Grundzüge der Spätphilosophie Husserls*, Freiburg-München 1962, p. 81.
[27] Cf. *Krisis*, Appendix XXVI, pp. 529–530.
[28] *Ibid.*, Appendix X, p. 448.
[29] Cf. H. Hohl, pp. 75–77 following.
[30] *Krisis*, p. 171.
[31] Cf. L. Landgrebe, *Phänomenologie u. Geschichte*, Italian translation by G. Forni, Bologna 1972, pp. 197–200.
[32] *Krisis*, p. 172.
[33] L. Landgrebe, *Phanomenologie und Geschichte*, p. 203.
[34] Cf. *Krisis*, pp. 143–159 following.
[35] Cf. E. Paci, *Tempo e Verità nella Fenomenologia di Husserl*, Bari 1961, pp. 244–250.
[36] E. Paci, *Bilancio della Fenomenologia e dell'Esistenzialismo*, p. 81.
[37] Cf. H. Hohl, p. 81.
[38] Cf. *Krisis*, pp. 206–207.
[39] H. Hohl, p. 81.
[40] *Krisis*, p. 99.
[41] "The historical consideration, which is our aim, regards our existence as philosophers and, correlatively, the existence of philosophy, that, in turn, exists, on the basis of our philosophical existence" *Krisis*, p. 538.
[42] *Ibid.*, p. 101.
[43] Cf. *Cartesianische Meditationen* . . . . , pp. 4, 48.
[44] Cf. *Krisis*, p. 466 and pp. 540–541.
[45] *Krisis*, Appendix XIII, p. 459.
[46] Husserl's interest in historiography is already present in 1923, when the grounds were

set down for a philosophical methodology of historiography as critical introduction to transcendental philosophy. See E. Husserl, *Erste Philosophie (1923/1924), Erster Teil, Kritische Ideengeschichte*, hrsg. von R. Bohm, Haag 1956.

47 Cf. *Krisis*, p. 288 and Appendix XIII, pp. 457–460; App. XIV pp. 474–478.

48 *Ibid.*, App. XXVIII, pp. 540–541.

49 *Ibid.*, App. XIII, p. 465. (The underlining is the commentator's.)

50 *The surrounding world that goes back to the primary foundation of philosophy and of philosophers' generation, is its living present.* In this sphere, the philosopher has his collaborators, his partners, in Aristotle, Plato, Descartes and Kant. *Krisis*, Appendix XXIV, p. 514.

51 *Krisis*, Appendix XIII, p. 467 (The underlining is the commentator's).

52 *Ibid.*, pp. 467–468.

53 *Ibid.*, p. 290 (The underlining is the commentator's).

54 Cf. K. Jaspers, *Philosophie*, vol. I, II.

55 K. Jaspers, *Vernunft und Existenz*, Italian translation by A. Lamacchia, Torino, 1971.

56 *Philosophie*, I, pp. XV–LV.

57 *Vernunft und Existenz*, pp. 49–62.

58 *Ibid.*, p. 63.

59 *Ibid.*, p. 122; pp. 130–149.

60 Cf. K. Jaspers, *Von der Wahrheit*, anthological translation edited by U. Galimberti, Brescia 1970, pp. 284–285.

61 *Ibid.*, p. 270.

62 Cf. L. Pareyson, 'Nuovi sviluppi del pensiero di Jaspers', in *Rivista di Filosofia* (1948/IV), p. 331.

63 Cf. W. Manasse, 'K. Jaspers und der Ursprung des Philosophierens' in *Merkur* XII/1958, p. 284.

64 Cf. K. Jaspers, *Descartes und die Philosophie*, French translation Paris 1938, pp. 85–86.

65 Cf. K. Jaspers, *Autobiografia filosofica*, italian translation by E. Pocar, Napoli 1969, p. 148. See G. Penzo, *Essere e Dio in K. Jaspers*, Firenze 1972, p. 217 on the connection between "History of Philosophy" – "Great" – "Philosophizing" – "Source".

66 The historiographical interest dates back to 1936 with *Nietzsche. Einfuhrung in das Verständnis seines Philosophierens*, goes on with *Descartes und die Philosophie* (Berlin 1937) and with *Schelling. Grösse und Verhängnis*, (München 1955), – after particular studies concerning Kierkegaard and Kant – to amplify with *Die grossen Philosophen*, München 1957. See H. Saner, 'Zu K. Jaspers' Nachlass (1969/1971) zu Geschichte der Philosophie', in *K. Jaspers in der Diskussion*, pp. 449–464 about the new critical prospects which can be given by the various unedited writings concerning this matter.

67 Cf. F. Costa, *La comprensione storico-essistenziale dei grandi Filosofi, Introduzione a Die grossen Philosophen*, Italian translation, Milano 1973, pp. 11–112.

68 K. Jaspers, 'Gibt es eine Weltgeschichte der Philosophie'? in *Aneignung und Polemik, Gesammelte Reden und Aufsätze zur Geschichte der Philosophie*, hrsg. von H. Saner, München 1968, pp. 9–13.

69 K. Jaspers, *Die grossen Philosophen*, pp. 188–191.

70 *Ibid.*, p. 117.

71 *Philosophie* I, pp. 282–287.

72 *Ibid.*, pp. 281–282.

[73] K. Jaspers, *Die grossen Philosophen*, p. 155.
[74] *Ibid.*, p. 349.
[75] *Ibid.*, pp. 1209–1210.
[76] K. Jaspers, 'Gibt es eine Weltgeschichte der Philosophie'? pp. 9–10.
[77] See *Ibid.*, pp. 10–13.
[78] The known and debated statement made by Husserl in his last years must be remembered: "Philosophy as a science, as a serious, rigorous, or indeed, apodictic science – the dream is over". *Krisis*, Appendix XXVIII, p. 535,
[79] Cf. L. Landgrebe, *Die Bedeutung der Phänomenologie* . . . . , p. 223.
[80] *Krisis*, p. 544.
[81] *Ibid.*, p. 99.
[82] *Ibid.*, p. 45.
[83] *Ibid.*, p. 99.
[84] *Ibid.*, p. 38.
[85] *Ibid.*, pp. 36–38.
[86] *Ibid.*, Appendix XV, p. 477.
[87] M. Sancipriano, *L'Ethos di Husserl*, Torino 1967, pp. 89–90.
[88] Cf. *Krisis*, p. 36; pp. 45–46.
[89] *Ibid.*, p. 46.
[90] *Ibid.*, pp. 206 following.
[91] *Krisis*, p. 511 (the underlining is the commentator's).
[92] Cf. for example – H. Hohl, *Lebenswelt und Geschichte*, pp. 78–79; E. Fink, *Welt und Geschichte*, pp. 164–168 following.
[93] K. Jaspers, *Die geistige Situation der Zeit*, Berlin 1931.
[94] *Philosophie I*, p. 273: "We've lost naivety".
[95] See *Philosophie I*, pp. 4–24.
[96] K. Jaspers, *Nietzsche und das Verständnis seines philosophierens*, cit.
[97] K. Jaspers, *Ursprung und Ziel der Geschichte*, Italian translation, Milano 1965, pp. 125–165.
[98] K. Jaspers, *Der philosophische Glaube angesichts der Offerbarung*, Italian translation by F. Costa, Milano 1970, p. 160.
[99] K. Jaspers, *Die Atombombe und die Zukunft der Menschen*, Italian translation edited by R. Cantoni, Milano 1961, pp. 285–286.
[100] U. Galimberti, *Heidegger, Jaspers e il tramonto dell'Occidente*, Torino 1975, p. 251.
[101] Cf. H. Saner *Zur K. Jaspers' Nachlass*, pp. 455–459.
[102] K. Jaspers, *Autobiografia filosofica*, pp. 149–150.
[103] K. Jaspers, 'Gibt es eine Weltgeschichte' . . . . , p. 12.
[104] K. Jaspers, *Der Philosophische Glaube*, Italian translation by U. Galimberti, Torino, 1973.
[105] Refer to his denouncement on the responsibility of German people for the coming of Nazism (*Die Schuldfrage*, Zurich 1946), his bitter considerations on the political destiny of the Federal Republic (*Wohin treibt die Bundesrepublik*?), München 1966. Afflicted by internal and external problems (*Freiheit und Wiedervereinigung*, München 1960). For a more precise analysis of Jaspers' political writings, I refer the reader to my paper 'Realtà politica e riflessione filosofica nel pensiero di Jaspers', in *La Nuova Critica, Quaderno* XXXIV, IX *Serie*, 1973/2, pp. 57–81.
[106] *Die Atombombe und die Zukunft der Menschen*, München, 1958.

[107] *Ibid.*, p. 123, see pp. 246–264.
[108] See *Ibid.*, pp. 30–99.
[109] *Ibid.*, pp. 417–481.
[110] *Ibid.*, p. 526 (the underlining is the commentators).

PIERRE TROTIGNON

THE END AND TIME

In Husserl's thought the concept of "telos" includes two essential strata, the first of which, alone, is not sufficient to necessitate a particular ontology of time.

In the first place, the form that teleology takes in the history of the human mind is the discovery of eidetic knowledge as an exigency of a universal demonstration of truth. The object of science could not be the lived-fact, even if the constancy of a lived-experience (*vécu*) confers an experiential generality on it; certainly, it can only come from possible idealities in which the laws of eidetic concatenation warrant its rational and systematic explication.[1] The ideal object is not given in lived-experience (*vécu*), it is intended (*visé*) through it, and the discipline of thought holds strictly to a "method", which is to say, to a finalized path. Platonism, which opposed the sensible essence of the phenomenon to the essential and eternal fixity of an idea, obscured its finalized character and replaced it by the assumption that the sensible *copies* the intelligible. The danger here is to miss the role of judgment, which is to say, the transcendental nucleus of thought. Judgment presupposes that the ordered activity of thought is implicitly led by a finality of knowledge and that, down to the passive syntheses, the pure being of ideality requires the possibility of its own recognition. Therefore, the sensible is not a faint shadow of the ideal, but is rather the essence which lays itself open to the gaze of thought within the hidden labor of the phenomenon. It is in this way that the teleology of science leads us back to a more radical, and more all-encompassing teleology which is revealed in first philosophy. The hidden labor which reflection will bring to full self-consciousness is a "subjective life" which is directed toward its own evidential confirmation.[2] This evidential confirmation, if its center is the certainty of the *cogito*, entails an objectivity (*Gegenständlichkeit*) which is given as an irreal immanence,[3] escaping from the immediate experience of time.

From this we move to a second essential point. The *telos* of knowledge is not only a transcendental meaning-intention (*visée*) of absolute certainty, but it is also the ultimate meaning-intention (*visée*) of human history which is directed toward the creation of a "pure community based on ideas".[4] Now, this teleology of humanity presupposes a precise theory of time, for if the

Tymieniecka (ed.), Analecta Husserliana, Vol. IX, 301–315.

history of the mind and of science is neither a series of accidents, nor a simple collection of facts, every instant of the conscious lives of humanity must be a possibility for an extension of the mind in which eidetic reflection unfolds the *eidos*.

In order to understand the intimate bond between teleology and time, we will take the exposition of the *Cartesian Meditations* as a guiding thread. This choice might seem strange. However we have seen the importance of the teleology of human history as it is analyzed in the *Crisis*. Now, the *Cartesian Meditations* and the *Crisis* form a significant couple. Husserl was always torn between Descartes and Kant who represented for him the two essential teleological meaning-intentions (*visées*) of knowledge. Now, Descartes represents the most profound meaning-intention (*visée*)[5] in that he presupposes that the knowing subject must be taken as an object for reflection, that it is there that science discovers its absolute *telos* and because, in the third place, philosophy appears there as the absolute teleological idea.[6] In this way the reference to the things-in-themselves and to the "transcendent" principle of the order of the world, which the harmonies and finalities of nature might suggest to our intelligence, are eliminated.[7] But, at the same time, an intentionality appears which initiates "stirrings" ("*amorces*") within conscious immanence, an ordered disposition of the "field", a "finality" of the meaning-intention (*visée*) such that the constitution of knowledge is to be thought of as the way in which the Same (*das Selbige*) becomes Other in order to recognize itself and to be intuitively seen. To put this in different terms, the universal category of all teleology is the *wish to see*, and even the *wish to be seen*, which is to be understood ontologically, and not as a simple psychological determination of man.

The teleological nucleus of this will to see is the Platonic concept of truth.[8] The meaning-intention (*visée*) of the rational form as model of all human existence, is immanent to the subject, so that if the knowing subject explicitly, and in a normal way, intends (*viser*) the determination of truths concerning the objects and the "world" of things, that act can have sense only because the very act of this finality is the bringing into operation of a more profound finality in which the originary rises to intuitivity through endless elucidations of the passive syntheses. True finality is thus without assignable limit. The endless opening (*ouverture*) of the field of history, as domain of originary intuitivity oriented toward its goal (*fin*), would only be a contradiction or a paradox if we forget that this assumption requires an overcoming of the natural attitude.[9] This overthrowing, in its turn, presupposes that the ego can be a "self" (*soi*) only through the repetition (*reprise*) of the teleology of truth within its most immediately intimate life. Natural consciousness must sense its own imperfections and lay itself open to the world as to the possibility of ceaselessly resuming its intuitive and intentional self-realization. At each instant the *world* is the form of the *telos*, however, this form can only be presented as an exploration of transcendental consciousness which exists in the mode of repetition and of eternal beginning. Husserl said that he found a "primitive" but "serious" sketch of this idea of the instantaneous repetition of the teleological meaning-intention (*visée*) in Berkeley.[10] Now, what he stresses, and what was clearly seen by Berkeley, is the bond between the relations among bodies, the theme of truth, the theme of the world, and the teleology of the intersubjective order.[11]

We are now ready to understand the lessons of the *Cartesian Meditations* on the relation between time and finality.

In the game of example and essence, every instant represents a possibility of reflection, a possibility for an extension of the mind. The constitution of phenomenological time, as Heidegger has so forcefully indicated, rests on the close tie between the intentionality and the temporality of consciousness. It is significant that Husserl refers to Saint Augustine and to the famous analysis of time in Book XI of the *Confessions* at the beginning of the *Phenomenology of Internal Time Consciousness*. In this theory, which Husserl believes has lost none of its great philosophical interest, the intuition of the instant in the present leads to an *extensio mentis* which is born from the out-rushing (*éclatement*) of life toward the *desideratum*, which is at once both desire and regret. Intentionality presupposes a finalized out-rushing (*éclatement*) of life toward the possibilities of the ego (*moi*), because each conscious state possesses a horizon composed of references to aspects which are "yet to come", "already passed", and "situated elsewhere". These are possibilities which are realizable by the ego (*moi*), what Husserl called "predelineated possibilities".[12] The play of the predelineated horizon shapes the intentional life of the ego (*moi*), and this potential activity which is uncertainly outlined by an overabundance of "Abschattungen", only has sense if the ego (*moi*) is able to relate itself to an intentional object which serves it as a noetic-noematic guide, to organize the passive givens of sensation.[13] The intentional object is a constitutive *a priori*; and we are able to perform a transcendental analysis of it because we are able to distinguish its essence through reflection. In this way, at the very center of intentional finality, the ego exists for itself, which implies that there must be a continuous evidence of the ego which constitutes it as continually existing within the finalized current of the *cogito*. The *instantaneous cogito* which *runs through time* thus presupposes two conditions:

a polar identity of the ego (*moi*), and a recognition of its identity through its successive syntheses. Intentional consciousness intends (*vise*) an object, a "cogitatum", which synthesizes the multiplicity; but, more profoundly, we are led back to subjective syntheses *through which the ego* (*moi*) *exists for itself*, within the "habitus" which I am able to recapture.[14] The unity formed by the polar ego (*moi*), which underlies the referential identity of the habitus *and* "that without which I could not concretely exist",[15] forms the monad, the *concrete ego in the full sense of the term*. Therefore, the self-consciousness of the intentional ego refers not only to objects, but also, *insofar as it is transcendental*, to the concrete monad.

Let us examine this monad. It can only appear through the effort of reduction in which consciousness identifies itself with the transcendental ego. Now, among the outlines which appear in transcendental objectivity, there is one which plays a singular role: *the other subject*, which is an object of the world endowed with the particularity of *being a physiologically animated body*, and is endowed with intentional behaviors. The other is an object which indicates a subjectivity within the world of my objects; it belongs to my sphere while transgressing the mode of being of my ego. The other's reflection, instead of returning to my monadic root, indicates another transcendental point of turning back (*rebroussement*), which, *nevertheless*, belongs to my sphere. In the interior of my being there are thus constituted numerous transcendental nodal points of which only one in always in concordance with "my" pure ego (*moi*). If I therefore put aside what is alien to me in order to retain only that which is "related to me alone" (*das Mir-Eigene*), I must not only eliminate animality, but also finalized objects like historical and cultural values, all of which are alien to me in the radical sense. However, a pole of pure transcendental finality of the ego, as self-conscious ego, will subsist; it is the *phenomenon of world* which is the transcendental moment of the relations between the ego and the monad. It is this which is the radical teleological orientation.

The single animate body is mine. This means that *in me*, in the relation of the ego to the monad, a world is laid open which is not only a transcendental opening (*ouverture*) in general, but is also the possibility of touching, seeing, and knowing a *nature*.[16] Perception, which presupposes an entire hidden architecture, is the grasping of a thing as "true", as identically achievable by *other egos*. In perceiving, the differences between sensations are abolished, and they are unified in the noematic form of the object, which is at once a common horizon for different egos and a common horizon for diverse types of sensible affections. This is so because perception is the experience of *nature*, which is to say, the examination of the ego by the monad within the transcendental opening (*ouverture*) of the world. Nature is the finality of the mind, a perspective turned back into the regressive road of the reduction, and fixed within the form of the present instant. It is clear that in this way Husserl indicates a kinship, a close affinity – at the very center of the phenomenological analysis of my animate body – between finality, reflection, intentionality, and time.

Let us examine the operation of "reduction to ownness" (*die Eigenheitlichkeit*). I take an object which has already been submitted to the eidetic reduction; I subtract from it everything which is not an exclusive property of the ego, which amounts to saying that I momentarily suspend the finalized and projective character of the opening (*ouverture*) of the world. If I "reduce to ownness" those "animate objects" which are "other men", I obtain "things". If I reduce my body to ownness, "so gewinne ich meinen *Leib*, und meine *Seele*, oder mich als psychophysische Einheit, in ihr mein personales Ich . . . ."[17] Why, in this same reduction, does the other become a thing, while I become a union of soul and body? Moreover, why does the elimination of everything which belongs to alien exteriority leave unchanged the relations between ego, other men, and the things? It is because the totality of the constitution of the world is inherent in me as *nature*, as inverted but ineradicable finality, through the reflexivity of my body in perception. Now, the reflexivity of my body in perception[18] is nothing other than the temporal extension of the now, or "primary memory"; for, at each moment of duration, a consciousness of the proximate past, and not consciousness of the present, is essential to the intuition of the time of being. This dilation of the instant in retention is completely different than the constancy of an *a priori* essence; it is not the permanence of an invariant, but a *continuous appearance*. Through its reflexivity, my body modulates the continuous appearance of the thing by introducing into the instantaneous aspect of the object, possibilities which are thought of as new possible teleologies for my gaze because they are already existent elsewhere for others or for me, but "before", or "after" this present instant. The order of eidetic variations is thus determined by reference to other reflective egos in the world, in such a way that the constitution of an *immanent unity of the temporal flux* presupposes a difference between the simultaneous and the successive, the foundation of which is the experience of the body reduced to ownness, and finalized toward the world.[19]

Starting from the reflexivity of the body in perception, a double genesis of temporality and intersubjectivity is possible; on the one hand, my body is the only original animate thing in the sphere of ownness, and it is through *assimilation* with it that I effect an *apperceptive transferrance* (*apperzeptive Uebertragung*) which leads me to conceive of the other body as another organism. This transferal is not an analogy, and it holds good for every object, since

every apperception in which we apprehend at a glance, and noticingly grasp, objects given beforehand – for example, the already given everyday world – every apperception in which we understand their sense and its horizons forthwith, points back to a 'primal instituting,' in which an object with a similar sense became constituted for the first time.[20]

But, whereas in the world of inert objects, every given can play the role of first foundation of a series, it is on the sole basis of my living, reflexive body that the interpretation of the other (*autrui*) as *alter ego* is possible. The particular form of apperceptive transfer is here that of an "originary pairing" (*ursprüngliche Paarung*),[21] which is a radical form of passive synthesis, that precisely in which the retentional constitution of time and the teleological meaning-intention (*visée*) coincide in the instant. On the side of the other ego, in effect, intentionality is led back from the perceived, alien body to my body, and the institution of the alter ego *leads back* to the duality *Leib/Seele* of my nature. On the side of time, intentionality is *led back* from the perceived object to the invariant and immanent ground of the noema and, in that way, creates an *interiority* within the reflective relation of the transcendental ego to itself. In the former, the stress is on the object which is thus given a place in intersubjectivity – an experience which founds *judgment* (*Urteil*); and in the latter, time brings the ego into existence, and gives it the form of a *conscience* (for *intérieur*) – an experience which is that of memory (*Erinnerung*). When judgment is supported by the interiorized memory of the world the teleology of transcendental consciousness is maintained.

Memory, understood as constitution of a past through a retentional extension of the instant, is at the very center of the relations between *eidos* and *telos*. Husserl's reference to Saint Augustine is not a simple encounter. Husserl seldom made citations. His references to Saint Augustine are laden with sense:

(a) in the *Cartesian Meditations*[22] he cites Descartes (the *cogito*), Socrates (Know thyself!), and Saint Augustine ("in te redi, in interiore homine habitat veritas") together to support a definition of philosophy as a movement which, after having lost the world through the suspension which yields universal self-consciousness, rediscovers that world in its truth on the foundation of the primary being-in-itself which is the community of monads.

(b) in *Erste Philosophie*, Husserl asserts[23] that Saint Augustine had, in effect, enunciated the *cogito*.

(c) in *Zeitbewusstsein* . . .[24] he stresses the importance of the *Confessions* in the history of the analysis of time.

It is important here to recall the consequences which were deduced from the Augustinian thesis on time, and see if they might help in clarifying Husserlian phenomenology. Time is a *distentio animi*[25] which radiates (*éclate*) toward a future and a past, and by which we measure the *affectiones*. The entire trajectory of affection presupposes an *exspectatio*, and comes to an end in a *memoria*.[26] However, this out-rushing (*éclatement*) of the human soul cannot be infinite since the whole life of man is destined, as is every partial unity of time, to be abolished in an end (*fin*) which, in the case of man, is called "death". *I* desire, *I* see, *I* conserve the memory of what is now present. But, for Augustine, it is God who desires, and who conserves the memory of my whole life, taken as a temporal unity.

"At ego in tempora dissilui, quorum ordinem nescio, et tumultuosis varietatibus dilaniantur cogitationes meae, intima viscera animae meae, donec in te confluam purgatus et liquidus igne amoris tui".[27]

The flux of time, the river of time is teleologically directed by the beatific eternity of god into which it rushes. The coming and going of things and men is, in fact, an entrance and an exit, "in tempora alia accedunt, alia succedunt".[28] In Husserl things are less clear; however, at the end of the *Cartesian Meditations* he says that phenomenology does not reject metaphysics, but that it takes up again the ethical and religious problems which the former metaphysics had expressed in a naive manner, "with absurd things-in-themselves".[29] And, in fact, Husserl was able to accept Augustine's analysis which shows that time cannot pre-exist the world. Without a creation of the world, and thus no "then" and no "now"; " . . . non enim est tunc, ubi non est tempus".[30] For Augustine, time belongs to the world of creatures, and therefore, it is and is not simultaneously. This essential mutability would be a pure chaos if it were not for the *contuitus* of our mind which, through the permanence of being which it contains, indicates how it is able to inclose the range of the past and the future. The mind is able to return to itself* (*redire ad cor*), examine its heart, which is to say, the source of the power of being. "Primo redi ad cor tuum".[31] then, "redi ad cor, et inde ad Deum. De proximo redis ad Deum, si redieris ad cor tuum,"[32] and finally, "intus Deus altus est".[33] This transcendent God, about which we could say that he is in me an *ascendent immanence*, is the light which makes all things visible. It is to this invisible light that we must return to see the truth. Now, what Saint Augustine is here forcefully stressing, in opposition to the Platonic doctrine of reminiscence, is that the vision of things in the verificatory light of God, like essential ideas, presupposes a reductive reflection which is focused on the "memory of the present".[34] Now, we encounter this same idea of a *Gegenwartserrinnerung*, of a memory of the present, in Husserl. I am able to represent something to myself as present simply on the basis of past experience, without having it before me "in the flesh", in such a way that I give to what is remembered a duration which reaches me in the now. This does not constitute a recall of the past, but a presence of the past in the present without a reproductive image. The recall of the past which is a remembrance (*Wiedererinnerung*), is not given as a moment of the originary presentative seeing (*das originär gebende Sehen*); however, in the same way, remembrance presupposes that the instantaneous phenomenal presence is finalized through a memorizing interiorization within the instantaneous, of that presence. This is the same formal structure as the relation of essence to its examples in the play of eidetic variations. This common form of eidetic variation and memorized presence shows the originary teleological impulsion (*vection*) from which rational consciousness emerges,[35] defined as a direct vision, the evident intuition of an *eidos*, in which position and motivation are one and the same. This means that the noetic and the noematic are strictly interchangeable. The end (*fin*) is identical to the origin; the movement of thought intends (*vise*) this initial overlapping at every instant, even if that instant is already the lived-through past. This presupposes that the *instant* is the only ontologically real time, the only time in which reality is given, but that it is a division between the origin and the end (*fin*) for a reductive and reconquering labor of thought. The supreme end (*fin*) is the instantaneous recapturing (*reprise*) of being in the truth of the instant.

The verificatory light of the instantaneous is the absolute end (*fin*). This is why it constitutes the proper domain of philosophy, "universale Wissenschaft von der Welt", a knowledge which determines the architecture, the eidetic relations, and the significations of the transcendental opening (*ouverture*) of the ego. This science cannot draw its impulse from some "immediate evidences" which deliver to me some being-in-itself or other. It is, from the first, a teleological position; the certainty of a final assertion which will be confirmed *through time*, by mastering the phenomenal appearance of being to the extent that we find the designative criteria for a "being fundamentally and finally valid in itself" ("*ein an sich endgültiges Seiendes*").[36] We thus understand that the crisis of philosophy is a crisis of European teleological ideas, manifesting itself as the impossibility of making sense of the originary experience of time, and giving rise to naturalism and psychologism through the *effacement of the interiority of thought.*[37]

What is this interiority of thought? A transposition of the memorizing interiorization will help us to understand it; it is the ascription of an *infinite end (fin)* to reason. The ascription of an end to reason rests on an originary historicity, which maintains a past and opens a future for all humanity through the potential belonging of every individual to this tradition. There is indeed a reciprocal affinity between the field which opens the culture of reason to men and the originary coupling of the monad with the alter ego in its sphere of ownness, and both, in their turn, are tied to the essence of time. The same profound essence of the *instant* founds the ego and opens history. Historicity, a process always already begun, is "*ein Allgemeines, das zum menschlichen Dasein gehört*".[38] This common lot is the ideal of a legitimation of that which is through the evident and principled vision of what could be.[39] How does this ideal, which is the absolute end (*fin*), belong in principle, to every man, at every instant? Here again we are going to see that it is temporality which originarily unifies the ego with the ontological and essential truth which it must, at every instant, elucidate and intend (*visé*) as an ultimate end (*fin*).

In fact, we are able to determine the essential belongingness of every phenomenon which allows it to be defined and situated. The domain of belongingnesses (*appurtenances*) is the domain of essences which are in founding-founded relations to one another. There is at least one essence when a given existent can be posited as belonging to the class which is homonymous with it, which is the experience of pure perception. Now, there is at least one being which has the property of referring to an essence which contains at least the possibility of the reflective existence of an exemplary, and this is the *ego.*[40] We are always able to compose between them essences belonging to the same domain. Let us call the *ego* '*a*', and examine the following objective sequence in reflection: *a*, (*a*), ((*a*)), etc. The simple presence of '*a*', which designates the ego is able to remove all uncertainty, for it is indubitable certainty.[41] The ego is thus never an empirical fact, and yet, as the first term in an absolutely certain series, it must be a fact. From the start, the ego stands in a founding relation with itself, and not one of simple belonging. This self-foundation of the ego within the life of the ego accompanies every act of belonging of a fact to an essence, thus opening "the form which necessarily includes all the *récus* of the pure ego,"[42] a gaze bearing on every operation of subjective life, an originary vision "which is the ultimate source of all legitimacy for every rational assertion".[43] This vision is neither immanent nor transcendent, it is a transcendence within immanence,[44] which, once again, repeats the ultimate structure of temporality and teleology.

Each instant is the *teleological opening* (*ouverture*) *of time* in the form of an *ego*, which is able to reflect the duality in its originary unity. The flux of lived-experience (*vécu*) is the experience of the opening (*ouverture*) of the instantaneous presence of time. This flux is not objective, and is not itself "in" time; "it is absolute subjectivity",[45] which we only metaphorically designate as 'flux', for the flux of time in its duration is the *image which designates the pure ego in its rational, teleological relationship to the essences*. This image accounts for the fact that the ontological opening (*ouverture*) is undergone as a splitting of the ego (*Ichspaltung*). The ego, as the foundation of the self, exists in the mode of *reflection*, which does not only modify the form of lived-experience, but also assigns to it a new significant end (*fin*). In this way, reflection does not constitute the lived-experiences of the ego into objects which the ego would then *have* at its disposal; but rather, in the objective life of the ego, it maintains the incessant renewal of the modes of being.[46] Through reflection, the ego comes to grasp the ontological sense of experience, through a modification of the finality of lived-experience, and not through a rending of its continuity. Thus the splitting of the ego presupposes the immediate belongingness of the split ego and the act of splitting; "in the living present (*in der lebendigen Gegenwart*), I have the split ego and the act of the splitting of the ego in a state of co-existence".[47] This co-existence applies to the ego as unity of its duality; and this unity which is intended (*visée*) as an ultimate end (*fin*), appears forthwith as the originary unity of being itself, without which the teleological purpose of humanity would have no sense. We are able to speak paradoxically of *the same ego* (*das selbe Ich*) at the very center of the *split* (*Spaltung*) and through the expedient of this split; "the active life of the ego is absolutely nothing other than this process of constant self-division in an active behavior (*das Sich-immerfort-in-tätigen-Verhalten-spalten*)".[48] The ego is the infinite explosive splintering (*èclatement*) of the self, which is ontologically identical to the presence of the thing itself in the present, as a perpetually renewed possibility to intend (*viser*) the ultimate end (*fin*) of the mind.

It is in this way that the profound signification of phenomenology as an awakening and as a conversion is confirmed. The reference to Saint Augustine and to Socrates at the end of the *Cartesian Meditations* is not a simple formality as we have already seen. The self-examination of consciousness is not a simple gaze at what we are, but the assumption of ends which intend (*vise*) the activity which, in constituting me, opens the manifestation of the things.

And, just as in Saint Augustine, where the self-examination of the soul is the movement toward the absolute light of god and his inexpressible presence, in transcendental phenomenology, reflection leads to the presence of the pure ego as a movement of ontological splitting, an origin without origin of all the forms which intend (*vise*) the spiritual teleology of humanity. In the Augustinian vision of man and the world, the human mind (*mens*) raises itself above the sensible world by going beyond itself and contemplating the divine ideas which constitute supreme Wisdom. This knowledge of the self and of the idea is not, however, identical with immutable being, but it is a simple participation in the divine nature.[49] The conversion itself, the inversion of thought which makes possible this moment toward the absolute, does not depend on the initiative of the human soul, but on a gift of divine grace which transforms possessive and prideful love into a submission of the soul to the act of god's love. The precept "redi in te ipsum" for this reason, must be placed in relation to that other Augustinian formulation, "te ipsum tibi reddam, quando te mihi redero".[50] Does this interplay of *redire* and *reddere* have an equivalent in Husserl? Most certainly. The return (*redire*) is the abandonment to the purely phenomenal, the approach to which through the eidetic reduction we saw in *Erste Philosophie*. However, this *return cannot be taught*, and the "absolutely constraining" motivation of phenomenology can have no other teaching authority than the *teleological sense of the ego in the instant* in which time opens and disperses the phenomenal field of the world. The reductive method does not signify an impoverishment of the phenomenal world, but rather it designates the suspension of the natural attitude which limits consciousness.[51] It is thus an unlimiting which, interiorized through the discovery of the pure ego as the root of all experience, is an *unlimiting of the ego* in reflection. The ego is the unlimited which opens sense in the instant. The conversion of thought to the unlimited is the *emergence* of sense-bestowing subjectivity (*die sinngebende Subjektivität*) and, through this emergence, there is the potentiality of an intersubjectivity extended to humanity as a whole acting as a teleological pole for thought. This infinite self-reflection (*Selbstbesinnung*) is the reflective return to the teleological essence of the ego.[52] We discover here the bond between the instantaneous decision of the conversion of the gaze, and the destiny of rational humanity. All of philosophy is a long self-meditation of humanity seeking self-understanding (*Selbstverständnis*),[53] and the interior light which enlightens man about himself (*Selbsterhellung*). Self-elucidation is the light of reason struggling against self-forgetfulness in which thought would no longer recognize that it is the center of the world. The supreme end (*fin*) of knowledge is to not forget.[54]

What, therefore, is radical finality? It is the meaning-intention (*visée*), which is a self-transcendence by which the divided consciousness brings being to light as instantaneous phenomenal reality. The being of consciousness, in its founding intentionality, is quite simply *being*, so that the initial act of the suspension of the "thesis of the world",[55] which sets off the absoluteness of consciousness, is the bringing to evidence of the functional character of consciousness, which is to say, the finalized character of temporality. *Being is in itself the absolute end* (*fin*), which clearly appears when we interpret intentionality not as the simple act of meaning (*visée*) of an object, but as the self-examination of thought which founds its out-rushing (*éclatement*) toward possible objects. For then, as Husserl tells us, we reverse the usual sense of the expression "being",[56] by supposing for good reason that the being of the ego is absolute – a proposition which is arrived at through the conversion of the ego in its splitting. To say that the being of the ego is absolute is not a psychological proposition, and still less is it an assertion of solipism, for in reality, it is a way of emphasizing that each *instant* is the opening (*ouverture*) of being in the form of a *transcendental ego* for which reality is the *goal* (*but*) of an investigation and of a conversion. Reason is the self-elucidation of being, it expresses the correlation between veritably real being and legitimatable being,[57] and in this sense, it is a "form of the universal and essential structure of transcendental subjectivity in general".[58] Philosophical self-exploration is the bond between the *originary opening* (*ouverture*) *of the ego in the present* and the *ideal of an absolute evidence of real being*. Rationality as self-examination (*Selbstbesinnung*) is an activity of constant reflection (*ständige Reflexivität*), which holds open the present and prevents it from falling back into forgetfullness. "It is only through this constant activity of reflection that a philosophy is a 'universal knowledge".[59] This movement of universal self-examination is a teleological history which defines man's being, because the very being which is taken up in the opening (*ouverture*) of the phenomenal, in the separation of being and sense,[60] is defined as a *Self*.

The *Self* of the reflective self-examination in the finalized history of knowledge is *being as such in its absolute essence*:

"The absolute polar idea, that of an absolute in a new sense, an absolute which is situated beyond the world, beyond man, beyond transcendental subjectivity, is the absolute *logos*, the absolute truth, as *unum, verum, bonum*."[61]

Absolute being opens itself freely as logos, which is to say, as a meaning-intention (*visée*) of an accord between sense and reality. Transcendental consciousness shows itself to be historical and teleological because, in the instantaneous present, it implicates the absolute Self as the transcendence of that which it reflectively achieves. For it, "to be itself" means to move through the opening of the phenomenal with the purpose of understanding it as a correspondence between the said and the given in a going beyond the present toward presence; "to be oneself" is to go beyond oneself toward the "Self". The existence of man forms a finalized history, even if each instant of it might well be entirely free of presuppositions if it is taken up by the gaze of the phenomenological conversion, because the will that we have to govern ourselves (*Selbstregierung*) presupposes that it is always possible for us to hold both the *archê* and the *telos*, the original and the final, in the unity of our gaze (*regard*):

"Jede Vorhabe, insbesondere die Mannigfaltigkeiten der Vermittlung intentional implizierende, bedarf der wiederholenden Besinnung, der Wiederholung als Erneuerung der Ursprünglichkeit . . . ."[62] The passive infinity of facts, of givens, of chance events arising from within the density of the temporal flux have, as a correlate, the reflective splitting of the ego through which passes the light into those shadows of the world.

We are now able to understand that the relations of temporality and teleology presuppose a particular metaphysics and a particular ontology.

The proper end (*fin*) of the mind is *intuition*, an act of possession of an object which fills the void of the signifying act. This intuition repeats the originary opening (*ouverture*) of being within phenomenality, it is at once its origin and its end (*fin*). Consciousness cannot understand the end (*fin*) which the signifying act which passes through it intends (*vise*), if it has not yet succeeded in intuiting it, or if it has not previously grasped the essence of signification and of intentionality.

Being is the fullness of the real, which can, however, be apprehended only within the eidetic game of essence and example. We can thus not unambigously qualify it in any language. We must say that it is at once passivity and activity, immediacy and distance, immanence and transcendence. However, the pure transcendent cannot be given. We begin from lived-immanence, and when we have that immanence, the transcendent stands out from it as intended end, an ultimate term in which every mind is able to recognize the truth. This cannot be achieved unless, in the same way, every consciousness apperceives in itself that the labor of knowledge brings to it the presence of the universal Self, the signifying and intentional aspect of which is formed through the splitting of the ego in thought. Through the successive supports which give it all the eidetic variations of which it is capable, consciousness advances toward the point where it will recognize "the thing in itself" ("die Sache selbst"). This presupposes that phenomenology is not a theory of subjectivity and of the ego, but rather, that it only analyzes subjectivity in order to abolish it within the pure presence of recognized reality. The teleology of spiritual and intellectual life intends (*vise*) the truth, understood as the presence of the thing itself, without any further reference to its unknown approach. It intends the point at which the transcendental ego will be able to utter the truth: "I am you yourself". If consciousness is so highly valued, if reflection is so highly esteemed a task, it is not because they secure for me a singular place in the world, but rather because, in them, every instant of time represents the renewed chance that the ultimate end (*fin*) might be attained. And this ultimate end is the paradox of a language which utters the truth: "You are my reality".

Université de Lille, III

Translated by Laurence E. Winters

NOTES

* Translators note from page 307; Not only does Augustine "play" with the double sense of "return to", or "turn back to", and "self-examination", or "self-inspection", but so does Trotignon – "retour sur soi-même" for example.

[1] Edmund Husserl, *Erste Philosophie*, I, 5. *Vorlesung*, in *Husserliana*, VII, 34–35.

[2] *Ibid.*, 4. *Vorlesung*, (VII, 30).

[3] *Ibid.*, 15. *Vorlesung*, (VII, 103–6).

[4] Edmund Husserl, *Die Krisis der europäischen Wissenschaften und des europäischen Menschentums und die Philosophie* in *Husserliana*, VI, 334 (*Krisis*).

[5] *Husserliana*, VI *Beilage* 10 and 15 – *Husserliana* VII, *Beilage* 18.

[6] *Erste Philosophie*, I, 9. *Vorlesung*, in *Husserliana* VII, 60–3.

[7] Edmund Husserl, *Ideen zu einer reinen Phänomenologie und phänomenologischen Philosophie*, I, §58, in *Husserliana* III, pp. 138–40 (*Ideen*).

[8] *Erste Philosophie*, I, 2. *Vorlesung*, in *Husserliana* VII, 15.

[9] *Ibid.*, 12. *Vorlesung*, in *Husserliana* VII, p. 81 sq.

[10] *Ibid.*, 21. *Vorlesung*, *Husserliana* VII, p. 150.

[11] *Ibid.*, VII pp. 151–2.

[12] Edmund Husserl, *Cartesianische Meditationen und Pariser Vorträge*, II, §19 in *Husserliana*, I. pp. 81–2 (*Cart. Med.*).

[13] *Ibid.*, II, §20 in *Husserliana*, I, p. 84.

[14] *Ibid.*, IV, §32 in *Husserliana*, I, p. 101.

[15] *Ibid.*, IV, §33 p. 102.

[16] *Ibid.*, V, §44 and the marginal note of Husserl.

[17] *Ibid.*, V, §44, in *Husserliana*, I, p. 128.

[18] *Ibid.*

[19] Edmund Husserl, *Zur Phänomenologie des inneren Zeitbewusstseins*, §§24–30, in *Husserliana*, X, pp. 52–63 (*Zeitbewusstsein*).

[20] *Cart. Med.*, V, § 50, in *Husserliana*, I, p. 141.
[21] *Ibid.*, V, § 51, in *Husserliana*, I, p. 142.
[22] *Cart. Med.*, V, § 64 (*Husserl.*, I, 182–183).
[23] *Erste Philosophie I*, 9. *Vorlesung*, in *Husserliana*, VII, 61.
[24] *Zeitbewusstsein* . . . . , Einleitung, in *Husserliana*, X, 3.
[25] Augustinus, *Confessiones*, XI, 26.
[26] *Ibid.*, XI, 28.
[27] Idem, *ibid.*, XI, 29.
[28] Augustinus, *Sermo 117*, 10.
[29] Husserl, *Cart. Med.*, V. 64, in *Husserliana*, I, 182.
[30] Augustinus, *Confessiones*, XI, 13.
[31] Augustinus, *In Johannis evangelium*, XVIII, 10.
[32] Augustinus, *Sermo, 311*, 13.
[33] Augustinus, *In Ps. 130*, 12.
[34] Augustinus, *Confessiones*, XI, 18.
[35] Husserl, *Ideen I,* § 136, in *Husserliana*, III, 333–6.
[36] *Krisis,* § 73, in *Husserliana*, VI, 269–76.
[37] ms. M. III 5 II b.
[38] ms. K III 9, cité in *Husserliana*, VI, 502.
[39] *Ideen*, I, § 136.
[40] *Ideen*, II, § 23 (*Husserliana*, IV, 101–6).
[41] *Ideen I,* § 46 (*Husserliana*, III, 106–10).
[42] *Ideen I,* § 82 (*Husserliana*, III, 200).
[43] *Ideen I,* § 19 (*Husserliana*, III, 44).
[44] *Ideen I,* § 57 (*Husserliana*, III, 137–8).
[45] *Zeitbewusstsein* . . . . , § 36 (in Husserliana, X, 75).
[46] *Ideen II*, Beilage VII, in *Husserliana*, III, 318.
[47] *Erste Philosophie II*, 40. *Vorlesung*, in *Husserliana*, VIII, 89.
[48] Ibidem, VIII, 91.
[49] Augustinus, *In Psalm., 41*, 7.
[50] Augustinus, *Sermo, 123*, 5.
[51] *Nachwort*, in *Husserliana*, V, 153.
[52] *Krisis*, Beilage XXVIII, in *Husserliana*, VI, 510.
[53] *Krisis,* § 73 in *Husserliana*, VI, 269 sq.
[54] *Krisis*, Beilage XXIV, in *Husserliana*, VI, 485.
[55] *Ideen I,* § 49, in *Husserliana*, III, 114–6.
[56] *Ideen I,* § 50, in *Husserliana*, III, 118.
[57] *Ideen I,* § 136, in *Husserliana*, III, 334.
[58] *Cart. Med.*, § 23, in *Husserliana*, I, 92.
[59] *Krisis*, in *Husserliana*, VI, 339.
[60] Ms K III 4, in *Husserliana*, VI, 489. Cf. *Ideen I,* § 49 (Husserliana, III, 112–6).
[61] Ms E III 4, p. 60, trans. A. Lowit and M. Colombié, in *Etudes Philosophiques*, 1954, p. 39.
[62] Ms K III 4, in *Husserliana*, VI, 485.

STEPHAN STRASSER

HISTORY, TELEOLOGY, AND GOD IN THE PHILOSOPHY OF HUSSERL

The historians of our epoch agree in admitting that Husserl is an important philosopher. Whatever their point of view, they could not deny that the phenomenological movement set in motion by him has contributed to the intellectual profile of Europe. They could not be blind to the fact that the philosophies of Scheler, of Heidegger, of Sartre, and of Merleau-Ponty presuppose certain directing ideas of Husserl's. Husserl is therefore a thinker whom we must study closely. But we know him little, we know him poorly, we know him superficially. We try to get rid of him by subsuming his case under ready-made categories. According to certain historians Husserl would be an "essentialist," an "intuitionist," a "Platonist"; according to others, an "immanentist," an "idealist," or a "subtle neo-Kantian." Once such a slogan has been found we have arrived: we have "understood" Husserlian thought by making it enter into a rubric.

But even those who have read some of these published works do not suspect the fact that *Husserl, the man and the philosopher, struggled all his life with the problem of God*. It is here a question of a trait which counters the idea of this thinker we generally give ourselves. It is typical that even a personal disciple, the German philosopher Wilhelm Schapp, knew nothing of it. In 1959, Schapp devoted to Husserl a commemorative article which ends with the characteristic remark:

> The arisal of philosophers from Thales to Kant makes one think of the battle of the giants who invaded or defended the heavens. This comparison may be understood more or less literally. The combat ended provisorily in the eighteenth century with Kant and his views on God, freedom, and immortality. *What would phenomenology have to say on this subject?*[1]

Well, the phenomenology of Husserl does have something to say on the subject of these problems. It is indeed interesting and instructive to analyze Husserl's thought from the point of view of metaphysics and the philosophy of religion.

This much being said, it is clear that we shall not speak of the man, but of the philosopher. In any case it will be a question of religious philosophy in the largest sense, and not of Husserl's attitude with regard to such and such

Tymieniecka (ed.), Analecta Husserliana, Vol IX, 317–333.

positive religion. *How does a phenomenological thinker arrive at posing the problem of God?* This is what will first interest us.

Let us start by noticing that time and again Husserl speaks of the problem of God. We could even say that his thought is haunted by the idea of the Divinity. From 1911,[2] we find in the published works, but above all in the unpublished works, passages which relate to the supreme Being. Husserl speaks of it explicitly and sometimes at length. Of course, he does not use equivocal expressions like "the transcendental being," "the absolute being," "the archetypal intelligence," or "the great unknowable." He speaks of God by naming him God. He always returns to it, if only after long intervals.

In what context does Husserl see himself confronted with the fascinating mystery? However uncertain and hesitant his attitude – we will have occasion to show it – on this point Husserl proves to be a mind of striking consistency. Contrary to the traditional doctrine God is not, according to him, the "cause" of the world. This is perfectly comprehensible. For Husserl, who had studied astronomy and mathematics, causality was a scientific concept designating a strictly defined function. By contrast, the idea of God was imposed upon him in another context, that of a *universal teleology*. As far as the manner in which Husserl characterizes this "telos" is concerned, we find considerable divergences. In the first book of the *Ideas* it is still a question of empirical facts which require a final explanation. Husserl names two of them: the evolution of the series of organisms up to the appearance of man, and, in the framework of the evolution of humanity itself, the awakening of culture with the treasures of the mind.[3]

For the phenomenologist another fact seems more significant, a fact which one could not explain empirically: the immanent teleology of our conscious life. It contrives that on the basis of processes which unroll in accordance with the laws of a blind causality we arrive at constituting a reasonably ordered world.[4] But that is not all. Husserl discovers teleological relationships in axiological consciousness,[5] gnoseological consciousness,[6] in the practical life of the subject,[7] and in the moral order of the world.[8] Evidently one would have to know exactly what the terms "telos" and "teleology" signify. We propose to speak of them later.

If it is true that since 1911 it is a question of the problem of God, it plays a role still more important in the later philosophy of Husserl. To understand the reasons for this we should examine above all two of his last works: the third volume of the *Phenomenology of Intersubjectivity*, containing writings dating from 1929 to 1935,[9] and *The Crisis of European Sciences and Transcendental Phenomenology.*[10] The texts concerning the phenomenology of intersubjectivity have only been published for three years. Their importance is capital. They contain several passages which have a metaphysical and religious import. We would like to cite two of them.

The first is taken from a letter which Husserl had addressed to E. Pearl Welch in 1933 and which the editor, Iso Kern, cites in his Introduction. In this letter Husserl writes the following:

> Philosophical problems are presented in their true sense as transcendental-phenomenological problems when one systematically follows an essential gradation. It appears then that religious-ethical problems belong to the highest degree. ... In the higher degrees of phenomenology, error, at the same time as the questions concerning the ethical life, a life authentically or inauthentically human, leading to an ultimate peace or ultimate unhappiness (to an individual and social harmony and disharmony), present themselves again as problems. It is there a question of a universal problematic which could just as well be designated by the name of teleology.... Therefore the problematic of phenomenological philosophy finally culminates in the question concerning the 'principle' of the teleology which has been rendered concretely accessible in its universal structures.[11]

This passage teaches us many things. First, Husserl assures us that he has not at all forgotten the questions concerning ethics, social life, religious life. Only, he holds to phenomenological philosophy's being systematically developed. He wants to begin at the beginning, and for him that means: with perception. If Husserl devoted many years to detailed descriptions of perceptual experiences it is because he wanted first to assure the basis of all his philosophy. However, following his program, the questions concerning social, generative, cultural, ethical, and religious life should not be neglected. If, in consequence, the problem of God figured at the highest level of this development, that does not mean that Husserl considered it to be a secondary problem.

The text we have just cited also allows us to understand the connection which, according to Husserl, exists between ethics and teleology. – The ethics that our philosopher professes is a humanistic ethics. According to him all men are called to become truly human, to organize a harmonious social life. They can respond authentically or inauthentically to their vocation. That vocation is nothing else but the expression of an immanent teleology. It is inherent not only to humanity, but to the entire reality of which humanity constitutes the supreme level. It is consequently rediscovered in all the structures of reality: in nature, psychic life, intellectual life. Thus, according to Husserl's program, this universal teleological principle should not be posed as a postulate in the style of idealistic philosophers, but should rather be demonstrated in phenomenological fashion thanks to concrete descriptions. – But what then is the nature of this principle? Are we authorized to say that for Husserl the Divinity itself *is* this principle?

Other passages permit us to respond affirmatively to this question. Here is a significant text: "God is not himself the monadic universe, but the entelechy which is immanent to it as the idea of a telos of an infinite evolution, of a humanity supported by absolute Reason which necessarily governs monadic being and governs it thanks to its own free decision."[12]

This text demands a certain interpretation. – The idea of a monadic universe had been presented by Husserl in the fifth of his *Cartesian Meditations.*[13] – But Husserl tells us clearly that, according to him, God is not identical to that community of monads; nor is he – as for Leibniz – the inventor and the builder of the harmonious monadic system. He is its *entelechy*. What does that mean? This technical term should be taken in the sense that Aristotle had given it in opposing it to "dynamics," to potency. We know that the Stagyrite had considered the psyche as entelechy, that is, as the perfecting act of a body which is in potency of living.[14] Analogously, God is, for Husserl, the immanent principle of perfection of the entire monadic universe. He is equally the principle which assures the advent of a humanity capable of realizing the reign of an absolute Reason. The divine principle therefore governs the whole life of the monads, but it does not govern it by imposing commandments and prohibitions "from without." The entelechy animates monadic being "from within." Consequently the individual monad itself will make its free decisions; but in making them it will, without suspecting it, obey the fundamental tendencies of its innate telos. – This "entelechy" is thus for Husserl nothing else but the divine teleology of which we have just spoken.

Let us now glance at the last work – unfinished work – of Husserl: *The Crisis of European Sciences. . . .* Interpreters agree in affirming that a certain infliction of Husserlian thought is to be sensed in the *Crisis*. What strikes them is that, for the first time, Husserl introduces the historical dimension into his research. "*Philosophy as science*, as serious, rigorous, indeed apodictically rigorous science – *the dream is over*," he remarks with regret; and in the same breath he exclaims: "Necessity of reflection. The reflection [must be] historical. *How* is history required?"[15] Of course, the need for historical reflection is accepted without further ado. It is a question only of the "how." How to introduce the historical point of view into phenomenological discourse; that is the question which preoccupies Husserl.

Why would it be so difficult to take historical perspectives into account? – Let us not forget that during a part of his life Husserl had been a man of science. He had studied exact disciplines in depth, such as mathematics, geometry, formal logic, and the theories precursor to what today we call logistics. The philosopher Husserl had long been fascinated by Leibniz's ideal of a "mathesis universalis," that is, of a formal, exact, and universally applicable science. Such a science would have the pretention of formulating apodictic, eternally valuable propositions "more geometrico." In 1935 Husserl felt he should renounce this dream. The necessity of taking the historical dimension into account was imposed. But the question of the "how" was all the more delicate.

Two possibilities were offered to Husserl in this epoch. The one would have consisted in a sort of historicism, such as Wilhelm Dilthey had established. According to Dilthey, reality was entirely and essentially historical. The pretention of philosophy and metaphysics to enunciate universally and eternally valuable truths would thus be vain. The history of philosophy would rather offer the spectacle of a series of world views (*Weltanschauungen*) belonging to definite types. Dilthey distinguished three of them: the materialistic-positivistic conception, that of subjective idealism, and that of objective idealism. According to him these philosophical views, each one opposed to the others, would always return. Impossible to grasp the nature of reality, of being, of truth as such. Philosophy thus could not be anything other than a typological doctrine concerning diverse conceptions of the world (*Weltanschauungslehre*).

Husserl could also have adopted another doctrine which, for different reasons, insisted on the relativity of truth. According to Heidegger all truth would depend upon human existence (*Dasein*). " 'There is' truth only in so far as *Dasein* is and so long as *Dasein* is," the author of *Being and Time* assures us; "That there are 'eternal truths' will not be adequately proved until someone has succeeded in demonstrating that *Dasein* has been and will be for all eternity."[16] But since human existence is unfolded in time, truth too should be historical. It would then be necessary to abandon the idea of a subsistent truth. According to Heidegger, truth (in the sense of the Greek word ἀλήθεια) would be produced uniquely thanks to a disclosure of beings. In other words: truth would be an event.

It is evident that Husserl feels attracted neither by Dilthey's historicism nor by Heidegger's existential analysis. He rejects the one's relativism and the other's attitude, which he considers to be a sort of irrationalism. He is convinced that both offend against the dignity of reason. For Husserl it is a question of introducing the historical element without making the slightest concession to the modern forms of skepticism.

We have said that the *Crisis* marks a turning point in Husserl's philosophical evolution. This is true in different connections. It also appears in this work that the *phenomenological metaphysics* which Husserl had announced in his *Cartesian Meditations*[17] *will be a metaphysics of mind and of reason*. This position evidently has consequences for his entire philosophy. "*Can reason and that-which-is be separated, where reason, as knowing, determines what is*?" This rhetorical question is found in the introduction to the *Crisis*. And the response follows without delay: "True being is *everywhere* an ideal goal, a task of *episteme* or 'reason,' as opposed to being which through *doxa* is merely thought to be, unquestioned and 'obvious.'"[18]

However – and this is the suprising turn that the later philosophy of Husserl takes – *the reasonable mind does not have an invariable form; its existence is a becoming*. Evidently that does not mean that the reasonable could ever become unreasonable or inversely the unreasonable reasonable. The movement in question is an "evolution" in the primitive sense of the Latin "e-volvere"; that is, it is a continual process of successive disengagements. It is a question of a constant transition from the implicit to the explicit, from that which is hidden to the manifest state, from that which is unconscious to the conscious state.

> Thus philosophy is nothing other than 'rationalism' through and through, but it is a rationalism differentiated within itself according to the different stages of the movement of intention and fulfillment; it is *ratio in the constant movement of self-elucidation*, begun with the first breakthrough of philosophy into mankind . . .[19]

If we adopt this point of view universal history moves into transcendental history. But a transcendental history could not be a sequence of fortuitous events. It should unfold following an immanent necessity, it should be produced under the form of necessary phases. The genesis of reasonable humanity is one of them, the eruption of philosophy another, the discovery of transcendental phenomenology a third.[20] "Here is universal and, at the same time, transcendental history," Husserl tells us in an important unpublished text dating from 1930:

> It is in all its phases disposition and realization, and as it is entirely development of itself by itself, containing in itself the sense of its evolution, that is, of its impetus and its accomplishment, we should also speak of disposition and of realization in relation to the whole, and the question of its 'telos' [of the 'telos' of all history] is raised . . . In order that there may be a world and a subjectivity which constitutes it . . . , it [transcendental history] should be produced under the form of essential degrees, passing from pre-being to being, and constituting in itself the reasonable man. Reason must already exist, and it must be able to reveal itself in a logical manner in the reasonable subject.[21]

It appears, therefore, that, following Husserl, reason should, in order to be able to manifest itself, produce the reasonable man, destined to discover reason and to bear witness to it.

If we consider events from this point of view *the history of philosophy also acquires a transcendental meaning.* It is true that if we consider the history of philosophy "from without" it is not at all distinct from a certain human cultural activity. As such it would simply be the chronicle of a social process which unfolds in space and time. But if we see it "from within" we perceive that philosophers are the supports of this evolving mind. The struggle of generations of philosophers thus becomes "the constant struggle of 'awakened' reason to come to itself, to an understanding of itself," Husserl remarks.[22] – The numerous historical considerations that we encounter in the *Crisis* and which, in their entirety, seem so little systematic, constitute a single effort on Husserl's part to show that transcendental phenomenology is the philosophy that all Western culture had unconsciously awaited and toward which the different philosophical currents had unsuspectingly converged.

Why this digression, one will ask. We see no relationship between Husserl's view of history and his idea of God. Yet this relationship exists. It becomes manifest as soon as we pose some questions concerning *the necessity with which this transcendental history unfolds.* How does it happen that reason tends with certitude toward its ultimate end, the light? Why *should* the reasonable man, the philosopher, the phenomenologist arise in the course of this evolution? It is impossible to account for this necessity by empirical research. That is because experience does not permit us to observe such a progress. Neither the concrete life of the individual man nor that of humanity gives place to the hypothesis of a constant aspiration to the clarity of reason. On the contrary: if we examine the lives of men we are often confronted with the symptoms of a confusion, of a wandering, of a mental illness. – This constitutes a serious difficulty for Husserl's transcendental phenomenology. For the illness which disorders individual consciousness can, in principle, become fatal for all consciousnesses at the same time. And if the phenomenon of a world existing objectively is the correlate of the subjective and intersubjective loans of reasonable monads, what would happen at the moment in which those monads start to rave? Would that not come down to the collapse of the world hitherto constituted in the unanimous way? Would not such a catastrophe signify an interruption, indeed, a failure of transcendental history? Would it not injure the manifestation of reason in the course of humanity's progressive evolution?

Husserl recognizes perfectly the importance of these arguments. He knows that he could not combat them by empirical research. But he refutes them as a phenomenologist. Intentional analysis makes us discover a finality inherent to the conscious life of the individual; it is that the most diverse functions and acts contribute to the constitution of a unique meaning. On the intersubjective level the most different men make efforts to fill in the gasp, to illuminate the obscurities, to stop the contradictions in their knowledge. In this manner they arrive at intersubjectively valuable judgments and decisions. That is why the constitution of an objective world, of a "world for us," is necessarily the work of a monadic community.[23] The idea of reason losing its way has therefore no transcendental import. It is true that there are abnormal subjects. But the very fact that we recognize them as such proves that we are aware of a norm. It is therefore impossible to deny in principle the existence of a finality inherent to the constitutive life of the individual monad, to that of the monadic universe and to the transcendental history of this universe.[24] – Very well, to that universal finality should correspond a universal end or, as Husserl prefers to say, a "telos." This universal telos is the motivating force of all historical development. It also offers an ultimate guarantee against a fatal deviation or a definitive stopping of that evolution. This supratemporal and transhistoric telos is therefore a metaphysical notion. *Husserl does not hesitate to identify it with the idea of God.*

"Is it possible that reason begins and ends in constituting experience?" Husserl asks himself;

> Is it possible that the constitutive process that it [reason] has freely set in motion be in vain, that being (and that is therein implied: a true, absolute value, rendering an ultimate existence possible) be a vain attempt which is arrested with the passing realization of myself, who am now by chance a reasonable subject, capable of being recognized and constituted as such: is it thinkable that reason not be active in a surtemporal manner and one belonging to all times? . . . It is here that *the idea of God* and the idea of a world teleology as principle of a possible totality of beings is imposed as a problem.[25]

In the same unpublished text God is also characterized as "the idea of an absolute pole"[26] toward which the flux of the intersubjective constitution of the world is directed. Again Husserl calls it the "absolute logos, the absolute truth in its full meaning of the *unum verum bonum* toward which every finite being is oriented, while being united by the aspiration which includes every finite being".[27] – One will now understand how Husserl could at once hold to the idea of a reason which changes its forms while developing, and reject historicism. According to him reason evolves in the course of history; but its evolution aims at an end which surpasses history. The absolute logos is supratemporal although it is manifested in the conscious lives which unfold in time. In short, it is thanks to the idea of God that Husserl's later philosophy is distinguished from a panhistoric and relativistic doctrine *à la* Dilthey.

This solution seems relatively clear. However it admits of doubtful aspects. Two questions of decisive importance should be discussed:

(1) Is the absolute logos of which Husserl speaks a finality immanent to cosmic evolution, or is it a transcendent and exterior pole in relation to that evolution? Is God himself the reason which develops, or is he the *finis ultimus* of that development? In other words: Is Husserl's religious philosophy pantheistic or theistic?

(2) If Husserl's God is a pole toward which cosmic evolution tends, but which does not itself evolve: how does such a conception agree with transcendental idealism? There, we encounter a considerable difficulty. Philosophically expressed, a transcendental God would have an objective, absolute existence, that is, an existence independent of subjective intentionality. At first glance such a thing seems to be incompatible with transcendental phenomenology. For that philosophy appears to be based upon the categories of a constituting subjectivity, on the one hand, and that of the products of its constituting activity, the real and ideal objects belonging to the phenomenon of the world, on the other hand. – We would thus have to choose between two possibilities. This God could be considered an ideal object, product of a subjective or intersubjective transcendental activity. He would then be a "mundane" phenomenon (in the phenomenological sense of this term). In this case it would be absurd to speak of a transcendental divine being. – The alternative would consist in the identification of God with subjective transcendental life. Then the Divinity would be identical to a reason seeking itself. We would have to do with a God who aspires to the light and to his full realization. In short, it would then be a question of a pantheistic conception, which would raise grave objections.

The question whether Husserl made note of these two obstacles, and how he tired to surmount them, is imposed.

But it is certain that Husserl was aware of the difficulties that we have just pointed out. That does not mean that his solutions were entirely satisfying. His attitude is rather hesitant. – According to the testimony of Ludwig Landgrebe, Husserl "attempted" to interpret transcendental subjectivity both in the sense of a pantheism and in the sense of a monadology.[28] The impression given us by the unpublished texts confirms this judgment. It is above all in the "Three Conferences on the Ideal of Humanity according to Fichte" that the notion of "telos" is determined in a pantheistic sense. "The I is absolutely autonomous, it bears its God in itself as the teleological idea which animates and directs all its pure actions," Husserl tells us.[29] In another unpublished text Husserl speaks of the possibility of a divine "supermonad" ("Uebermonade").[30] He discusses this idea, he does not adopt it. Thus it is intentionally that Landgrebe speaks of different *attempts*. Moreover, there are passages in the unpublished texts and in the published works which are opposed to a pantheistic interpretation.

Let us first return to the first book of the *Ideas*. God is there characterized as transcendent not only in relation to the world, but also in relation to "absolute consciousness." It is "an *'Absolute' in a totally different sense from the Absolute of Consciousness*" and it is "a *transcendent in a totally different sense* from the transcendent in the sense of the world".[31] Whatever Husserl thought of it before and afterward, at the moment of the conception of the *Ideas* God was for him not at all inherent to the transcendental *ego* or to transcendental intersubjectivity. – In Husserl's later philosophy God is most often described as the telos, the teleological idea, the final principle of evolving reason. It is clear that these formulae may be interpreted just as well in a pantheistic as in a theistic sense; for the possibility that, according to Husserl, this teleological idea, this final principle *inhabits* reason in the process of developing could not be neglected. But Husserl also says that "to absolute subjectivity corresponds, as an infinite path which permits it to arrive in evolving at its true being ... an absolute, supreme idea".[32] In the same unpublished text Husserl speaks of an "absolute ideal pole" to which "corresponds the universal transcendental subjectivity which lives in the flux of the constitution in common with a valuable world".[33] That which is found as ideal pole *facing* transcendental subjectivity could not be immanent to that subjectivity, nor coincide with it. That is the reason why we interpret such passages in a theistic sense.

Let us now examine the second question we have just formulated. We have observed that the God who would be an ideal pole for evolving reason should exist in an absolute fashion, that is, independent of subjective or intersubjective reason. Would such a conception be compatible with the fundamental convictions of transcendental phenomenology? We have already formulated the alternative: this God should either be a phenomenon constituted by transcendental consciousness, or inhabit that consciousness. We do not see any way of escaping these alternatives. However, it is uncertain whether Husserl considered the idea of an absolute being absurd.

One thing is certain: This problematic should have been developed in the *Crisis*: Husserl announces it clearly. From the introductory paragraphs he poses in a fairly brusque manner the question of the absolute. He poses it under the form characteristic for him: it is first a question of the possibility of an absolute *knowledge*. It appears, however, that, for our philosopher, absolute knowledge is at the same time knowledge of the absolute, that is, of God. "The problem of God clearly contains the problem of an 'absolute' reason . . . ," Husserl tells us.[34] The problematic therefore is not of an epistemological nature, it is of a metaphysical nature. – This question concerning the absolute becomes for the old Husserl a crucial question. Thus he asks himself in an unquiet, indeed tormented tone: "How can thinking achieve anything but relative truth? . . . "[35] Is not the idea of an absolute truth, of a truth in itself, of an objective truth meaningless?

Let us first ask what the word "objective" could mean in this context. It does not mean, as for Kant, "intersubjectively valuable." Husserl has no illusions about that sort of objectivity: "Experience . . . as communal experience and reciprocal correction . . . does not change the relativity of experience; even as communal experience it is relative . . . "[36] The objectivity in question here is certainly not that to which the sciences of nature lay claim. Husserl always rejected the opposition we readily make between exact statements and subjective statements; in our text, he affirms once more that this opposition is "absurd".[37]

How does Husserl resolve this problem? Does he offer us a solution? In the epilogue of the *Crisis* which constitutes so to speak a provisional proof posed over the unfinished edifice we find a passage which gives us an idea of the manner in which Husserl meant to surmount this difficulty. Here is what he says of it:

> He [man] verifies, argues, and decides in a rational way – but does the whole idea of 'truth in itself' have a meaning for him? Is this [idea], together with its correlate, what exists in itself, not a philosophical invention? But it is not a fiction, not a dispensable invention without significance, but one which raises – or is called to raise – man to a new level in a new historical development [*Historizität*] of human life, a historical development whose entelechy is this new idea . . . "[38]

This text permits us to understand the metaphysical ground of the later Husserlian philosophy. It teaches us some important things:

(1) The "truth in itself" is an idea.

(2) To the idea of a "truth in itself" corresponds that of a "being in itself." Husserl does not find these notions to be absurd.

(3) The ideas of a "truth in itself" and of a "being in itself" are inventions of the philosopher, but they are not vain and arbitrary fictions. In other words: this philosophical invention does not have the character of a paralogism; it

does not issue from an inevitable transcendental illusion which the critical philosopher would have to unmask.

(4) The discovery of the idea, "truth in itself," marks in the history of humanity the beginning of a new era.

(5) The idea, "truth in itself," is at the same time an entelechy which moves and animates the evolution of a humanity which has just entered a new historical phase.

The key word in this whole passage is undoubtedly the word *idea*. It allows Husserl to escape the alternatives we have just formulated. *The idea is neither constitutive consciousness nor constituted phenomenon*; it belongs neither to transcendental subjectivity nor to the mundane horizon. It is rather that which guarantees the ultimate unity and finality of all the constituting loans. Without being something of this world it constitutes "an absolute, ultimate reality." It is the telos of the reason whose evolution unfolds in different phases: but it is not identical with the reason which still seeks itself. On the contrary, it is the motor of that evolution; it makes reason realize itself at successive, always more elevated levels.

In Husserl's thought this ideal telos is the Divinity itself. All the texts plead in favor of such an interpretation: God is almost always called "idea," "ideal end," "ideal telos," "entelechy," etc. Let us cite another one of those philosophical monologues to which we listen every time we read Husserl's unpublished works:

> God as idea, as idea of the most perfect being, as idea of the most perfect life, which constitutes of itself a perfect world, [of a life] which creatively develops from itself a universe of perfect spiritual beings in relation to a nature perfect in the highest degree. Philosophy as correlate of the idea of God, as science concerning the Being existing in an absolute way.[39]

It is true that there is in philosophy no technical term as equivocal as the term "idea." In the context of Husserlian thought one should take two possible meanings into account: the Platonic ἰδέα which the spiritual soul can apperceive thanks to an intellectual intuition, a concept akin to the notions of "eidos," "essence," and "ideal prototype." But one must also think of Kant for whom the idea is a concept in relation to which one could not find in the sensible world a corresponding object.[40] It is certain that Husserl was familiar with both meanings.[41] – But the notion of "telos" is equally ambiguous. To a person who has had an Aristotelian-Thomistic formation the idea of a God who would be the final cause of cosmic evolution appears perfectly familiar. However, it is certain that Husserl did not think of such a final causality. He examines this possibility and rejects it.

But would we not then arrive at Aristotelian metaphysics, that is, at ποιοῦν, at the final idea which animates all world history and which, at the same time, should be conceived as a Platonic idea?" he asks; "Yet no, in so far as we would not truly have posed an ideal end (following the schema of the mature tree which is the end and the limit of development).[42]

This passage is particularly instructive. Why aren't Aristotelian entelechy and the Platonic idea truly ideal ends? The example of the tree implies the answer: because it is a question of a determinate end, because such an end can be attained, because it can be the object of an intuition. The idea, the truly ideal end, by contrast, could not be realized in a practical way, nor grasped theoretically. Husserl said it formally: the idea of God does not correspond to the highest axiological degree we could imagine, it corresponds to an *infinite and unlimited augmentation* of values. This conception brings Kant to mind in so far as the idea is transcendent, in so far as one could not find an object which would conform to it. However, Husserl distinguishes himself from Kant because for him the idea is not an "heuristic fiction," valuable uniquely for the regulative use of the understanding. It is for him an ultimate reality, an absolute reality. – We can therefore conclude by remarking that *God is for Husserl from the speculative point of view an "ideal"* (in the sense we have just determined), *from the practical point of view an ideal telos.*

That is what the historian of philosophy and interpreter of Husserl could observe after an examination in depth. May I be permitted to add some critical remarks. (It goes without saying that, however severe it may appear to be, my critique does not impair the admiration that Husserl's effort to arrive at a religious philosophy inspires in me.) I would like to limit myself to three considerations:

The first remark has reference to the *idea of history* such as Husserl proposes it to us. One wonders if he doesn't rather speak of a cosmic development whose outcome is fixed in advance. It is true that, according to Husserl, the process of humanity's perfectioning could be impeded and retarded in a passing way, but it is certain that it could not fail. But that view of history goes against history as we live it. Precisely in the framework of a phenomenological philosophy this perspective does not satisfy. Maurice Merleau-Ponty, struggling against the conception of history which is at once idealistic and Marxist, protests: " ... history is impotent to achieve anything without the consciousnesses which take it up and so decide of it ... "[43] We could reverse the terms of this characterization by saying: a situation where something can either succeed or abort is an historical situation; a series of events whose outcome depends upon our efforts, our courage, our intelligent decisions, that is history. A cosmic process which, in any case, will arrive at its ends may not merit the name of history.

A second consideration attaches to the first. It relates to *ethics*. We recall that Husserl told us that "absolute Reason necessarily governs monadic being and governs it thanks to its own free decision." The formula seems rather vague; but those who are familiar with Husserl's thought understand it. Our philosopher means that regulation by absolute Reason is not imposed upon the monadic being from the outside. There is no question of heteronomy. The consciousness of the individual monad makes its decisions in perfect liberty. It governs its own conduct autonomously. But, as we have indicated above, even while acting freely, consciousness unsuspectingly obeys the fundamental tendencies of absolute Reason. – All that makes us think of the "ruse of Reason" of which Hegel spoke. But in our epoch, in the epoch of totalitarian regimes, everyone knows that that absolute Reason which beguiles individual consciousness is nothing other than a disguised tyrant. It is an absolute power which, thanks to cunning and subtly camouflaged influences, hinders and impedes the true ethical decision. – It is doubtful that Husserl knew this modern problematic; but it is certain that his attempt to reconcile the idea of autonomy and that of heteronomy does not reach a satisfying solution.

To close, let us say a word about Husserl's religious philosophy. It is intentionally that Husserl always repeats that God is an idea, a principle, an ideal telos. These are diverse ways of saying that, in the framework of his religious philosophy, God is not a person. It would therefore be useless, perhaps even absurd, to implore him, to venerate him, to pray to him. Once again we have to do with a "God of the philosophers," far removed from the God of Abraham, Isaac, and Jacob (to speak like Pascal).

Despite all these objections it will have to be recognized that Husserl made a far-reaching effort to resolve the ultimate questions of ontology and metaphysics. Whatever certain historians, certain critics, and even certain disciples have thought, Husserl was not the philosopher who exhausted himself describing the appearance of an extended thing to a spectator or the sonorous impressions a prolonged tone made upon an auditor. He was not afraid to pose the most fundamental questions. It is thanks to this radicalism, to this philosophical curiosity, to this metaphysical disquietude that Husserl deserves the name of a great philosopher.

University of Nijmegen *Translated by Philip Lawton*

NOTES

[1] Erinnerungen an Husserl,' in *Edmund Husserl, 1859–1959*, Nijhoff, La Haye, 1959, pp. 12–25, see p. 25; my italics, S.S.

[2] *Idee der 'philosophischen Disziplin,' Idee der Philosophie*, manuscript dating from 1911, initial F I 14.

[3] *Ideen zu einer reinen Phänomenologie und phänomenologischen Philosophie*, erstes Buch, Husserliana III, La Haye, Nijhoff, 1950, p. 139; *Ideas: General Introduction to Pure Phenomenology*, trans. W. R. Boyce Gibson, Collier, New York, 1962, – abbreviation: *Id I* – §58, p. 158.

[4] Ibid. and ms. F I 14, p. 42.

[5] Ms. F I 14, p. 40, 42, 47.

[6] Loc. cit., p. 46.

[7] *Erneuerung als individualethisches Problem*, ms. dated 1922, initial F II 7, p. 148, and *Drei Vorlesungen über Fichtes Menschheitsideal*, ms. dated 1917, F I 22, p. 19, 20.

[8] Ms. F I 22, p. 22.

[9] *Zur Phänomenologie der Intersubjektivität*, dritter Teil, Husserliana XV, Nijhoff, La Haye, 1973, abbreviation: *Int III.*

[10] *Die Krisis der europäischen Wissenschaften und die transzendentale Phänomenologie*, Husserliana VI, Nijhoff, La Haye, 1954, abbreviation K; *The Crisis of European Sciences and Transcendental Phenomenology*, trans. David Carr, Northwestern University Press, Evanston, 1970.

[11] In all the cases where there is no recognized translation we will let the German text follow in the notes. "Die philosophischen Probleme erschliessen sich mit ihrem echten Sinn als transzendental-phänomenologische in einer wesensmäszigen systematischen Stufenfolge. Es zeigt sich dabei, dass die religiös-ethischen Probleme solche der höchsten Stufe sind . . . . In der höheren Stufe der Phänomenologie wird der Irrtum in eins mit den Fragen des ethischen Lebens, des Lebens in echter oder unechter Menschlichkeit, in letztlicher Befriedigung oder Unseligkeit (einer individuellen oder sozialen Harmonie und Disharmonie) von neuem Problem. Es handelt sich um die allumfassende Problematik, die auch unter dem Titel der universalen Teleologie angesprochen werden kann . . . . So ist der oberste Abschluss für die Problematik der phänomenologischen Philosophie die Frage nach dem 'Prinzip' der in ihren universalen Strukturen konkret erschlossenen Teleologie"; *Int III*, pp. LVIII–LIX.

[12] "Gott ist das Monadenall nicht selbst, sondern die in ihm liegende Entelechie als Idee des unendlichen Entwicklungsethos, des der 'Menschheit' aus absoluter Vernunft, als notwendig das monadische Sein regelnd, und regelnd aus eigener freier Entscheidung"; *Int III*, appendix XLVI, p. 610.

[13] *Cartesianische Meditationen*, Husserliana I, Nijhoff, La Haye, 2d edition, 1973, pp. 121–183; *Cartesian Meditations: An Introduction to Phenomenology*, trans. Dorion Cairns, Nijhoff, The Hague, 1960, abbreviation CM, pp. 89–151.

[14] *On the Soul*, II, 2, 412a

[15] "Philosophie als Wissenschaft, als ernstliche, strenge, ja apodiktische Wissenschaft – der Traum ist ausgeträumt . . . ." "Notwendigkeit der Besinnung – die Besinnung historisch – *wie* bedarf es der Geschichte?"; *K*, appendix XXVIII, p. 508; trans. Carr, p. 389.

[16] "Wahrheit 'gibt es' nur, sofern und solange Dasein ist . . . Dass es ewige Wahrheiten

gibt, wird erst dann zureichend bewiesen werden, wenn der Nachweis gelungen ist, dass in alle Ewigkeit Dasein war und sein wird". *Sein und Zeit*, 7th edition, Niemeyer, Tübingen, 1953, pp. 226 and 227; *Being and Time*, trans. John Macquarrie and Edward Robinson, Harper & Row, New York, 1962, §44c, pp. 269–270.

17 *CM*, §60.

18 "Ist Vernunft und Seiendes zu trennen, wo erkennende Vernunft bestimmt, was Seiendes ist?"; "Ueberall ist wahres Sein ein ideales Ziel, eine Aufgabe der Episteme, der 'Vernunft,' gegenübergesetzt dem in der Doxa fraglos 'selbstverständlichen,' bloss vermeintlichen Sein." *K*, p. 9 and p. 11; trans. Carr, p. 11 and p. 13.

19 "So ist Philosophie nichts anderes als "Rationalismus," durch und durch, aber nach den verschiedenen Stufen der Bewegung von Intention und Erfüllung in sich unterschiedener Rationalismus, die ratio in der ständigen Bewegung der Selbsterhellung, angefangen von dem ersten Einbruch der Philosophie in die Menschheit . . ."; *K*, p. 273; trans. Carr, p. 338.

20 See *K*, appendix XXVI, p. 503.

21 "Das ist die universale und dabei transzendentale Historie. Sie ist in allen Stadien Anlage und Verwirklichung, und da sie als Ganzes Selbstentwicklung ist und in sich Entwicklungssinn, also Aufstieg, Erfüllungsgang birgt, so ist auch für das Ganze von Anlage und Verwirklichung zu sprechen und die Frage nach ihrem Telos tritt auf . . . Damit die Welt und die sie konstituierende Subjektivität sein kann (. . .), muss sie, in Wesensstufen verlaufend, auch den Vernunftmenschen in sich konstituieren. Es muss Vernunft schon sein und muss sich zur logischen Selbstenthüllung in dem vernünftigen Subjekt bringen können." *Anfangende Vernunft – Teleologie – Gott*, manuscript dating from 1930, initial E III 4, p. 19a.

22 " . . . im ständigen Ringen der 'erwachten' Vernunft zu sich selbst, zu ihrem Selbstverständnis zu kommen . . . "; *K*, p. 273; trans. Carr, p. 339.

23 See *CM*, §36, 37.

24 See *Int. III*, text 14, p. 210, 211, and appendix XXVI, p. 438.

25 "Kann Vernunft anfangen und aufhören im konstituierten Dasein? Kann der konstitutive Prozess, den sie schliesslich freitätig ins Werk gesetzt hat, vergeblich sein, kann Sein (und darin beschlossen wahrer, absoluter Wert, der letztes Sein möglich macht) in dieser Weise ein leeres Absehen sein, das stecken bleibt in der vorübergehenden Verwirklichung von mir, der ich jetzt gerade Vernunftsubjekt bin, als das erkennbar, konstituierbar: kann es anders sein, als dass Vernunft überzeitlich und allzeitlich am Werke ist? . . . Hier muss die Gottesidee und die Idee der Weltteleologie als Prinzip einer möglichen Seinstotalität zum Problem werden"; ms. E III 14, p. 30, 31.

26 "absolute ideale Polidee"; loc. cit., p. 36a.

27 den "absoluten Logos, die absolute Wahrheit im vollen und ganzen Sinn, als das unum verum, bonum, auf die alles endliche Seiende in der Einheit [des] jedes endlich Seiende umfangende Strebens hingerichtet ist"; loc. cit., p. 36b.

28 *Phänomenologie und Metaphysik*, Hamburg, 1949, p. 190.

29 "Das Ich ist absolut autonom, es trägt seinen Gott in sich als die reinen Tathandlungen beseelende und leitende Zweckidee"; *Drei Vorlesungen über Fichtes Menschheitsideal*, manuscript dating from 1917, initial F I 22, p. 25, 26.

30 *Substanz und Monade*, manuscript dating from the years 1921–1923, initial E I 17; see also volume two of *Zur Phänomenologie der Intersubjektivität*, Husserliana XIV, appendix XLI, pp. 300–302, *Möglichkeit der Verschmelzung von Monaden, Möglichkeit*

einer (göttlichen) Uebermonade.

[31] *Id I*, §58, p. 142, Husserl's italics; trans. Gibson, p. 158.

[32] "der absoluten Subjektivität, als der unendliche Weg, zu ihrem wahren Sein sich zu entwickein . . . eine absolute oberste Idee entspricht"; ms. E III 4, p. 36a.

[33] " . . . eine absolut ideale Polidee" der "die transzendentale Allsubjektivität im unendlichen Strom der vergemeinschafteten Konstitution der Weltgeltung . . . lebend entspricht"; *ibid.*

[34] "Das Gottesproblem enthält offenbar das Problem der 'absoluten' Vernunft . . . " *K*, §3, p. 7; trans. Carr, p. 9.

[35] 'Wie kann Denken anderes leisten als relative Wahrheiten?"; *K*, §73, p. 270; trans. Carr, p. 336.

[36] "Die Erfahrung, die Gemeinschaftserfahrung und wechselseitige Korrektur . . . ändert nichts an der Relativität der Erfahrung, sie ist auch als Gemeinschaftserfahrung relativ"; *ibid.*

[37] " . . . ein Widersinn"; *K*, p. 271; trans. Carr, p. 337.

[38] " . . . er [der Mensch] bewährt, argumentiert und entscheidet sich vernünftig – aber hat für ihn die ganze Idee 'Wahrheit an sich' einen Sinn? Ist das, und korrelativ an sich Seiendes, nicht eine philosophische Erfindung? Aber doch nicht eine entbehrliche und bedeutungslose Erfindung, sondern eine solche, welche den Menschen auf eine neue Stufe erhebt, bezw. zu erheben berufen ist in einer neuen Historizität menschheitlichen Lebens, deren Entelechie diese neue Idee ist . . . . "; *K*, p. 270, 271; trans. Carr, p. 336.

[39] "Gott als Idee, als Idee des allervollkommensten Seins, als Idee des allervollkommensten Lebens, das aus sich die allervollkommenste 'Welt' konstituiert, das aus sich die allervollkommenste Geisterwelt in Bezug auf eine allervollkommenste Natur schöpferisch entwickelt. Die Philosophie als Idee, als Korrelat der Gottesidee . . . . als Wissenschaft vom absolut existierenden Sein"; ms. F I 14, p. 43.

[40] *Critique of Pure Reason*, Transcendental Dialectic, first book, second section.

[41] See *Id I*, p. 9, 10.

[42] "Aber kämen wir dann nicht zur Aristotelischen Metaphysik, als sum ποιοῦν als der beseelenden Zweckidee der ganzen Weltentwicklung, die zugleich als Plantonische Idee gefasst werden müsste? Doch dies insoferne nicht, als nicht eigentlich ein ideales Ziel gesetzt wäre (nach dem Schema des vollendeten Baumes als Endgrenze der Entwicklung)"; ms. F I 14, p. 43, 44.

[43] *Phénoménologie de la perception*, Gallimard, Paris, 1945, p. 512.

Section II: Eschatology and the "Form of all Forms"

ANGELA ALES BELLO

TELEOLOGY AS "THE FORM OF ALL FORMS" AND THE INEXHAUSTIBILITY OF RESEARCH

Husserl's assertion that teleology is "the form of all forms" suggests the need for a reexamination of his analyses in this perspective, which may be considered fundamental for discovering a *leitmotiv* in phenomenological investigation itself.

Teleological research seems to be bound in a particular manner to the ultimate significance of phenomenological inquiry, an inquiry that consists of the perennial question "What are things like?" and in the last resort concerns the possibility of man's understanding the life-world, not only for speculative purposes but with a view of finding criteria for conduct.

Although, as Husserl himself recognized, phenomenology in this respect remains within the furrow of the theoretical inquiry characteristic of Western culture, I yet think that, going along Husserl's intentions, imposes a need for radicalism surpassing all traditional philosophical systems. The *πρώτη φιλοσοφία* is not a theory among other theories; it is not a point of view side by side with others already proposed in the history of speculation, but rather constitutes an inquiry that goes "beyond" the theorizations, below the "crust of facts" and outside of the bounds of conventional schemes in order to highlight the originary.

I. TELEOLOGY AND PRECATEGORICAL LEVEL

The historical and present importance of phenomenological inquiry is therefore due to the radical change it proposes with respect to past philosophies, and to the fact that it brings to the fore the ultimate and deep dimension, the precategorical one that underlies everything that is theorized and therefore made categorical. If this was once an arrival point of phenomenological analysis, it has now become an obligatory starting point for those who want to continue to use and refine the means of investigation offered by Husserl.

It is at this level that teleology as the "form of all forms" makes its appearance in a primary and also lasting manner. It reveals itself in the dimension of originariness that phenomenology was the first to clearly indicate and, even though it is a constitutive element, it does not emerge immediately but has to be sought for.

Tymieniecka (ed.), Analecta Husserliana, Vol. IX, 337–351.

By means of regressive inquiry (*Rückfrage*), the originary structures of subjectivity and therefore of the world can be examined in their connection-distinction right back to the *Urkinästhesen, Urgefühle, Urinstinkte.*[1] One thus realizes that in "every fact" and even in subjectivity (for subjectivity, above all, is a fact)[2] there is a form of unity (*Einheitsform*):

... die Wesenform ... von der Weltlichkeit. ... Damit scheint schon 'instinktiv' die Konstitution der ganzen Welt für mich vorgezeichnet, wobei die ermoglichenden Funktionen selbst ihr Wesens-ABC, ihre Wesensgrammatik im voraus haben. Also im Faktum liegt es, dass im voraus eine Teleologie statthat.[3]

If a teleology is discovered in the "fact," a full "ontology," inasmuch as it brings out the fundamental and constitutive element of reality, must be identical with it.

What is meant by the statement that the constitution of the entire world is "instinctively" pregiven in a form of unity, that is, an essence (*Wesensform*)? What does this predetermination indicate? Are all things already known in advance, so that the dynamic aspect of knowledge is eliminated by a strict determinism?

It is only at the precategorical level that we discover the full meaning of knowledge itself in its dynamic tension. Indeed, if it is true that we can arrive at the essential structure (*Wesensstruktur*) of the world in an apodictic manner, then the entire gnoseological process must remain an "open" process.[4]

The validity of this assertion is confirmed by the analysis of evidence. Evidence consists of "... Selbsterschauung des Gemeinten bzw. erfüllende Verwirklichung in der Selbsthabe, als Synthese der Identitätsdeckung bloss vergreifender Meinung mit dem erfüllenden Selbst."[5] Given the identity thus discovered, the fulfillment (*Erfüllung*) of the *Leervorstellung* substantially indicates an uncovering (*Enthüllung*) of what is already given.[6] *Selbsthabe* as subjective element has its counterpart in *Selbstgeben* as objective givenness; evidence is therefore nothing other than *adaequatio intellectus et rei*: "... danach ist also bewährende Evidenz nichts anderes als in der Ursprünglichkeit der Selbsthabe vollzogenes Bewusstsein der Adaequatio rei et intellectus."[7]

One may therefore ask whether evidence gives truth in a full sense. If the answer is yes, then an apodictic, adequate, and definitive knowledge would be possible. But Husserl's answers to this question are full of vacillations and changes of mind that in the final analysis are connected with the meaning he attributes to "reduction"; one must therefore go back to the beginning and ask oneself "what one wants to obtain" and what one "effectively obtains" by means of reduction.

If "reduction," understood as suspension of world existence, has transcendental subjectivity as its residue, then one must ask oneself whether such a "residuum" is apodictic and, further, whether knowledge of the ego itself is adequate.

If adequation is the same as apodicticity, as Husserl holds in *First Philosophy,*[8] then the ego is not only apodictic, but self-knowledge must also be adequate; however, if ". . . *das Ichdenke . . . nicht adäquat erkennbar ist*,"[9] as is argued elsewhere in the same book, then it would seem that the very apodicticity of the ego is called into doubt, even though it is affirmed in *Cartesian Meditations.*[10]

Subjectivity as an absolute and unquestionable residue is the result of the "Cartesian way" that seeks a beginning on which the building of knowledge can be erected. But the other ways of reduction seem to reject this perspective, especially the one defined by I. Kern as the *Weg über die Ontologie,*[11] which results in the apodictic beginning being called into doubt. If one still speaks of apodicticity, the term is intended to refer – as I have said before – to an essential structure and not a beginning.[12]

The discovery of this essential structure takes the form of an analysis of transcendental subjectivity, but in its temporal dimension, and this is why one can speak of doubt and error. In other words, unquestionability would involve absolute staticity, and therefore also a knowledge defined once and for all; but questionability of the radical starting point would open the way to skepticism of relativism, and the entire problem therefore needs further consideration.

In what sense can one speak of "apodicticity" of the ego, of the world, or of evidence?

We have already seen that the apodicticity of the ego is bound up with a certain way of understanding phenomenological reduction. What, then, is the answer as regards the world? Assuming that reduction does not take us to a "residuum,"[13] but will lead us back to a fundamental, the basis of the precategorical life-world, the apodicticity of the being of the world must clearly be "relative."[14] But relative to what? Once again, this would seem to refer us to the absolute apodicticity of the "I"; indeed, what other than the *ego* could be apodictic? Apparently there is no way out of this dilemma.

I think that this can be clarified precisely by examining the evidence. Granted that evidence is *adaequatio intellectus et rei*, we have to see whether we are really faced with an *Endgültigkeit*. In discussing this problem, Husserl makes a distinction between mathematical truths and those depending on experience, and holds that in this field it is impossible for truth to be determined in an absolute manner; but it is possible to realize this fact and to describe the modalities of the relationship between *intellectus* and *res* without determining its contents once and for all, i.e., in an absolute manner. In the last resort, after all, the entire discussion is concerned with the problem of truth possession as regards the "I" and the world.[15]

If the process of knowledge is a process of *Bewahrheitung,*[16] truth remains a limit idea; and this justifies the teleological structure of evidence as described in *Formal and Transcendental Logic* and is related to the problem of how the given is given:

> Alle Selbstgebung, und zwar alle, die schon begrenzt wird auf wirkliche Präsentation, steht unter Gesetzen der Gradualität der Klarheit, worunter wir allgemeinst verstehen eine Gradualität der Selbstgebung, die ihren idealen Limes hat in einer absoluten oder 'reinen' Selbstgebung, wofür absolute Klarheit also nur ein anderes Wort ist.[17]

And therefore:

> Ein immanentes Selbst, das zur Urstiftung gekommen ist, is für das aktive Ich auf Grund möglicher Wiedererinnerungen ein beständiges mögliches Telos, oder vielmehr es ist Mitglied eines Reiches von Seiendem an sich, das ein universales mögliches Telos für dieses Ich ist.[18]

While the straining toward understanding is justified by the temporal structure itself, which prevents a full adequation and apodicticity,[19] this selfsame temporal structure also suggests the possibility of overcoming dispersion, and this precisely through rememoration (*Wiedererinnerung*). Such an effort toward understanding does not mean "dispersion," but rather the possibility of highlighting the inexhaustibility of research and, in addition, the acquisition of some firm points. These latter are represented by the function of recollection peculiar to the rememoration, which enables us to form some basic concepts like that of "object," and thus to overcome immediacy in the conquest of a firm knowledge. But of what does this conquest consist? Analysis of evidence causes the precategorical dimension to emerge as a level on which the two poles of the "I" and the "thing" are present in their interconnection and, consequently, it is possible to speak of "constitution." But even if we grasp the very structure of constitution, have we really solved the problem of the full possession of truth once and for all?

The teleological structure of evidence, which is simultaneously the teleological structure of consciousness, prevents this ultimate conquest. In that case, can one really answer the question regarding the meaning of the "I" and the world? In a certain sense an answer can be given, and in another sense it cannot, at least not a full one.

The inexhaustibility of research is connected precisely with the emergence of the essential structures;[20] the seeming paradox of this statement resolves itself it we bear in mind that it is possible to understand and indicate the meaning of evidence, i.e., its teleological structure as revealed at a precategorical level, without thereby solving the problem of the possession of truth once and for all.

One might note, indeed, that an affirmative answer to the question proposed above could have a twofold result: first, that of adequacy and apodicticity, by virtue of which true knowledge regards the "I" (and this is a theme that emerges in the Husserlian analyses of the Cartesian way of reduction); second, another and very different result in which transcendental reduction is "primordial,"[21] in the sense that the analysis of transcendental subjectivity is the ground on which the relation between the "I" and the world manifests itself. Otherwise stated, we can say that at this point there emerges the Husserlian paradox of the "I" that is both subject and object. Subjectivity is the point at issue here.

The criterion of evidence (understood as *adaequatio*) seems to involve a dualism between the *intellectus et res*, a dualism that seems to be in contrast with evidence as self-givenness, in which the distinction between the two terms should disappear, and consequently also subjectivity as a preferential element. Nevertheless, even though it is true that the theme of transcendental subjectivity retains its validity also when one follows the way of the life-world, as W. Biemel holds,[22] one has to examine whether subjectivity is or is not intended as a "residue" and therefore in an apodictic sense.

The critique of subjectivity, that formulated by D. Conci[23] being a case in point, tends to consider it as a categorical "residue," which is in contrast with the theme of self-givenness.

The point at issue can therefore be formulated as follows: Are the subjective and objective poles categorical products or do they merely emerge at a categorical level, reflections of something that is already constituted at the precategorical level?

In reality the moment of *Selbsthabe* and that of *Selbstgeben* seem to be delineated already at the precategorical level as the ultimate foundation of the gnoseological perspective brought out by the phenomenological inquiry; in fact, to whom is the given given?

The subjective pole is originarily delineated and subsequently emerges as a moment of recognition to the extent to which it can grasp the meaning of its originary self-givenness through reflection; but it also grasps the indefinite character of givenness itself, albeit in its teleological trend toward adequacy. Fullness, situated at infinity, refers us back to a visualized but not completely attainable end. It is in this sense that one can comment on the passage in *Krisis* where Husserl, accepting and developing a fragment of Heraclitus

("Der Seele Grenzen wirst du nie ausfinden und ob du auch jegleichen Strasse abschrittest; so tiefen Grund hat sie"), recognizes that ". . . das unendliche Ganze in seiner Unendlichkeit strömender Bewegung auf Einheit eines Sinnes gerichtet ist, aber freilich nicht so, als ob wir ihn ohne weiteres ganz erfassen und verstehen könnten."[24]

This means that the "dualism" is not that of *intellectus* and *res*, but is rather related to the infinity of the adequacy process and, in the last resort, to the transcendence of originariness. Indeed, evidence as *adaequatio intellectus et rei* must not be understood as a congruence of two unrelated elements, but rather as a process of verification: ". . . Uns eine Vorstellung evident zu machen, das ist doch, sie zu ursprünglich erfüllender Bewahrheitung bringen."[25]

From a certain point of view, therefore, we have a belief that all can be clarified, from another we find the admission that this belief has a merely regulating function. This is possible because the apodicticity and the adequacy of the *ego* are questionable, otherwise a valid and univocal knowledge would be attainable.

It is from this perspective that one must see the criticism that Husserl levels against the interpretation of phenomenology as "Cartesianism," defining it as a ridiculous misunderstanding: ". . . als ob ihr 'ego cogito' eine Prämisse oder Prämissensphere wäre, um aus ihr die übrigen Erkenntnisse . . . in absoluter 'Sicherung' zu deduzieren."[26]

If the analysis of the life-world, as a result of the discovery of transcendental subjectivity, shows that ". . . die Welt, die für uns ist, nach Sosein und Sein unsere Welt ist, ganz und gar aus unserem intentionalen Leben ihren Seinsinn schöpft, in einer aufweisbaren apriorischen Typik von Leistungen,"[27] it also makes it clear that evidence ". . . gibt gleich Beispiele für die Gradualitäten der Klarheit und für die daraus zu schöpfende Idee (eine Idee!) vollkommener Klarheit, der ich mich annähern kann."[28]

II. TELEOLOGY AND THE HISTORICAL WORLD

The reflections outlined above serve as a basis for examining the sociohistorical world. Many of the analyses of *Krisis* are bound up with teleology understood as an interpretative key of the historical process, which is a process of the realization of reason.

If we want to understand why Husserl summons reason to his aid and explain the peculiarity of this position, which seems to résumé the well-known arguments already proposed in the past, we have to refer to the analyses of *Ideas* and remember that

ein spezifischer Vernunftcharakter ist aber dem Setzungscharakter zu eigen als eine Auszeichnung, die ihn wesensmässig dann und nur dann zukommt, wenn er Setzung auf Grund eines erfüllten, originär gebenden Sinnes und nicht überhaupt eines Sinnes ist.[29]

Rational is equivalent to "fulfilled," and a rationally motivated position has ". . . in der originären gegebenheit ihren ursprünglichen Rechtsgrund,"[30] and that is why evidence is ". . . die Einheit einer Vernunftsetzung mit dem sie wesensmässig Motiverenden."[31] It is clear, Husserl goes on to say, that what is originarily given is adequate only in an eidetic sense, while, as far as a real thing is concerned, ". . . keine auf solch einer inadäquat gebenden Erscheinung beruhende Vernunftsetzung 'endgültig,' keine 'unüberwindlich' sein kann."[32]

Here the precategorical level does not yet appear, even though a few pages later Husserl asks himself whether one can speak of the *Urdoxa* having an *Urvernunftcharakter.*[33] Nevertheless, the connection between rationality and an originarily offering givenness is established, and thus also the difference between the eidetic sphere (which is referred to as the *Wesensstruktur*) and knowledge of the contents.

How can this view be made to harmonize with the "theoretical" reason that Husserl refers to and which he seems to regard as an instrument? In actual fact, however, reason has no instrumental value; it is not a capacity but rather the realization of an *adaequatio*, a realization of which one can become aware at the reflective level. That is why theoretical reason is "doxic" and truth and evidence are "doxological" and "theoretical."[34]

Since Husserl here speaks of a realization, an achievement, reason is the end toward which man tends, but an end that is never attainable, a having to be that would imply a life lived in the sense of apodicticity, as the arrival point of human existence.

The consequences for the sociohistorical world are very important; teleology as the "form of all forms" inevitably manifests itself also at this level, but solely because underlying it all there are the considerations we made in the first part of these notes. The reflections about the historical development of humanity, even though they did not form an intimate part of phenomenological analysis from its beginnings, find a natural framework precisely in the particular manner of understanding the life of consciousness as a flowing in the temporal sense and therefore possessing both a historical dimension and an internal teleology. By enlarging the scope of this reflection to encompass humanity, therefore, one can understand the development of humanity in the historical sense, as well as in the teleological one. The history of humanity is not committed to mere "chance," but is connected with a definite "task."

What is the purpose of humanity's task, and what are its characteristics? Who can discover this task or define it? It is revealed preeminently as a result of philosophical reflection which, ever since the days of ancient Greece, has impressed a characteristic mark upon Western speculation. Is this, we may wonder, a stereotype of the German mentality, a residue of the idealistic view that holds Western civilization to be superior to its Oriental counterpart? There may well be such a residue in Husserl's mind, but his justification for this assertion is very different from the one given by the idealists: in Western culture he recognizes the emergence of a philosophic attitude tending to radicality.

One might object that, in Husserlian terms, Western philosophy has to be considered as a categorical product, and that he had not understood this fact, even though he had been the first to discover the categorical dimension. In other words, he had made the same mistake as the one for which he reproached Galileo, i.e., to have discovered its secret and concealed it at one and the same time. In his case the concealment is said to consist of his having continued along the lines of the "intellectualistic" or "rationalistic" tradition of Western philosophy, the discovery of his assertion that every "construction," everything that is elaborated, rests on a precategorical foundation.

All the same, the use Husserl makes of the term "reason" can throw some light on what he means by theoretical research and the role of the philosopher. His continuation of past traditions does not consist of a renewal of intellectualistic and rationalistic views; the *episteme* to which he refers is similar only by analogy to that of Greek or Cartesian philosophy; and what it has in common with its Greek or Cartesian counterpart is more a need for radicality than any similarity of contents.

What has just been said brings out, first, the "novelty" of phenomenological research as compared with the illuministic or idealistic approach – with which it seems to have some contact – and second, its relevance and validity in connection with the contemporary debate about the interpretation of history and reality.

If we want to make the first of these points clear, we cannot but comment on Husserl's proposal that the philosopher or, better still, the phenomenologist, should assume the attitude of a disinterested spectator. One of the easiest mistakes it is possible to make in connection with the interpretation of phenomenology is to believe that "disinterestedness" is equal to "neutrality," i.e., taking no definite position in judging things. On the contrary, a "disinterested spectator" is one who endeavors to free himself of conditioning ideologies and personal motivations and to adopt a research attitude that aims at discovering the true meaning of the authentically "pregiven."

Disinterestedness does not therefore mean "disengagement," but rather a maximum effort to read and understand reality, something that leads to an "engagement" for an authentic development of mankind. But before we can really appreciate this constructive aspect, we must also note to what extent this theoretical and rational attitude differentiates itself from the illuministic and idealistic one.

From the phenomenological point of view, indeed, one might say that the former makes a wrong use of reason, which is employed in a purely formal manner in the various fields of human research. Phenomenology's relationship of connection/distinction with respect to Kant is emblematic in this respect: while from a certain point of view Husserlian inquiry remains within the ambit of Kantian transcendentalism, Husserl also goes beyond Kant and deeper than Kant by virtue of the former's recognition of the precategorical. Kant's failure to arrive at this level prevented him from solving the problem of the connection between the scientific and finalistic interpretations of nature, which are juxtaposed but not connected, being due to different gnoseological instruments: *Verstand* and *Urteilskraft*.

Although Husserl accepts the formalization peculiar to the scientific approach, he not only does not regard scientific knowledge as the only valid one, but discovers a wider dimension to which it must be traced back, i.e., the precategorical level and its teleological structure. We can thus understand his critique of scientism and the unmasking of its claims.

Husserl's criticism is even more incisive when it comes to the concept of reason as understood by the idealists, because Husserl refuses to accept it as an absolute.

As a result of this twofold critique it becomes possible to expound the development of Western culture, highlighting both its positive and its negative aspects, without yielding to the temptation of either an irrational interpretation as a reaction to the pretense of reason to be absolute or a materialistic one on the rebound from an idealistic trend.

The rational aspect, inasmuch as it has claimed to be all-embracing, has acted as an element of concealment, giving rise to idealistic metaphysics and the positivistic absolutization of scientific knowledge. Husserl has always fought against these positions. But one might still object that this entire line of argument concerns the critical function of phenomenology with respect to nineteenth-century philosophy, a function that is no longer effective in our own times following the rejection of both idealism and positivism.

In fact, however, in the positions of Kant and Hegel one can trace a problem that seems to be everlasting and has been insistently reproposed throughout the history of Western speculation; in our own times, for example, it has come to the fore in the polemic between T. Adorno and K. Popper. Faith in the method of scientific research seems to go hand in hand with a liberal political view that considers tolerance to be of the utmost importance. Criticism of science's pretense to have an absolute value, on the other hand, leads one to conceive a type of society with absolute and univocal characteristics. What we are interested in here is not so much the political or sociological controversy, but rather the fundamental contrast of approach, the difference in what is considered to be valuable and true. Indeed, I am convinced that the nucleus of the discussion is concerned with the very idea of truth, and therefore also with the choice of what is valuable and the criteria for this choice.

A liberal position taken to its extreme consequences involves not only the danger to which Adorno, Marcuse, and others have drawn attention, i.e., a merely apparent tolerance, but something that affects the very possibility of having a criterion of value and is therefore far more profound and serious: if all things, all attitudes, are of equal value, then there is nothing that is really valuable. Here we come face to face with an indifference that justifies everything, but does not provide any guidance.

As regards the intolerance of the opposite position, this is the direct result of the conviction that one is in substantial possession of truth; this truth has been grasped once and for all, and it must therefore be made to prevail in its full validity. The clash arises because a position that accepts many truths, and thus denigrates its contents, comes face to face with one that absolutizes its own content and sees but one truth.

The basic problem of our own time is therefore once again the search for truth.

When we examine the situation of man and his history from a phenomnological point of view, we note a constant endeavor to understand the meaning of things, a meaning that is closely linked to their "telos." What is man's goal? The answer to this question also clarifies the one regarding the nature of man. Is it possible to establish criteria of choice and orientation that have an intersubjective character, but not a conventional one, and are determinant for the purposes of an authentic development of humanity? Or, to come back to the theme we have just discussed, can one avoid both dogmatism and skepticism? It is precisely through teleology, which points out the infinity of the task but also provides us with a structural element, that this problem can be solved.

Infinite tension must not be understood in the Fichtian sense, i.e., as a unity continually reproposed but never attained as the result of a laceration, but as a perennial shortcoming of man's possibilities vis-à-vis the full comprehension of the truth.

Does this assertion indicate a pessimistic point of view? No, I should rather call it a realistic position that highlights the need for finding value criteria and structural elements, but prevents the search from being discontinued on the pretext that all has been conquered.

It also clarifies the meaning that has been attributed to progress, which must not be understood as a linear development on the way to an automatic achievement of conditions of well-being or a full comprehension of reality.

At first sight there is a contrast between this statement and the teleological view, but this dissolves as soon as one bears in mind that this view is connected with the task of delving deeply into something, just as the realization of what has been understood is closely connected with responsibility, doing what ought to be done. This is the "new kind of praxis" that Husserl talks about; it can and ought to be established, but man can also fail. This is not a fatalistic view of the course of history, but involves the responsibility of all men.[35]

The individual will must conform to the universal one as a will of life, and this not just in the primary sense of a simple biological reproduction or conservation, but as a trend toward the "true human being" characterized both by the conquest of full self-consciousness and by the practical realization of a "creative" resolution of the contradictions. The creativity that Husserl has in mind here[36] is not a creation from "nothing," but an attempt to reach true being, and therefore also its fullness, by successive steps and partial realizations.

That is why Husserl, when explaining the meaning of teleology, presents it in two ways, i.e., as the self-understanding of humanity and as the conquest of truth. It seems that these two goals do not coincide; in fact, while the former is attainable and will be reached when man understands that his task is that of conquering truth, the latter is situated at infinity and is therefore unattainable. Self-understanding does not mean that one has reached the truth, but merely that one has become aware of the true and ultimate purpose of the historical process. For this reason Husserl holds that the "final" form is also the "initial" one,[37] in the sense that humanity – even though it understood the value of research for the first time in ancient Greece – could not complete the process of realization because it tends toward infinity.

In this way one can explain the history of philosophy, which is a continuous search for something that is situated at infinity (truth), while each individual philosopher is aware of what constitutes the authentic goal of the task (self-understanding).[38]

This duplicity involves the subdivision of the ultimate purpose into an immanent and a transcendent perspective. Indeed, inasmuch as reflection leads humanity to realize the principal problems connected with its existence, the self-consciousness aspect may be defined as immanent; but the true meaning of this existence has to be continuously sought after, so that there is a perennial trend toward the attainment of truth and the realization task is truly infinite: for truth is the pole that is unattainable and yet has to be postulated; and in this sense it is therefore transcendent. In other words, in history we may discover an inexhaustible trend toward the attainment of truth, which in the last resort must be considered as metahistorical and metatemporal.

III. TELEOLOGY AND THEOLOGY

Another problem arises at this point, because the inexhaustibility of research involves an infinite tension between what man can understand and what things actually are. That is why teleology brings us back to theology, a term that I use here in the etymological sense, i.e., in a discussion about God. Husserl's published works contain no organic treatment of this subject, but one can find a number of significant starting points that show that he was interested in the matter; indeed, the internal formulation of the problem is already such, at least in my opinion, as to open the road to a more far-reaching analysis.

First of all, one may note that in Husserl's thinking the motive of teleology is invariably connected with that of theology, although the opposite is not always true; moreover, both of them assume a particular configuration that always depends on the actual trend of the search.

The reflection about consciousness, which reveals the presence of the acts and experiences in the stream of consciousness, and the transcendence of things with respect to this consciousness, the "I," lead Husserl to face up to the problem of the transcendence of God.[39] Similarly, the treatment of empathy leads him to postulate the existence of a consciousness that thoroughly knows all the others,[40] and he returns to discussing this question in connection with the interpretation of subjectivity as a *monad*, asking himself whether a supermonad (*Übermonad*) is conceivable. The need for an absolute and primary element, which is bound up with the search for an ultimate justification, is connected in most cases with the discovery of a finality present in consciousness, in things, in the human world (understood as moral world).

It is only logical, therefore, that the quest for originariness as precategorical, which appears as the arrival point of Husserl's speculation, should have its counterpart in the need of a primordial necessity (*Urnotwendigkeit*) as the ultimate and profound justification. No longer God as idea, but as a foundation that is not only beyond but also below all things. And Husserl arrives at this basis as a result of his reflection on teleology contained in a text written in 1931.[41] In fact, he there treats not only the moral problem, considering God's will as "absolute and universal will that lives in all transcendental subjects," but formulates another starting point: while searching for the primordial structure (*Urstruktur*) of reality by means of questioning flash backs, he underscores, as we have already seen, the teleological trend present in reality or, better, in factuality.[42] The ascertainment of the presence of teleology in primordial facticity (*Urfaktizität*) thus raises the question whether it may have its justification in God: "Wir kommen auf letzte 'Tatsachen' – Urtatsachen, auf letzte Notwendigkeiten, die Urnotwendigkeiten."[43]

The different ways of formulating the problem of God, corresponding to the various modalities of approach to the reality of phenomenological research and the trend of the problems that are tackled by it, all converge toward the recognition of a level of transcendence, of absolute knowledge, of an originary basis, and all of them are attempts to define God.

For the purposes of our present analysis of teleology, however, the really interesting feature is constituted by the fact that this infinite tension may find its full realization only in a new dimension, a metahistorical one, whose presence is clearly indicated by phenomenological inquiry, but which Husserl does not want to utilize or develop further.

In the aforementioned text[44] Husserl once again expresses his intention of bracketing this line of thought, a statement that can already be found in *Ideas.*[45] Here the egological aspect of phenomenological research undoubtedly makes itself felt in all its relevance. Having discovered the teleological structure within it, and notwithstanding the fact that it points to the existence of a deeper level, Husserl yet prefers to reduce the originary to the "originary operative performances" or, more precisely, to stress that it is the ego that makes the process evident and that all research must therefore start from this latter.

However, one cannot avoid the problem that a deeper study of the ego leads not only to the "originary operative performances," but also to a primordial necessity that one cannot ignore if one accepts the validity of the previously made remarks regarding reduction and the role subjectivity plays vis-à-vis it. Rather, the very fact that the problem of the ultimate basis emerges through reduction shows that this operation, always provided that it is carried out correctly and right through to the end, is truly a means of authentic discovery. It permits teleology to be thrown into full relief, and it is through teleology that the *Wesensstruktur* of reality is revealed. Thus we have here the emergence of a "leitmotiv" that, via a discussion of gnoseological problems, leads to historical ones and eventually to God, enabling us to take the first steps along the endless road of the search for truth.

Centro Italiano di Fenomenologia,
Rome

NOTES

[1] E. Husserl, 'Zur Phänomenologie der Intersubjektivität', *Husserliana* Band XV, M. Nijhoff, The Hague, Text Nr. 22, p. 380.
[2] *Ibid.*, p. 385.
[3] *Ibid.*, p. 385.
[4] E. Husserl, 'Erste Philosophie', *Husserliana* Bd. VIII, p. 400.
[5] E. Husserl, 'Analysen zur Passiven Synthesis', *Husserliana* Bd. XI, p. 102.
[6] *Ibid.*, p. 71.
[7] *Ibid.*, p. 102.
[8] 'Erste Philosophie', p. 35.
[9] *Ibid.*, p. 397.
[10] E. Husserl, 'Cartesianische Meditationen', *Husserliana* Bd. I, par. 9.
[11] I. Kern, 'Die drei Wege zur transzendental-phänomenologischen Reduktion in der Philosophie Edmund Husserls', *Tijdschrift voor filosofie* nr. 2, 1962.
[12] 'Erste Philosophie', p. 406.
[13] *Ibid.*, p. 432.
[14] *Ibid.*, p. 397.
[15] 'Analysen zur Passiven Synthesis', p. 103.
[16] *Ibid.*, p. 69.
[17] *Ibid.*, p. 202.
[18] *Ibid.*, p. 203.
[19] 'Erste Philosophie', p. 398.
[20] *Ibid.*, p. 406.
[21] *Ibid.*, Beilage XX, p. 438.

[22] W. Biemel, 'Reflexionen zur Lebenswelt-Thematic' – *Die Phänomenologie Heute*, M. Nijhoff, 1972, pp. 49–77.
[23] D. Conci, *Prolegomeni ad una femonenologia del profondo*, vol. I, Rome, 1970.
[24] E. Husserl, 'Die Krisis der europäischen Wissenschaft und die transzendentale Phänomenologie', *Husserliana* Bd. VI, p. 173.
[25] 'Analysen zur Passiven Synthesis', p. 66.
[26] 'Die Krisis der europaischen Wissenschaften', p. 193.
[27] *Ibid.*, p. 184.
[28] E. Husserl, *Formale und transzendentale Logik*, Halle, 1929, p. 253.
[29] E. Husserl, *Ideen zur Phänomenologie und phänomenologischen Philosophie*, III, p. 335.
[30] *Ibid.*, p. 335.
[31] *Ibid.*, p. 336.
[32] *Ibid.*, pp. 338–39.
[33] *Ibid.*, p. 342.
[34] *Ibid.*, p. 343.
[35] 'Die Krisis der europäischen Wissenschaften', pp. 275–76.
[36] 'Zur Phänomenologie der Intersubjektivität', XV, 380.
[37] 'Die Krisis der europäischen Wissenschaften', p. 274.
[38] *Ibid.*, p. 443.
[39] 'Ideen zur Phänomenologie', vol. III, par. 38.
[40] 'Zur Phänomenologie der Intersubjektivität', XIII, 9.
[41] *Ibid.*, XV, Text Nr. 22.
[42] *Ibid.*, p. 385.
[43] *Ibid.*, p. 385.
[44] *Ibid.*, XV, Text Nr. 22.
[45] 'Ideen zur Phänomenologie', vol. III, par. 38.

MARIO SANCIPRIANO

TELEOLOGY AND THE CONSTITUTION OF SPIRITUAL FORMS

I. TELEOLOGY AND CONSTITUTION

When Husserl published his *Logical Investigations* (1900–1901), he clearly already had an end, as he aimed to supply an evolutionary explanation of psychological constitution ("die Teleologien der geistigen Konstitution") on the analogy of the teleologies of natural constitution.[1] In his later works, with a significant first hint appearing in *The Idea of Phenomenology* (1907), he tried to go deeper into subjective activity, not in the evolutionary sense of biology, but rather in its transcendental implication, by way of his theory of the constitution of objects in consciousness.

As compared to empirical-critical analysis and psychological studies, by that time still affected by biology as shown in the sentence cited above from *Logical Investigations*, this transcendental turn allowed Husserl – although still close to the concept of the empirical ego – to adopt in his *Ideas I* different types of teleological models, which were no longer those of the relationship "individual-species," nor the coordination between natural causes and effects, but rather the models characteristic of the structures of pure consciousness oriented in their intentional prescriptions. Further, dating from the second part of *Formal and Transcendental Logic* (1929), it is possible to notice a deepening of subjective constitution as the genesis of the original acts of the transcendental subject.[2]

It is our aim in this paper to proceed to an analysis of subjective constitution as expressed by the goals of the human mind, and to outline the particular acts of human consciousness that – insofar as they belong to the dynamic structures of the pure ego – are liable to be mediated with the symbols and intuitive contents of spiritual life. In fact, while on the one hand natural constitution is conditioned by biological causes which, once they are put into existence, will unfailingly produce their effects with the passing of time, even though with some differential approximation, on the other hand transcendental constitution "prescribes" future acts creatively and remains open to various possibilities depending on the extent to which interior time consciousness conforms to the law of free choice. This law is internal to lived experiences (*Erlebnisse*) and precisely connotes them as "spiritual" ones.

Tymieniecka (ed.), Analecta Husserliana, Vol, IX, 353–365.

We would not go so far as to maintain that the egological structures that form the pure ego have to be qualified in themselves as "spiritual" ones; yet it is true that they include, in their dynamics, all the necessary conditions to take on a content of interest, willful or voluntary acts, moral values, and ends that can be connoted as such.[3] We shall not expect phenomenological description to provide a new type of *spiritualism* but simply a *science of the spirit* that takes into consideration the primary constitution of spiritual forms in the evidence of their self-donation and development, with regard to their specific goals.

The phenomenological process of reduction, which determines the *epoché* not only on natural attitudes, but also on the very forms of the mind in order to radically trace back the structures of pure consciousness, was not enough to prevent Husserl from maintaining that "das reine Ego eine Unendlichkeit von Formen in sich schliesst".[4] Such an infinity of possible forms, because of its very infiniteness, finds more adequate explication in the spiritual realm, even if it is faced in its various determinations with all the limits of human enterprises. One might share Mr. Broekman's view and say that, in its most genuine approach, Husserl's phenomenology is aimed at the purity of egological structures which, in their origins, do not possess a spiritual content of their own, nor include the moment of will;[5] the same structures rather form the a priori features that influence experience, and are liable to mediate with the symbols and images of spiritual life. It is spiritual life, in the very split between subject and object, or man and thing, that reproduces the distinction internal to the relationship between *noesis* and *noema* in terms of pure consciousness, by enriching it with its meaning. The a priori egological structures, because of their intentional purpose of mediating with the contents, both ideal and effectual, of human experience, are then more akin to what we normally mean by "mind" than to what we mean by "body". The *facts* of life are included in them with their continuity with *ideas* along a line of effectual-ideal development, even if the outcome of the egological activity can be measured more properly in the mind than in the animal nature, which possesses tangible and factual features but does not constitute what belongs more specifically to human teleology.

At this stage, before proceeding to an analysis of the particular accomplishment of the multiple forms of the mind, we have to go back from the concept of the constitution of such multiple forms, to the concept of constitution in itself, or *self-constitution* in transcendental subjectivity. In a manuscript on teleology, Husserl deals with "die Selbstkonstitution der transzendentalen Subjectivität als ins Unendliche auf Vollkommenheit, auf wahre Selbsterhaltung

Gerichtetsein".[6] By establishing a relationship between *arché* and telos, he limits the pretense to know and accomplish, in a system of the Absolute, the subjective being that is out of time and always what it used to be: *τὸ τί ἦν εἶναι*.[7] There is a tension between *arché* and *telos*, but also a distance and therefore a suspension: perfection (*Vollkommenheit*) is not completely given in the beginning, it is rather an intentional object. The subjective being, at least in his human side, appears to be always striving to reach for a goal: he can never own himself as a pure act, nor does he own his goals altogether, so that he might come to know completely his identity. In particular, his spiritual acts have a *history* of their own originating from the *self-constitution*, and a *pre-history* originating form natural constitution.

When we affirm that the structures of the pure ego are already oriented towards spiritual life and are analogous to it in their transcendental constitution, we are not – contrary to the idealists – claiming to make an absolute deduction of the spiritual world from the transcendental subject. Far from that, we are simply suggesting a correlation of that world with the subject: a correlation that is above all a coordination in terms of a rational teleology. It thus becomes possible to discover the adequacy of spiritual life to achieve completion (*Vollkommenheit*) of any intellectual, aesthetic, social, moral forms, prearranged in the same originary structures. Further, we are not claiming to make a deduction of the *Ich-Mensch* and its ideal formations from a hypothetical absolute subject considered as *causa sui*, but we intend to ascribe to the transcendental subject the faculty of establishing infinitely possible forms, for "von Akt zu Akt fortgeht".[8] We therefore attribute to the *Ich-Mensch*, first-person bearer of a transcendental activity, the quality of being the centre of an activity aimed at objects, that is, more eminently, at the objects of spiritual life. Such an eminent position of this category is due to its prerogative to act in the infinite space of human yearnings, even if there exists a limited choice susceptible of practical realization. It is also eminent because it carries the *Erfüllung* of the idea of perfection or *completion* already existing in the *cogito* as a formal principle, and confers on it its multiple intuitive contents in the various forms of spiritual creation.

II. CONSTITUTION OF SPIRITUAL FORMS

It is now high time to proceed to an accurate analysis of the originary acts of consciousness that form themselves as being spiritually oriented. We will have to trace in those acts the particular forms that are related to the various ontological regions of the spiritual being, with their respective teleologies. We will limit ourselves here to considering the fulfillment (*Erfüllung*) of the originary structures as from their teleologic moment of formation, to their development into accomplished forms, according to the following moment: (1) theoretical, (2) ethical, and (3) religious.

1. Theoretical Form or Contemplative Life (Vita Contemplativa)

A teleology is internal to the logical processes that take place on a perceptive and categorical level, so that the constituting consciousness establishes – in the various degrees and modes of knowledge – its relationship with objective reality and finds in it the noematic completion of its acts. The acts of consciousness that have their goals within subjective life are, however, also objective ones, in so far as they reveal its contents and permanent forms, both on the level of a "pure psychology" that delineates the individual ways of theoretical and practical learning, and on the level of those forms that more properly belong to the "spiritual" sphere, with their aesthetic, moral, social, and religious objects.

Already in the beginning of his *Formal and Transcendental Logic*, Husserl assigns to logic the task of radically examining its own teleological sense. He himself carries out such an examination by noticing the structures that open up, on a perceptive level, a stable relationship with a given object, while representing an ever-increasing conquest of its various glimpses (*Abschattungen*). This work fixes an important moment in Husserl's ideal evolution, when he passed from *static* to *dynamic* constitution or, if we are to use his own words, *genetic* constitution, according to which consciousness has the capacity of conferring meanings on things in an active way. Among these meanings, those pertaining to spiritual life – thanks to their creative originality – seem to be the nearest to the a priori structures from which develops genetic constitution, which is based, in this case, on the experiences of interior life.

In a theoretical environment, intentionality is no isolated fact, nor simply a subjective one; it rather permits a synthetic unity of reason, which teleologically links the "I think" with the events of spiritual life and their objective ends.[9] Universal *mathesis* – in spite of its descriptiveness – contains, at least implicitly, the theoretical elements and characters peculiar to contemplative life. When we pass from a formal description of logic to the recognition of *logos* and transcendental egology, the theoretic interest of phenomenology is increased, even though such a process – from the point of view of realistic, or historicistic, or spiritualistic interpretations – goes beyond the intentions of Husserl's methodology, which are always of a reductive nature.[10]

Phenomenology always reduces *cogitata* (conscious acts) to the basic structures of the *cogito* in pure intentional consciousness; and yet, in this phenomenological reduction, there is evidence of the factors that are helpful in characterizing the forms of spiritual life. Conscious acts can have an empirical content; but, when they are regarded as "*qua cogitata*", they can but have an ideal content and therefore more properly belong to the realm of mind than to the realm of material events. In fact, thetic consciousness – *constituted* as a *constituting* agent – is the center of developing acts, intended most certainly not in the sense of natural evolution but rather as a teleological predisposition towards a completion of man's being, that is, of an eminently spiritual being. Now, it is a known fact that in man's being, because of the unity and continuity of effectual and ideal life, there remains a fundamental analogy between spiritual and natural activity, at least in a twofold sense: (a) consciousness is receptive in "*passive genesis*", since it takes on, in its anthropological extension, customs and trends (passive synthesis); (b) consciousness is productive of new and original acts in "*active genesis*" (active, or creative synthesis). Thus, originality and creativity are characteristics peculiar to any event that somehow negates a pre-existing series, and yet they appear to be more specifically connected with spiritual life.

In the first of the above senses, that is, passive synthesis, in so far as consciousness has a synthetic nature, there opens up a horizon of infinite possibilities (*protentionale Horizont*); also, in so far as it is conditioned by the stream of its own lived experiences, it forms itself in a passive way, at least to a certain extent: "Das strömende Bewusstsein wirft zwar beständig vorwärts einen protentionalen Horizont aber es ist nicht abzusehen, dass diese Vorzeichnung eine so bestimmte und eine so zwingende sein sollte, dass wir sagen dürften für die Zukunft sei eine Norm wahren Seins als eine bindende festgelegt".[11]

According to the aims of a science of the spirit, it is necessary to point out the indeterminacy of the "protentions", which are motivated but not strictly predetermined. In the spiritual realm, which is after all the realm of the ends, there exists orientation rather than rigorous prescription. The adoption of the teleological principle affords, from the beginning of all free and conscious acts, a certain anticipation of "act" over "power", that is, of the future over the past, by virtue of the expectation and choice of feasible actualities. In such a way the given possibilities, from the very moment of their passive genesis, leave enough room for the initiative of anticipating the future in its various possible solutions, with particular regard to those most suitable for the construction of the spiritual world. According to Husserl, teleological consciousness is the form of all the forms of possible worlds, which is the realm of indeterminacy ("Reich der Unbestimmtheit"): a realm dominated by the ego through the forces of spirit to the extent in which it succeeds in making it the seat of its choices and initiatives.

Now, in the realm of the mind, from the point of view of passive genesis, the constitution of spiritual forms already has a preliminary task of its own, represented by the formation of sedimented habits that most aptly answer the questions raised by intellectual or moral life, make such a life easier, and contribute to the formation of a *moral nature*. It is in such a "nature" that are to be found the residues of lived experiences and the conditions that prearrange the proper human behavior, determined in conformity with moral ends. In the depth of this "nature", starting from its unconscious layer, originate those motivations that are channeled by human will towards the production of spiritual forms. Then it is possible that such logical and moral activities, having as their pre-possible space the indeterminacy of the future, achieve their fulfilment. Already in passive genesis spirit emphasizes the "genetic" much more than the "passive" aspect. Sedimented habits (*Habitualität*) are most certainly connected with the Latin *habitio* (Gr: ἕξις), rather than with *habitus*. In fact, *habitio* (*nomen actionis*) denotes a possession, activity, prerogative pertaining to a spiritual man who, as such, is always owner and never owned; while *habitus* (*nomen acti*) is, at least partly, a passive condition.

In further stages of *active genesis* we find an even greater space in which we can practice those mental activities peculiar to the various regions belonging to spiritual life. Although genesis is active, at first, in the order of genealogy of pure logic, our genetic knowledge can widen in scope to consider the development of the social–historical world. Logical invention, from the sphere of the "I think", moves to the spheres of volition, capacity, and doing, yet the unavoidable *contaminatio* with the empirical ego cannot remove from it its reducibility to the essential structures, from which it is always possible to go back to a more clear and conscious understanding of the human world.

In *active genesis* the answers of the ego find their expression, which is not only *I think*, but also *I move, I do.*[12] From the individual body originates kinesthesis, that is, the dynamical movements inherent to the life of the ego ("ichliche Beweglichkeit") and to the progressive unveiling of all sides of the objects.[13] The kinesthetic movement has its first manifestation in bodily life, in the "active egological functioning" ("ichtätige Fungieren des Leibes");[14] but it permeates all subsequent subjective life and historical development. We might add that there is not a moment of physical kinesthesis and perception that is not taken on in full awareness by the spiritual being, unless it is permanently buried in its unconscious. In this connection we might here remember Rosmini's "sentimento fondamentale", which is a way of feeling alive in the energy and extension of the body, and is also a predisposition to "percezione intellettiva" (fundamental intellectual feeling).

From perception we pass on to the general horizon of the world as unity of the objects ("einheitliches Universum der seienden Objekte").[15] Now, although such a unity implies only the assumption of a cosmology – and Husserl abhorred ideological categorizations – it is a sure fact that, in the human appraisal, it can be considered a world view (*Weltanschauung*), that is, an evaluation sometimes serene, sometimes tragic of our being in the world.

Perspective, by delineating the lineaments of objects, makes it possible for us to perceive them as a succession of "adumbrations". These are at first supposed to be a component of the immanent unity of every single object, as defined by its usual name; then they are understood as included in a general evaluation, that is, in their relationship with the "construction" of the spiritual being. It is then the spiritual being that agrees with things or rejects them, depending on his own exigencies, by indefinitely multiplying the perspective adumbrations, or by limiting or rejecting them as dictated by his own judgment or moral exigencies. In this connection Chapter IV in the second section of *Formal and Transcendental Logic* is of fundamental importance. There the active genesis of appraisal is thoroughly examined with a research into its furthest motives in pre-predicative evidence, to reach an apex in the "conviction" that it is to be found at the top of reasoning, that is, every categorial relationship. For our purposes in this paper we wish to emphasize that the genealogy of logic reproduces the processes in the formation of appraisal while looking for a genesis of sense (*Sinnesgenesis*) beyond the stiff categorial support. Such a genesis permeates the knowledge of ever new initiatives. It thus becomes possible, while trying to give shape to a creative teleology, to promote spiritual life as well, by using the intellectual tools of a functioning intentionality, that produces to a certain extent an increase of the significant being and an expansion towards freedom, that is, autonomy of reasoning.

In the light of such a progress from a philosophy of life to a philosophy of the mind, are to be interpreted the writings of the *Crisis*, in which the world appears to be a theatre for the thoughts and actions of every human being: "Alle unsere theoretischen und praktischen Themen . . . liegen immer in der normalen Einheitlichkeit des Lebenshorizonts 'Welt'. Welt ist das Universalfeld, in das alle unsere Akte erfahrende, erkennende, handelnde, hineingerichtet sind."[16] An understanding of the world, even the most objective one, suffers from our inability to give it a meaning coherent with our experiences, our thoughts, and our spiritual evaluations.

2. *Practical Form or Active Life (vita activa)*

Of no less importance for the formation of a science of the mind is the "a priori constructive teleology" that considers the construction of a moral world as defined in Husserl's courses on ethics. We will here limit ourselves to commenting on a few passages that may help to clarify our assumptions in joining *alpha* and *omega* in the constitution of spiritual acts oriented to their specific ends. As it was to appear more clearly in *Experience and Judgment* (1939, but posthumously edited by Landgrebe in 1954), voluntary intentions and speculative intentions both obey an identical dynamic logic: the practical intentions of will, as well as those of intelligence, are teleologically oriented and move from a pre-possible or possible, that is, virtual, state of things towards the achievement of given goals. Besides, already in the fundamental distinction between subject and object and the double intentionality there existing (from subject to object and, to a certain extent, from object to subject), the moral world represents the *analogon rationis* of the intellectual world.

It is true that in the speculative trend we do not find, in the achievement of its goals – that is, its intellectual results – the actuation of external objects in the most specific sense of *doing*. Such external goals are the stated purpose of will. Here the forms of the *I think* remain within the cognitive sphere, where there exists a stricter homogeneity between the *cogito* and the *cogitatum qua cogitatum*, while the *cogitatum* is heterogeneous in so far as it is existence, external world, goal of the action. The analogy existing between knowledgeable and practicable trends is, however, inherent in their process with the contribution of a material *hyle* to the intentional *morphè* in the various types of intuition and thought both to knowledge and morale. Thus is realised the progressive fulfillment of the functioning intentionality in the usage of cognitive and operative means, as well as intermediate ends, up to the achievement of their respective goals and, also, in the possibility of coming back to contents already attained and left behind, by reproducing intuitions and actions in the form of distant memories or recollections of the past in present times.

There exists, no doubt, an analogy – explicitly stated in Husserl's courses on ethics – between intellectual and voluntary forms. Consequently there is a correspondence between the doctrine of "pure practice" and the theoretical doctrine, from which it is to be evinced that, when human experience was originally formed in its a priori features, knowledge and will must have had a mysterious constitution in common. If cognitive intentionality and willful intentionality have a parallel analogous genealogy (*active genesis*) and a tension objectively oriented to their specific ends, at the roots of such a genealogy there is a single constitution that belongs, in its various stages, to the basic structures of consciousness. Even if, in practice, will and feeling realize themselves in action overstepping the sphere of interior life, there is no interruption occurring between *idea, will,* and *action*. The courses on ethics shows the way to overcome the logicist prejudice, as Paul Ricoeur recognizes in Husserl's phenomenology. Moral action, because of its intuitive contents, its obligation to acknowledge in others the quality of *subjects* and *centers*, as active as we are, *of a spiritual life*, offers a *via regia* to transcendence and assigns to intentionality its most realistic goal.[17]

Also in *Experience and Judgment*, with the development of a genealogy of logic that ensures the possession of objects – from the capacity of forming associations and synthesis, to culminate in the taking up of a position of the ego in the predicative judgment and in "conviction" – we recognize the tension towards what is true and can be available for practical ends in so far as it spreads to the sphere of action, to the ways of *being able* and *doing*. What Husserl says about conviction (*Überzeugung*), considered as the highest possible logical act, can immediately be applied to the realm of spiritual values thanks to the meta-empirical meaning assumed, in that context, by "persuasion", which does not always require an immediate contact with experience.

It is possible, then, to find a certain parallelism existing between desirable and cognitive objects, that is, between *voluntary intention* – in its temporal "protention" and in its creative power – and *logical intention*, with the significant activity of consciousness. Intellect and will, in their mutual interaction, come out from what a scholastic would call an originary "indifferentia propensionis", peculiar to a consciousness reduced to its primitive indetermination. They acquire, however, a determination in their paying "attention to ..." and in their "interest for ...", to achieve a preferential choice of values. Also the energy internal to will comes out, in an intentional activity, from the originary dispositions (*ursprünglichen Dispositionen*), to achieve a fulfillment of the acts. Every virtue is thus a habit in the active sense (once again *habitio* rather than *habitus*): "*Tugenden sind* ἕξεις".[18] The moment of freedom appears in the initiative and rational determination of a virtuous behavior: freedom as *choice*, and freedom as *reason.*[19]

It is the creative energy of will that comes to full evidence in the *Grund probleme der Ethik* (*insbesondere der formalen Ethik*). Thetic will stretches out in the time formed by consciousness, that is, duration – in the Bergsonian sense of the progressive enrichment of the acts of the mind – to reach its climax in action. Action is the natural outcome (*entquillt*) of a process that includes "the creative phase of constituting" in a progressive horizon of ever new acts of will.[20] In the ethical field one could say that the opposition, between "constitution" and "creation" is overcome or, at least, attenuated in Husserl's philosophy.[21] What is "constituted" in the *innere Zeitbewusstsein* projects the character of creative continuity in further possibilities. In an eminent way, spiritual acts as opposed to the processes that take place according to natural causality, are constituted as creative acts, because of their absolute novelty and their origin from a void, in freedom. From them has its origin a variety of forms in aesthetic, gnoseological, religious, and political phenomenology.

There thus takes shape a doctrine of "formal practice" referred to the creative energy of will, which stretches out from the point of the initial "*fiat*" to that of fulfillment. "*Es ist vollbracht*":[22] it is fullfilled and, at the same time, it is over, so that it may be possible to start anew for ever and ever (*immer wieder*). In fact, in formal practice, the doctrine of constitution enunciates "the creative power inherent to will", even if it limits itself to connecting the intention expressed by the practical instance ("*fragende Intention*") with the fulfillment of the action ("*Erfüllung*"), thus confirming a general structure of constituting consciousness, rather than expanding a doctrine of freedom that aims to create ever new instances.[23] Husserl's doctrine of freedom – by all means to be read in the light of his manuscript papers relating to the constitution of time – is susceptible of further developments based on a creative teleology and on the grounds of a more explicit mediation of the general structures of consciousness with the particular figures of spiritual activity.

The task that confronts phenomenology as a method for ontology is accomplished where that of metaphysics begins. To the latter it can offer, besides the rigor of its concepts and its instruments of research, quite a few operative guidelines. Intentional teleological models are dominant in the finalities – devoid of all practical ends – characteristic of the fine arts, which prefer *fruition* of their objects rather than *usage*. In economics there is a predominance of ends that aim towards what is *useful* or *vital*; in politics one might well wonder what is the goal of a society globally considered, not so much to impose on everybody a single idea but to make individual interests agree with the *common weal*. In religious aspirations, lastly, we see that *prevision*, with the enrichment of the contents of faith through revelation, becomes *prophecy*.

3. Religious Form (vita mystica)

Religious experience is certainly very rich with spiritual forms. Our theme leads us to consider religious experience as an originary phenomenon of human life and to study its predispositions in intentional consciousness. In the religious relationship we can find the extremes of the duality subject–object, since the *cogito* is confronted with an infinite object: the very distance between the origin of the intentional function and its objective goal is an infinite one. Besides, the *cogito* presents itself, in its origin, as a form devoid of historical existence before the contents of its actuation in the *res cogitans*, that is, the spiritual substance.

The originary religious predisposition has therefore a twofold appearance: (a) as acknowledgement of human inadequacy and the invitation to transcend into the ineffable; (b) as intentional predisposition towards perfection and, thus, as development of positive values already present in man. Religious experience, then, does not aim solely at filling the gap in spiritual life; it contains the essence of a positive reference to an absolute *completion* that transcends the human condition, perfect up to a certain point, to accomplish it in perfection. During such a process all the yearnings that were *formally* implicit in the intentions of the pure ego and in the Cartesian idea of perfection find their accomplishment. Such an accomplishment becomes possible thanks to the mediation of the symbols characteristic of faith.

The forms of the mind that actuate themselves in religion, that is, in Hegel's terminology, in the objective spirit of an infinite "representation", are more than anything else a confirmation of the symbolic mediation of intuitive contents with the originary structures of consciousness. The greater the distance from the alpha of subjective constitution to the omega of its farthest ends, the more urgent it is to reach a mediation. From the point of view of religion, it seems that Man-God is the most representative figure.[24]

Religious consciousness is somewhat appalled at the emptiness inside it; an emptiness that may take on the most diverse contents, even those of negative sign. Nevertheless it finds a comfort when it discovers within it an infinite intentional purpose to live the fullness of the forms in the union with God (*theopoiesis*). It is, however, because of the unmeasurability of the final goal with the originary act of consciousness that we must assume as necessary the existence of a principle of transcendence.

If God, according to Husserl, is the remotely far man ("*unendliche ferne Mensch*"), just because of this infinite remoteness, He is capable of transcending man.[25] Should the distance between man and God be a finite one, and the two terms be measurable, there would not exist a transcendence but a continuity between the two. Whereas, if our God is the God of an infinite remoteness – and no distance is greater than that between human existence and the *Esse per se ipsum subsistens* – then the very *Selbstkonstitution* is aimed at the other, the all-other, that is, transcendence. It is in the immanency of *objective inwardness*, that is, in the Augustinian *veritas*, that transcendence reveals itself, whenever the idea – both theological and teleological – arises to the divine perfection.

University of Siena *Translated by Roberto Birindelli*

NOTES

[1] E. Husserl, *Logische Untersuchungen*, I, 9, §53, *Husserliana*, Vol. XVIII, 1975.
[2] E. Husserl, *Formale und transzendentale Logik, Versuch einer Kritik der logischen Vernunft*, Niemeyer, Halle, 1929, sect. II.
[3] On the independence of the structures of consciousness, cf. J. M. Broekman, *Phänomenologie und Egologie*, Nijhoff, The Hague, 1963, p. 190.
[4] Husserl, *Cartesianische Meditationen*, p. 108.
[5] J. M. Broekman, *Phänomenologie und Egologie*, p. 190: "Keine seelischen Inhalten wie etwa Willensmomente." According to A.-T. Tymieniecka, Husserl believed that "the permanent features acquired by the pure ego remain also transcendent to the acts and the stream of consciousness in which they occurred"; she also quotes Ingarden's efforts toward the concreteness of the human being, thus supporting also "the living ego of the concrete living person" (A.-T. Tymieniecka, 'Beyond Ingarden's Idealism-Realism Controversy with Husserl', *Ingardeniana, Analecta Husserliana*, Vol. VI, Reidel, Dordrecht, 1976, p. 300).
[6] E. Husserl, 'Teleologie. Die Implication des Eidos transzendentale Intersubjektivität im Eidos transzendentales Ich (15 November 1931)', *Husserliana*, XV, p. 378.
[7] Aristotle, *Metaph*. VI, 1, 1028a.
[8] E. Husserl, *Ideen II*, p. 98.
[9] E. Husserl, *Formale und transzendentale Logik*, p. 232.
[10] For further discussion, see M. Sancipriano, *Il Logos di Husserl*, Bottega d'Erasmo, Turin, 1962.
[11] E. Husserl, *Analysen zur passiven Synthesis* (1918–1926).
[12] E. Paci, 'La fenomenologia come scienza nuova', *Aut aut*, no. 60 (1960), p. 350.
[13] E. Husserl, *Krisis*, section 28.
[14] *Ibid.*
[15] *Ibid.*, section 2.
[16] *Ibid.*, section 38.

[17] Roman Ingarden stresses the idealistic interpretation of Husserl's thought and attributes to what Husserl called *intention* the impossibility of transcending the presumption of the world: "Die Realität . . . nur als Vermeintes 'und darüber hinaus Nichts' gefasst wird" (Ingarden's letter to Husserl, 1918). This does not mean that, in the development of Husserl's thought, there has not loomed possible, on the part of consciousness, an experience of the real (cf. A. Półtawski, 'Constitutive Phenomenology and Intentional Objects', *Analecta Husserliana*, II, 1972, p. 90).
[18] E. Husserl, 'Beilagen zu den Grundproblemen der Ethik (insbesondere der formalen Ethik)', Winter 1908–1909, Summer 1911, p. 76 in the MS (Husserl Archive, Louvain).
[19] *Ibid.*, p. 77.
[20] *Ibid.*, pp. 142ff.
[21] Regarding this opposition, see also R. Sokolowski, *The Formation of Husserl's Concept of Constitution*, Nijhoff, The Hague, 1964: consciousness, while constituting the meaning of things, does not create them but is only the necessary presupposition. Cf. A.-T. Tymieniecka, *Eros et Logos*, Nauwelaerts, Louvain-Paris, 1972.
[22] E. Husserl, 'Beilagen', p. 144.
[23] In the development of Husserl's thought, from the synthesis of *Ursprungsetzung* in *Ideas I* – interpreted by Paul Ricoeur as "un fiat createur du moi pur" (E. Husserl, *Idées directrices pour une phénoménologie*, Gallimard, Paris, 1950, p. 413) – we pass to the *Urschöpferische Akte* in MS K III 11, 1935, where the reference to a proto-creation is made explicit.
[24] See also our contribution to the 30th Symposium of the Center for Philosophic Studies in Gallarate, Italy: M. Sancipriano, *Homo-Deus. Considerazioni di fenomenologia della religione*, Brescia, 1976. On the twofold position expounded above, that is on the meaning of the ineffable at the origin of religious experience (A. J. Heschel) and on the originary religious intentions (M. Scheler), cf. A. Babolin, 'Metodo critico-fenomenologico-metafisico nella filosofia della religione, off-print from *Il metodo della filosofia della religione*, La Garangola, Padua, 1975, pp. 322, 328.
[25] E. Husserl, *Krisis*, section 12.

ELIO COSTANTINI

METAPHYSICS OF BEGINNING AND METAPHYSICS OF FOUNDATION

I. INTRODUCTION

We may summarize the starting point of our study in the following three points: (1) Man's intellect by its very nature aims at the recognition of truth. His will by its very nature is oriented toward the good. Thus man in his entirety, in both his existential structure and his ontological status, basically has being as his end. Man's orientation toward these ends emerges from the history of philosophy, even though interpretations have been varied, sometimes contrasting and sometimes even negative.

We may extrapolate the assumption that future philosophy will maintain the same finality. If this were not so, philosophizing itself would cease to have any meaning. But the contrary is true, since philosophizing is an immanent need of man and his philosophical search will continue for as long as he exists; the basic ends will always remain the same: truth, the good, being.

(2) Husserl's phenomenology, which appears as a new philosophy, as a new method of research, is also open to the same ends. Phenomenology, based as it is upon intentionality, is open to the being of things, it is open to the world of life, it is open to a total view of the world. This opening however, is, no longer that of a classic attitude opposing man to the world but that of consciousness which grasps the phenomena of the world as they present themselves spontaneously and immediately in the structures of consciousness itself.

(3) The reductive method used by Husserl leads to the emergence of the transcendental Ego. The latter is not the Cartesian ego which is still considered to be a congenial part of the world, but the pure Ego in which the natural world is mirrored and can be described in its essential structures which conform to the immanent structures of transcendental consciousness.

After delineating, in subsequent sections, the initial position of Husserl by contrasting it with those of Thomas Aquinas and Kant, we expect to arrive at the formulation of our argument.

II. RESOLUTION OF ONTOLOGY INTO A METHODOLOGY

(A) The mind's orientation toward understanding being in an ideal form is

Tymieniecka (ed.), Analecta Husserliana, Vol. IX, 367–379.

served by a reductive process by Husserl. This process, which is taken as far as possible, reveals the absolute transcendental Ego, as the foundation of all apodictic knowledge and a constituent element of all reality. The reasons why thereby an ontology of consciousness replaces an ontology of being are that: (1) the thinking subject and the object of thought are no longer considered as realities that are already constituted and intrinsically purposive; and (2) the relation between the intending ego and the object intended is a dialectical one that goes from one pole to the other within consciousness with no chance of getting outside it in order to grasp reality in its natural being.

1. First Observation

The Ego constitutes the object (first intentionality) and constitutes itself (second intentionality). This is a single intentional act, as the ego cannot apprehend the object without apprehending itself in relation to this object. This activity lends meaning to the object, but as the ego is simultaneously extrapolating the meaning – bestowing it upon itself – one is dealing with the meaning of meaning, as the ego attributes meaning to the function itself attributing meaning to an object. In other words, the significant and the significate, the meant, are one and the same thing.

2. Second Observation

The Ego consists of an activity which gives meaning to everything. If the Ego has this capacity, it must be in virtue of a principle.

The question thus arises: if this principle is an objective one, a point of reference that is neither the signifying consciousness, nor the knowing thought, it directs the search toward real being; if the principle is subjective, i.e., if the Ego is its own principle, we remain closed within the realm of immanence; no other alternative is possible.

(B) *Thomas Aquinas – Kant – Husserl.* Husserl rejects gnoseologism, which flourished with Kant, as is demonstrated by Bontadini (in *Studi di filosofia moderna*), but he also rejects Aquinas' gnoseology which leads to an ontology of the real being and a consequent realistic metaphysics (Aquinas, *De Veritate*; Husserl, *Erste Philosophie*).

Husserl suggests a third solution for the theory of knowledge which has the advantage of sublating gnoseologism but the drawback of rejecting realism.

For the three mentioned philosophers the gnoseological construction takes on the following form: For St. Thomas, the formal *a priori*, i.e., the *objectum formale quo*, is directed toward the real being whose presence is revealed by sense data; for Kant, the *a priori* of the intellect functions only when filled with the phenomenal data of sensibility; while for Husserl, intentionality as an *a priori* element is achieved only in sense intuitions. Thus for all three of them there is an *a priori* element and an *a posteriori* one.

In spite of this the differences between them are considerable: For Thomas Aquinas the *a priori* is a "detegens" one, i.e., one that reveals the being of the entity, whose thematization opens up the way to the metaphysics of foundation; while for Kant the *a priori* is an "abscondens" *a priori*, i.e., it does not grasp the being of the entity and thus precludes any possibility of metaphysics. Finally, for Husserl, the intentional *a priori* grasps logical objectivities as justified by sense data, but these objectivities are only formal and do not allow one to reach the real being or substance in itself, which is the true object of the cognitive process. Thus for Husserl the *a priori* is neither *detegens* nor *abscondens*; although open to being, it does not allow the metaphysics of being to be retained. For him absolute knowledge is still inaccessible, hidden not by the phenomena as Kant holds, but by the logical objectivities that are valid only for a transcendental consciousness.

III. INTENTIONALITY AND REDUCTION

(C) Intentionality is an inner experience of opening toward being; it is a formal presence, the knowledge of being conscious of something that does not belong to consciousness. In this sense it is a modality, a typical structure of consciousness itself.

Phenomenological reduction is not a spontaneous event; it is a deliberate operation imposed by man against his natural tendency to experience the world as a preexisting reality. It also involves man's experience of himself as a physical and spiritual reality.

Nevertheless, the possibility of performing the phenomenological reduction teaches that intellect can carry out operations of its own that are not linked to the body. It also demonstrates, however, according to Merleau-Ponty, the "impossibility of a complete reduction," i.e., impossibility of eliminating the "natural-corporeal" element.[1]

These views, however, are foreign to Husserl, for whom the intellect is the absolute transcendental Ego and not the concrete personal ego of any individual man. But this conception of Husserl is challenged by Merleau-Ponty, for whom the corporeal element is something irrelative, that cannot be grasped rationally.

We hope to have indicated the major points upon which phenomenology moves away from Aquinas' and Kant's philosophy; it remains an open field, and Merleau-Ponty is right to say that phenomenology "to the very extent to which it keeps faith with its intention, will never know where it is going."[2]

(D) *Noetic activity of the Ego*. The activity of the intellect is in the phenomenological conception an intentional activity for the purpose of knowing and has its own temporal form. This activity is thus an *actus cognoscendi*. It is directed toward the constitution of noematic objectivities receiving logical meaning from the cognitive act itself, and also having their own temporal form, which differs from that of the things lived flowing with the consciousness.

Thus the act of knowing must be founded on the *actus essendi*, i.e., in being. Therefore, we would venture to say that the act of thought presupposes the act of being. But as Husserl does not attempt to start with the assumption that man is basically a being endowed with a specific ontological foundation (e.g., that of matter and form), it follows that the act of constitution, instead of being founded on being, is founded on the absolute Ego, which is essentially a source and an instrument of knowing.

We reach here then a crucial difference of attitudes between the scholastic approach to philosophy and that of phenomenology. The first one seems to seek for its start a foundation in Being. The second expects to formulate and solve all the great questions concerning man and universe by pursuing their origin in cognition.

The central issue of our study is the status of teleology in both of these philosophical approaches. Therefore we have to envisage it in the above perspective.

IV. BEING AND FOUNDATION

In his works Husserl gave an incomparably extensive treatment to the problem of cognition. The problem of cognition for him was a primary and fundamental one; he returned to it persistently, and it would not be wrong to say that its difficulties were a spur to his entire research work. The solution to the problems of cognition is thus seen as the premise or starting point for further research.

In any case, both the validity of the ontological experience and the possibility of a metaphysical experience, as well as the possibility of proceeding speculatively toward the metaphysical foundation of the finite beings and things, depend on the way in which these problems are solved.

Husserl gave no exhaustive treatment to the teleologies nor any systematic one to the problem of history. His works contain, here and there, a few references to history. However, in view of the Husserlian emphasis upon cognition, these references must be related back to the solution Husserl gives to the problem of cognition. Therefore, to inquire into the teleological problem in Husserl, we have first to survey the gnoseological problem and to verify whether the solution he suggests is a valid one.

1. Concept of Experience and Being

What is meant by experience? The problem was first raised by Parmenides and was still being debated by Kant. The two alternatives were immanence or transcendence of being? In short, does experience reveal being or conceal it?

Giannini gives an interpretation of the concept of experience which is in line with Thomist thought, especially with the *De veritate* of St. Thomas.[3] For Giannini experience means an unconditional opening to the given. However, the question arises about the meaning of this definition. Unconditional opening seems to mean, first of all, experiencing experience; then it means opening toward being, in the light of which experience of the finite real entity must be considered. The datum is actually the real entity or entities which constitute the world. However, an entity subsists, but it is contingent; if it is contingent it "has" being but it is not being itself.

Therefore, in Giannini's perspective, the being of the entity founds the entity, but demands in turn to be founded. In other words, the entity given to me in sense experience is an entity insofar as it is itself a being, but for the entity to be an entity it is not enough to be its own being "but [it] demands that the being of the entity should also be founded by the subsistent Being."

In the scholastic approach as interpreted by Giannini the being of the entity is not the subsistent Being, but the being of the entity means "necessary relation to the subsistent Being," in the sense, however, of "transcendental relation in the narrow meaning of the term as, by its very essence as a finite entity, it means relation to the subsistent Being."

Giannini quotes C. Fabro[4] and points out that the philosophical problems regarding the theory of knowledge must grasp

> not only the way in which man opens up to his intuition of objects (psychology), not only that relation (contact) that the subject sets up with the object but also and mainly the absolute content (being) toward which the intrinsically operative intentional act of knowledge tends.[5]

He thus concludes that modern philosophy, after embracing as its precondition a method in which a distinction is made between inner and outer experience, ends up, as regards the given object, by considerably "weakening its organic structure"; so that both empiricism and rationalism lose their "contact with reality" wherever "synthetic judgment" is opposed to "analytic judgment."[6]

Giannini claims that, in spite of the reduction carried out by Husserl to reality in itself, i.e., reality that is neither denied as inaccessible nor created by the subject, and that, in spite of the intentional opening of the intellect toward it, phenomenology does not seem to be able to free itself from the "grip of transcendentality."[7]

2. Concept of Experience in Kant

Turning to Kant we see that the concept of "experience" is no longer interpreted as an unconditional opening, but as a closure in the face of the being of the given.

This means that the Kantian *a priori* does not characterize experience as an opening toward being but, as it is a psychological function and its scope is restricted to phenomenal contents which are also subjective, it is "walled in," as Vanni Rovighi says, by the sense contents, and there is no longer any chance of reaching the being of the given or the given in itself.

This amounts to depriving man of ontological and metaphysical experience.

3. Concept of Experience in Husserl

In Husserl the concept of experience may still be defined as an opening toward the given, although this opening is no longer unconditional. We have to ask, what is it conditioned by?

It seems in fact that Husserl relativizes the given as a datum in itself to the datum as a given through consciousness. In this way being is grasped, but it is a being objectified by and through consciousness and created by the noetic activity of the transcendental Ego. Consequently the constitutive function of consciousness is the condition of experience. Being takes its origin and remains within the bounds of consciousness and to speak of a given-in-itself is meaningless. Constitutive acts give being a meaning, even though the given in itself or the being of the datum differs from the noetic process and is not exhausted by the latter's activity.

4. Husserl's Texts

A number of his texts reveal how Husserl clarified his concept of experience after 1891, i.e., the year he published *Philosophy of Arithmetic*. In this work Husserl asserts that the theory of knowledge is present in it in a general sense and that it is subordinate to the predominant psychological interest. Husserl's work sets out to give "a philosophical clarification of pure mathematics." Therefore, it is not the datum of experience (which is the *real* datum), but mathematical objectivity understood as logical truth even though it is not thought by any mind.

To start with, Husserl eliminates in this work the distinction between outer and inner experience maintained by Descartes and Locke, although this elimination favors inner experience in which the world is built up as an object by the subject. In fact, Husserl says: "It is only a question of the independence of the material entity when speaking of actual consciousness."[8]

In this case the real entity in itself becomes a content of conscious acts and the content of conscious acts forms an indivisible whole together with consciousness. While the concept of experience as an opening toward the real being is resolved into a closure.

If the task of phenomenology is to describe and analyze "experiences of representation, of judgment, of knowledge" for the purpose of attaining "clarity and distinction for a critical understanding of pure logic,"[9] all that is not sufficient for an adequate concept of experience.

In point of fact, if the essence or the being of the entity is separated from the real Entity with which it is one, and becomes a correlative object of consciousness; if correlation dominates the being of the entity and the being of consciousness; then the finite being in its concreteness is no longer thematizable and the experience of the entity *qua* entity is no longer valid.

5. *Concept of Being*

Like Kant, for Husserl "being is not a real predicate." "I can see the color, but not the being colored." He wonders if it is "in the real sense" that we can speak about being "a character," and answers in the negative. Husserl therefore concludes: "being is absolutely nothing perceptible."[10] We cannot then conceive of experience as an opening to being.

6. *Contents of Consciousness*

Hence for Husserl the object to be experienced is no longer the "real" entity, but the object of consciousness, i.e., the eidetic essence. Perception is the vision or intellectual intuition of the same datum as an ideal object. It has the advantage of apodicticity: "we are certain about this objectivity, according to evidence peculiar to it, in a way that is absolutely as original as that which makes us certain about the objectivity perceived through experience."[11] But a question that is "difficult to explore" is raised, viz., the capital distinction between sensory perception and categorial perception remains to be explained.

7. Perception

For Husserl sensory perception is by its very nature incomplete. Only ideal objects enjoy absolute adequation. Husserl maintains that insofar as the connection between intellect and sensible or ideal things perceived is achieved by intentionality, "the sense of its intentional object, i.e., of the thing insofar as it is given in perception, means, that it, the object, is in principle perceptible only through . . . adumbrating . . . perceptions."[12]

Thus sensory perception is inadequate and does not provide a certain and indubitable starting point. As a result "if verbal meanings are carried out on the foundation of acts that achieve only an obscure and confused consciousness, the rational nature of the evidence must necessarily be lacking."[13]

At this stage Husserl differentiates and clarifies his concept of experience: the real entity emerges cognitively from a relation with consciousness; this relation resides in the noema/noesis structure of the rational act; the rational act is correlated to the progressive apparition of the thing;

> but the apparition can only be unilateral and imperfect; now, what appears to consciousness is not only what it really appears to be but simply the thing in its entirety, in its overall sense, which can be intuited only unilaterally and, in many respects, remains indeterminate.[14]

The perceiver thus perceives the perceived thing, but the perceived thing is never grasped in its complete real structure or actual existence, i.e., as an entity existing in its own right. Therefore there seems to be left out from perception so conceived a residual being, and thus a gnoseological transcendence sets in. We may then ask, what is this residual being? It is the being of the entity; it is the entity that subsists in itself.

For Husserl, as it should be clear, the being of the entity is not thematizable in an ontological sense, as extraneous to consciousness. Thus no metaphysics of ontologico-metaphysical foundation is possible, only a (consciousness that we could call) "metaphysics of beginning."

V. IS IT POSSIBLE FOR PHENOMENOLOGY TO DEVELOP IN THE DIRECTION OF A METAPHYSICS UNDERSTOOD AS THE ULTIMATE FOUNDATION OF BEING?

After we have critically assessed the shortcomings of Husserl's position we propose to examine whether they could not be overcome.

(A) In fact, Husserl states that phenomenology has three preconditions; (1) it is open to all legitimate contributions; (2) it is open to all possible developments at the level of critically interpreted research; and (3) it shuns all dogmatism, as it does not present itself as a closed and definitive system, as Hegel's philosophy does.

Thus, phenomenology does not favor either a realistic or an idealistic position, but claims to be a new way of philosophizing based upon the absolute certainty of its specific procedure. Could we then with this procedure first reach the individual being and then proceed speculatively toward the metaphysical foundation of its finitude?

The opening toward being of which Husserl speaks, is an initial opening, but it does not go as far as to grasp the being of the entity as such. The given *qua* object of the intentional telos of the intellect is not the "datum" as such, but is given as a noematic correlate of the noetic act of the subject living it.[15]

(B) *Positive Evaluation of Phenomenology*. If the noetic act conferring meaning denotes the act of acting and the noema denotes the meaningful content, then it is carried out; if the purpose of Husserlian methodology is "the detection of this bipolarity which basically betrays a relativity of both subject and object," a first positive value to be assigned to the "new teleological interpretation of *Wesenschauen* and *Weltanschauung*" lies in a determined effort to sublate "gnoseologism" implying gnoseological immanence and, consequently, metaphysical immanence.

But we may wonder whether the phenomenological concept of experience is successful in opening the way to gnoseological transcendence. Gnoseological and metaphysical transcendence cannot be seen as "presuppositions" which, as such, are left aside by Husserl, but are conditions which are "critical of an authentic way of philosophizing that seeks unity of being in the multiplicity of the entities."[16]

Let us now consider a second positive value emerging from Husserlian reflection. Experience means opening toward being (this has already been said); but to say "opening" means to say "ontological participation" in the act of cognition having as object the act of being. Moreover, Husserl does not deny the ontological nature of the act of thinking.

Experience as an act of the consciousness thus implies a constant purposiveness of the mind itself toward a being that is not consciousness itself: a being that the mind itself does not create or posit. This ineliminable ontological implication of the act of experience was observed by Heidegger who, at least initially, oriented his research toward a reappraisal of being. Therefore, both the sublation of gnoseologism and the ontological nature of the act of experiencing are decisive elements in the phenomenological structure and allow one to speak of a metaphysical beginning in phenomenology.

It still remains to be seen whether it is possible after starting off from these elements, to direct research toward a metaphysics of being or of foundation, or, as Luigi Stefanini would say, a "metaphysics of fulfillment."

(C) *The Problem of Sublation.* This is a problem that has tormented modern thinkers, and still torments modern philosophers. Husserl's followers, Adolf Reinach, Hans Theodor Conrad, Hedwig Conrad-Martius, Edith Stein, Max Scheler, and Roman Ingarden, on the one side, and Eugen Fink and Ludwig Landgrebe, on the other, are divided on the issue whether phenomenology of consciousness is capable of exfoliating the total universe of cognition or whether to accomplish it we have to make recourse to metaphysics.

If thinking is thinking of being, if thinking and being are the same thing, not insofar as being is resolved into thinking but insofar as being is the object of thinking (Parmenides), then noetic activity cannot simply be restricted to grasping abstract objectivities or, as Franz Brentano would say, "illusory" objectivities. The experiential opening must be total. Only in this way, according to Antonelli, does Husserlian "revealing descriptibility make a world emerge objectively in a concrete matrix of dynamic participation."[17]

Thought is always found to be intentionality directed toward being "so the ontological structure of thinking activity does not end in nothing because of the fact that it emerges as consciousness from an experience of. . . ." It is thus clear that thought is "being manifesting itself as intelligible, and thus it is emerging from being."[18]

Being measures itself against being, i.e., being refers to being through the mediation of thinking.

Thinking is thus measured by being and so thinking itself is revelatory of being. Husserl speaks of opening to being in the infinite intentional possibility that has the grasping being as its end.[19]

This proposal for a sublation of the problem and a development of phenomenology toward an ontology of being, although implicit in "intentionality," which is the main factor of the phenomenological construction of the philosophical discourse, seems to lie outside the construction itself because of the strictly defined limits within which phenomenological inquiry takes place.

In spite of this, we have seen that a positive appraisal of Husserl's phenomenology allows us to speak of the possibility of a metaphysical beginning or, if one prefers, of a metaphysics of consciousness approached in its foundation.

It is not possible, however, to speak of such a metaphysics of foundation without going beyond the constituted objects and concentrating upon individual consciousness considered in its indivisible bipolarity of subject and object itself.

Therefore, while it may be possible to conjecture from where the teleologies intrinsic to the phenomenologically approached universe of discourse originate, it is not possible to forsee where they tend.

In this sense we may interpret our initial remark that phenomenology, insofar as it appears according to Merleau-Ponty as a "dialogue" or "an infinite mediation," "to the extent to which it keeps faith with its intention, will never know where it is going."

VI. THEME OF THE TELEOLOGIES

1. The Purpose of Intentionality

The sharp distinction set by Husserl between real entity and ideal entity, between fact and essence, complicates the theme of the teleology of the intentional life of consciousness.

Man, says Landgrebe, is a teleological being. This is true, but he is such to the extent that he is seen as having a finite ontic and ontological structure which is also perfect in its degree of being. Nevertheless, man is further open to being and thinks and acts always in the direction of being. Thus, in him, theoretical activity is distinct from, but not separable from, practical activity.

The theoretical and practical implications of this are stressed by Paul Ricoeur in his *Philosophie de la volonté*, where he says that the will must also be included among the theoretical modalities of intentionality. We have then to take up again the issue of the object.

Let us recall that the object of the intentional act is not the real entity, which has been left aside by the epoché; it is the idea reconstructed by consciousness. The intentional telos of the constituting Ego charges the fruit of the constitutive process with meaning, and it receives its temporal form in the conscious process. Thus the object originates as a noematic objectivity whose subsistence, as it were, is due to its relation with the subject. In this case what subsists is neither a concrete being nor an idea but the subject thinking it. In contrast, in the real world what subsists is the real entity and not the rational being of the entity alone.

Therefore, the terminal stage of the process of cognition is not the entity from which one starts off in order to explain it as a function of the being that founds it; it is rather some result of the constructive rationality inherent in the structure of consciousness.

The same may be said of the intentionality of the will. This is why Ricoeur speaks with reference to will of a residual irrationality that escapes the grasp of thought.

2. The Teleology of History

As we have outlined above, Husserl reverses the classic relation: the mundane entity is no longer considered in its essential relation with the being that founds it; the cognizing entity is no longer seen in its dialectical tension, in ontological dependence upon the subsistent Being that founds it; man's freedom of choice is no longer set face to face with transcending values.

For Husserl, the mundane entity is resolved into the essence or noematic structure constituted by the activity of the subject; the essence is thus related to the subject; the person is understood in relation to universal consciousness; freedom of choice is relative to the transcendental functioning of the absolute Ego. Furthermore, Husserl stresses the sharp distinction made between the factual historical moment and the theoretical moment of its interpretation.

This distinction has been sublated by Hegelian idealism according to which the activity of the spirit is given as both historical and philosophical development; it is sublated by historicism, which considers as valid only the historical judgment and resolves the entire historical process into praxis. But Husserl rejects both solutions of this distinction.

Angela Ales Bello gives us the following interpretation: "Quite the contrary, Husserl simultaneously takes into account immediacy and mediation, maintaining the validity of the theoretical judgment as a judgment acknowledging the meaning of each fact."[20] Thus in its explication the cognitive activity affects both the experiential fact and the explicative act of reflection.

Nevertheless, Bello admits that, in this connection, "transcendentality actually piles up obstacles from the gnoseological point of view." And yet, a situation of coexistence exists within phenomenology between the "conscientialistic" view "which entails a closure in the subject as a source of all knowledge" and the subjectivistic view "which posits the metaphysical problem beginning from an analysis of consciousness." Furthermore, she recognizes the "duplicity between subject and object and between the experiential moment and the moment of explanation."[21]

Historicity, therefore, of man and his world, and not historicism, is purported by Husserl; yet it is conceived as a transcendental historicity. That is, man exercises freedom with reference to values, but they are presented by the transcendental Ego. Thus, although a telos is inherent to history, historical teleology is lacking a metaphysical foundation.

University of Viterbo

NOTES

1 Merleau-Ponty, *Fenomenologia della percezione*, Milan, 1965, p. 23.
2 *Ibid.*, 31.
3 Giannini, *La tematica della transcendenza*, Desclée, Rome, 1964.
4 *Percepzione e pensiero*, Brescia, 1962.
5 *Ibid.*, p. 87.
6 *Ibid.*, p. 89.
7 *Ibid.*, p. 90.
8 *Richerche Logiche*, II, 43.
9 *Richerche Logiche*, II, 4.
10 *Richerche Logiche*, II, 608.
11 *Logica formale e trascendentale*, p. 150.
12 *Ideen*, I, 80.
13 *Ideen*, I, 284.
14 *Ideen*, I, 286.
15 Cf. M. T. Antonelli in *Atti dell'XI Convegno di Studi Filosofici di Gallarate* on the subject of phenomenology, Brescia, 1956, pp. 232–33.
16 *Ibid.*, pp. 235ff.
17 *Op. cit.*, p. 238.
18 *Ibid.*, p. 238.
19 *Ibid.*, pp. 238–39.
20 *Edmund Husserl e la storia*, Parma, 1972, p. 127.
21 *Ibid.*, p. 128.

A. L. KELKEL

HISTORY AS TELEOLOGY AND ESCHATOLOGY: HUSSERL AND HEIDEGGER

Die Geschichte ist das grosse Faktum des absoluten Seins; und die letzten Fragen, die letztmetaphysischen und -teleologischen Fragen, sind eins mit den Fragen nach dem absoluten Sinn der Geschichte.

Edmund Husserl

The dialogue between Husserl and Heidegger is generally approached from one point of view, namely, by describing the debt of the younger toward the elder, to whom the former is related by a sense of friendly admiration as much as by a real *symphilosophein*. One thus attempts to discern in Heidegger's thought the truly Husserlian phenomenological elements which are thought of as now determining, now insignificant, and to which one opposes everything that separates Husserlian phenomenology from Heideggerian ontology. Little attention has been given to discovering the points of convergence of the two thinkers; and there has been still less consideration given to the Heideggerian ideas which Husserl might have accepted, even if unwillingly.

But rather than engaging in the fruitless exercise of attempting to reveal and describe – to catalog – the influences they might have had upon each other, our purpose here is the more modest one of initiating a dialogue which seems to have been lacking during Husserl's lifetime. But what was received and what given is now difficult to determine – particularly in view of the silence of each man concerning the influences upon him. Of course we do not wish to deny that they might have influenced each other. On the contrary, is it not "the exclusive privilege of the greatest thinkers" to allow themselves to be influenced?[1] We prefer, however, to examine the philosophical development of each instead of their mutual influence.

Nothing could better illustrate the communion of thought between these men than their interest in the history of thought – an interest found in the later Husserl (without being totally absent in his previous work) and throughout the work of Heidegger. Paradoxically, what for want of a better term one resigns oneself to calling their "philosophy of history" develops with Husserl through a singular drawing together of his own teleological vision and the rational teleology brilliantly developed by Hegel in the last century – despite

Tymieniecka (ed.), Analecta Husserliana, Vol, IX, 381–411.

the former's hostility toward Hegel's metaphysics of history, to which Husserl attributes the responsibility not only for the rise of skeptical historicism (that distortion of any true thought of history) but also for the weakening of faith in philosophy.[2] With Heidegger, who early came under the influence of Hegelian thought, that influence was exhibited in a conception of history which increasingly rejected the teleological myth of the progress of Reason.

There is accordingly a twofold approach to the intellectual relationship between Husserl and Heidegger. First, to what extent there is (or is not) in both Husserl and Heidegger a vision of history which is indissolubly teleological and eschatological – even if Husserl never uses the latter term nor Heidegger the former (except in a critical manner). In other words, we shall concentrate primarily on discovering how their overall views converge. Second, by a more discriminating approach we shall attempt to discern the characteristic features of their respective conceptions, such that what is *teleology of reason* for the one will appear as *eschatology of being* for the other. These two approaches will reveal as many convergences as divergences in their vision of history – both equally significant and illuminating for the idea of philosophy which they were each developing. It is therefore essential to refrain from ignoring these divergences or regarding them as negligible nuances; for it is these "apparently negligible nuances" which, as Husserl remarks, decide the destiny of philosophy.[3] Nor is it a question of attempting to erase the distances, which will appear insurmountable only to those who forget that true proximity is frequently disclosed by a seeming distance.[4]

I

While Husserl may once have had the reputation of being an ahistorical, even antihistorical thinker, we now know that as the idea of phenomenology became more precise for him, so the history of philosophical thought became inseparable from the very concept of philosophy (a development which occurred as early as the pre-*Crisis* period). No doubt, the philosopher was himself responsible for the fallacious image of "an ahistorical Husserl," since he continued to insist that phenomenology defines itself as much in terms of its origin as by its fundamental project, that is, both by its antihistoricism and its antipsychologism. Since it is descriptive phenomenology, feeding itself on the organic source of eidetic intuition and clarity, its method radically excludes all recourse to historical facticity. No longer can philosophical, or mathematical, truth be subordinated to psychological or historical contingencies, announced the author of *Philosophy as a Rigorous Science*; for the eidetic reduction is, in itself, a reduction of all historical facticity.[5] Thus Husserl argued against compromising with a species of historicist thought which had originated in Hegelianism – a mode of thought responsible both for the withering away of the scientific impulse of philosophy and for the loss of all faith in philosophy (which, in Hegelianism, saw itself reduced to the level of a mere *Weltanschauung*). Husserl reacted to the threat of skeptical relativism contained in historicism by asserting that one does not need to borrow from the history of philosophy to become a philosopher, since "the idea of philosophy, as that of truth, is supratemporal".[6]

At the same time, while he opposed empirical history serving as a basis of validity or of verification of scientific or philosophical truths and theories, he recognized the "immense value" of history for the philosopher who, he admits, "needs history [but] not however, in the manner of the historian" who loses himself in the complicated details of the philosophical systems of the past.[7] The history of philosophy will hereafter be recommended to the philosopher as a fertile ground of inspiration. Therefore, if it is true that the phenomenologist owes nothing to the history of philosophy insofar as it is history, the understanding of the meaning of his own project requires that he understand his own historicity and the historic rootedness of the idea of philosophy that he tends to realize.

The lectures Husserl gave in 1923–1924 under the title *Erste Philosophie* expose systematically, for the first time, this "eidetic history" of philosophy from Socrates and Plato, to Descartes, Locke, and Leibniz, and from Berkeley and Hume to Kant, and literally outline the paths and wanderings of the idea of philosophy – the first of all sciences and which phenomenology wants to revive.[8] It is this idea which brought Husserl to the concept of history which was to be an integral part of his understanding of his own philosophical position before his *critical* reflection on history became clearly directed by the *crisis* of thought – which revealed to him the general crisis to which European humanity was prey; a lucid glance at the teleology of European history sufficed, as the philosopher insisted as early as *Cartesian Meditations* and confirmed in the period of the *Crisis*, to bring to light the historical origins of the contemporary crisis.

It follows that it is no longer possible to see the conjunction of phenomenology with the history of philosophy as an incidental encounter, that is, as fortuitous and due to the painful (but external to phenomenological thought) experience that Husserl had of the irruption of real history into his private life. Such an explanation is all the less plausible for, even in the period of descriptive phenomenology, pure subjectivity discovers itself in a temporal dimension; and the Husserlian *ego cogito*, which in this respect differs from the Cartesian *cogito* – which closes itself up on the instant – conceals a temporal plenitude which will progressively orient phenomenology toward what will be named the genetic analysis. Now, if temporality constitutes the essence of individual subjectivity, is not history the temporality of intersubjectivity of which *First Philosophy* rediscovers the crucial function? Just as intersubjectivity outlines the horizon of transcendental subjectivity, so history inscribes itself as the horizon of intelligibility in the vision of the philosopher's developing self-consciousness.[9] It is the ground which carries both our present and our future and – just as my own past is buried in an "in itself" which escapes us just as we escape from it – so history is not at an inaccessible distance outside of us, precisely because we are situated in it by the very historicity of our existence. However, what is in it of concern to the philosopher are not the contingent facts of objective empirical history; rather it is the meaning, the *telos* which animates it and which acts like an historical *a priori* supporting all historical facticity, which is of philosophical significance. It is because it conveys to us this meaning that it appears in its essence as tradition.[10] For all cultural form, if it has its own historicity, inscribes itself through the community in the universal historicity, and thus serves as the foundation for the meaning of all history of fact.[11] Now in this sense, philosophy, more than any other cultural domain, is the transmission of the idea, that is to say, of the telos through which it constitutes the continuity and unity of a traditionality. But far from being a simple conceptual continuity and unity, it institutes the unity of meaning of the universal historical horizon, and more than any other type of knowledge it, in itself, promotes a radical self-consciousness and self-comprehension.

In Husserl's approaches in *First Philosophy*, the history of philosophy is, in part, the uninterrupted effort of philosophy to say what it is in its essence. Beyond its own historical reality, it tells of the hidden meaning of the history of European humanity in general. The reduction of empirical history allows us to draw out what was originally present in the form of the ideal of a rational universal science; the history of philosophy teaches us that the creation of an authentic rational science is "the supreme condition of the possibility of a cultural development" of humanity and of its accession to an authentic rational humanity.[12] It can only be a science animated by a vision of ultimate rationality on condition that it be carried out on a basis of absolute justification and self-responsibility. That is why the universal science of philosophy enjoys such an eminent position in the history of humanity in general, for the latter can only reach perfect self-mastery if it progresses under the aegis of "the absolute Idea of philosophy".[13] This explains its universality and absolute rationality, at the service of which the philosopher will fulfill his noblest mission, that of being the "civil servant of humanity".[14]

Thus, if one would only patiently examine the texts, one would discover the hidden tension which exists, at least since *Ideen I*, between the Husserl who, on the one hand, rebelled against the teachings of history – convinced of the necessity for a radical reduction of all historical and philosophical facticity, and a Husserl, who, on the other hand, divined (as early as 1911) "the immense value" of history for the self-consciousness of the philosopher in his own living present. A more careful examination cannot fail to reveal that the "ahistorical Husserl," as the author himself made clear in a letter to Georg Misch in 1930, stood away from history only temporarily (without, however, losing sight of it) in order to refine his method of scientific questioning directed toward history.[15] It remains true that the relationship between phenomenology, which aspires to be the "absolute archeological science" (*Urquellenwissenschaft*), and "historical phenomenology," which takes the history of thought as a fertile ground from which all authentic philosophical *Selbstbesinnung* can draw, will long remain ambiguous. While Husserl began by taking a sort of "vow of historical poverty", later on he recognized the extent to which he himself was "the spiritual inheritor" of a philosophical past that occupied the whole of the historical horizon at which his present interests and concerns originated.[16] In so doing, he limited himself (he was convinced) to answering the call of history itself, since he discovered within it the idea of an absolutely and ultimately foundational archeological science, which, like an *immanent telos*, had guided the whole of the history of thought.[17] Yet history has still a further function; for everything continues to happen as if, as a record of bare fact, history was restricted to showing merely the possibility and the will to realize such an absolute science; what is missing is historical reality itself.

In Heidegger the history of philosophical thought is present right from the start, and he inaugurated the dialogue – which was never to end – with the philosophical *tradition* in his thesis of 1916 (on scholastic thought); even before this he had articulated the original meaning of a "historical deconstruction," which was to be the fundamental movement of all future ontology. Heidegger, more than any other philosopher, defined the task of philosophy as a dialogue with the initiators and promoters of the great philosophical tradition. To philosophize authentically is, for him, to follow, step by step, the many paths of thought since the dawn of Greek philosophy. The cause (*Sache*) of thought could not better manifest itself than in this dialogue with the history of its slow and patient unfolding; it is therefore necessary for the thinker to reach out to those who preceded him in order to recapture the authentic witness of thought and to take upon himself the "provenance" of thought – not, however, in order that it should serve us as an anachronism but rather to draw attention to the signs of the approach of true thought. If he wants to evade all misunderstanding, he will have to examine, therefore, the "law" of this dialogue with the history of philosophy, about which (according to Heidegger) Hegel was the first to discover the necessity for philosophy itself.[18] Far from being the sign of a simple historical curiosity, the dialogue with tradition will attempt to penetrate what gave its force to past thought; but it will search for it, less "in what has already been thought than in the not thought from where thought receives all its impulse".[19] Hence, what is required of the philosopher is not a hypothetical "step forward" but rather "the step which retrocedes" (*Schritt zurück*), and which has nothing to do with an illusory attempt to resuscitate a past irrevocably past; on the contrary, it appears to be the only step capable of carrying him toward what still remains of our future.[20]

Let us clearly understand that the history of philosophy and its "recapture" in a thinking dialogue, even if it were to resemble the patient work of the historiographer and his doxographic method, can only be fertile, in Heidegger's eyes, as it was for Husserl, on condition that it is not "pure history," that is, historian's history (*Historie*), conceived as the empirical science of past facts and events. History must cease to be the faithful narrative of a factual past and become a stimulant to philosophical thought. It is then that the history of philosophy will be the patient "repetition" (*Wiederholung*) of the cause of thought.

In the same way, *Sein und Zeit* defines the preliminary paradoxical task which imposes itself upon the thinker confronted by the fundamental question of metaphysics, which "today has fallen into oblivion": whoever intends to pose the question must first reject everything which in the history of thought has contributed to oblivion; such a task involves the "destruction of the history of ontology" which the author poses as a condition for the creation of all new ontology.[21] It is because the human being in his essence is rooted in a fundamental historicity, that he is enabled to comprehend himself and all things through the interpretation handed down to him by history and that therefore his philosophical questioning is itself through and through historical. Now if the human "being-there" descends to tradition, if it tends of itself to lose itself within it, it abandons the responsibility of questioning and answering which binds its existence.[22] By asserting itself of itself, tradition is, as it were, an integral part of the existential facticity of the *Dasein*. But instead of revealing what it transmits, it prevents, in reality, all effective understanding by masking it with the veil of obviousness. While it is supposed to root men in their proper historicity, tradition finishes by uprooting them from their original historicity by hiding behind a keen interest for the past, the most flagrant expression for which is *historicism* (or historism). Hence Heidegger, like Husserl, is extremely harsh on the denatured or denaturing thought of history which is known under the name of historicism and in which he sees, in the end, the enemy of all true "historial" thought (*geschichtlichen*); for it is the systematization of the attitude of the objectification of history which "calculates the future from the images of the past as they are defined to it by the present." Today, therefore, the "domination of historism" appears to him nowhere more manifest than in the hold of the mass-media on world opinion.[23]

It is clear that the philosopher does not escape this servitude of all tradition. The history of philosophy as an ontotheological tradition appears literally as the progressive covering up of its own origins and as the oblivion of the originating question of its own project. Thus the task prescribed to philosophical thought by *Sein und Zeit* enjoins with the phenomenological reduction of all historical facticity. Yet, paradoxically, ontology or metaphysics continues to live off this tradition to which it has been degraded, and under the form of which it has imposed itself upon Western philosophy.

However, let us beware of taking the "phenomenological destruction" of the history of ontology in a purely negative sense. Clearly it enjoins us to destroy all the concealments and misrepresentations undergone by initial ontology in order to "discover" the original experiences and to reintroduce thought in its living traditionality. "Destruction" by no means signifies the simple rejection of tradition, the single-minded will to get rid of it, to abolish a past which weighs down heavily on present thought.[24] If it involves a *critique*, it is a critique of the present of this ontological past whose intention, which is positive, is in no way a will to bury the past in nothingness.[25] Thus it would be better to talk from now on of the "deconstruction" of the history of philosophy rather than of its "Destruction." Heidegger explains later what "the critique" aims at; he stresses that the path to take does not involve a "break with history"; if it invites us to deconstruct the "purely historizing statements of the history of philosophy," it is in no way a "betrayal of history"; on the contrary, it requires "appropriation" and metamorphosis of what tradition offers.[26] The phenomenological deconstruction is fundamentally "liberation," Heidegger explained, and it "renders free what tradition delivers".[27]

If it is true, then, that we always think in the atmosphere of a tradition which is, indeed, the transmission and "donation" of thought through history: "only a true experience of tradition will bestow upon us the present".[28] Here is unequivocally affirmed the eminent, ineluctable function of tradition as much as its deep-seated ambiguity: it consists in directing our thought by tracing in advance the paths to explore – for only the "already thought" can indicate to us what beckons to be thought – without at the same time surrendering us to our irrevocable past; on the contrary, it frees us from a servile submission to a lifeless past and leads us, insofar as it is a thought which precedes us, toward the liberty of the dialogue with what was.[29] Taken in its deepest essence, the authentic tradition, far from being a shackle and "a millstone around [our] neck ... frees [us] in urging us toward what awaits us".[30]

No doubt Heidegger does not follow Hegel completely in his audacious identification of philosophy with its own historic unfolding; for him we only truly philosophize "when we enter into dialogue with the philosophers" without becoming pure historians of philosophy who merely assemble and describe in an historizing manner other men's philosophical opinions and doctrines. The historial, rather than historical, dialogue to which Heidegger invites us is a way of considering the meditations of previous philosophers of the past in an effort to understand the end toward which philosophy is moving: for "that in the direction of which the tradition of philosophy leads us, it delivers us".[31]

Sein und Zeit brought to light the historical rootedness of philosophical questioning and thus stressed the necessity for the "repetition of the question of Being" because it is, in its essence, an "integrally historic question" – not in the banal sense in which ontological thought, in search of the truth of being, should illustrate and confirm itself by the empirical history of ontological doctrines, but in the deeper sense in which the thought of Being is like the "Memorial of Being," and as such belongs to the history of Being as its destination.[32] One can guess the profound change in Heidegger's concept of history as early as 1935. "Geschichte" (and not "history") does not signify something past – that which is no longer to come – nor is it the pure present of today which is not to come either, but simply "occurs"; history as "coming-from" (*Geschehen*) determined as from the future is the way in which men, in assuming what was, act and undergo their present.[33] Philosophy is thus through and through historial, because it articulates questions in the light of history; that is to say, in its essential relationships to the totality of being philosophy discloses what is to come while relating it back to its own commencement.[34]

The way in which the philosopher is invited to assume his own philosophical historicity could be qualified as "archeology," provided that one sees in this word something other than the desperate attempt to resuscitate an irretrievable past: in this original (and primordial) sense "archeology" would thus be the patient and unflagging effort of self-communion of thought on that which, since the dawn of Occidental philosophy, has not ceased to serve as the dominant function of thought.[35] In the last analysis, Heideggerian archeology, like Husserlian archeology, conceals an historial significance which transcends any sort of purely historical enterprise. Both for Husserl and for Heidegger (even if the former reaches this conclusion more slowly) philosophical thinking and the *Selbstbesinnung* of the philosopher necessarily pass through a *Historische Selbstbesinnung* (according to Husserl's expression), of which, assuredly, each philosopher, in his own way, refines the original meaning, both, however, refusing to confuse it with the banal investigation of the historian of philosophy.

II

There are further similarities in the thought of Husserl and Heidegger. Starting from their vision of philosophy and its relationship to the history of philosophy, there begins, bit by bit in both, to take shape what we are forced to call – by an expression certainly suspect to both of them – a "philosophy of history" based on the preeminence and originative role ascribed to the history of thought. Husserl convinced himself that philosophy is not one cultural form among others, for it is not comparable to any other form, for example, art; nor is it a craft, although laying claim to a little more seriousness. On the contrary, philosophy is characterized by a universality and rationality which transcends all other cultural forms; so that one would be mistaken in limiting its aim to that of a purely theoretical knowledge. "Philosophy," wrote Husserl, as early as 1924, "even if it relates itself to the universal domain of the values of knowledge alone ... possesses a significance extending to all pure culture and has, thereby, correlatively, a significance for all of humanity and for all rational universal humanity".[36] We must not believe that this is simply to establish "an historical fact," at the end of a long history, or which we should record for a century marked by its philosophical interests. It is, on the contrary, to express a necessity of essence. Although, constituting apparently only a part of the cultural totality of humanity, it appears, however, that by its position of universality and rational radicalism, philosophy is called upon to fulfill an infinitely more important mission than that customarily attributed to it: for Husserl it has truly the mission of serving as a guide to cultural humanity in order to lift it "to a universal self-consciousness and to an ultimate rationality".[37] For the demand of a rational justification of its existence does not address itself to the single individual called upon to govern his life according to the norms of reason – instead of allowing himself to be dominated by a blind organic development. Philosophy is concerned instead with the establishment of all human community. Husserl praises Plato for having been the first to understand this eminently "humanist" function of philosophy: "for mankind to succeed in elevating itself to the level of a true and authentic humanity, it is necessary to first develop an authentic science ... in which those who are called upon to guide humanity (the archons) draw the truths according to which they will organize rationally the life of the community".[38] Far from it being a cultural form among others, a type of theoretical activity comparable to any other kind of knowledge, philosophy is "the rational foundation, the principle and condition of the possibility of an authentic community," and thus of an authentic humanity.[39] This is why Husserl feels that there is but one valid way of considering the history of philosophy, namely, that one which consists in approaching it

> from the point of view of the grandiose function which it assumes in the history of humanity, having, as its mission, to elevate it to universal self-consciousness and to ultimate rationality, thanks to which it will be able to usher it to an authentic humanity.[40]

It is precisely the history of philosophy which teaches us that the creation of an authentic rational science is "the supreme condition for the possibility of the cultural development" of humanity and for its accession to a rational humanity – which Husserl opposes to the blind development of a humanity remaining captive to its organic animality.[41]

> Authentic humanity means a humanity arrived at true maturity and perfect self-mastery, rendering it at all times able to direct its life with full responsibility, an ability which only philosophy, the science animated by a goal of ultimate and universal rationality, is capable of confering upon it. In other words, the only authentically human culture is that founded on the rationality of absolute norms and principles justified by an absolutely autonomous and universal science.[42]

Such is the mission which the *Crisis* assigns to the philosopher – the "civil servant of humanity" – thus burdening him with a responsibility far transcending that of his private existence, that of his theoretical thought; in effect, the philosopher has to account for the meaning which humanity has for itself and thereafter for what humanity is and will be in a future which philosophy itself must forge by means of both the consciousness it achieves of its task and the responsibility it takes in the destiny of humanity.[43]

Such is the task of all future philosophy aspiring to the title of a science; and such was the idea of philosophy that guided it throughout its history, since its beginnings in ancient Greece – "cradle of the European spirit" – where, for the first time in the history of humanity, there arose the ideas of one absolutely rational, universal science and of a new humanity and human culture, founded on philosophical Reason, that is, on a Reason conscious and ultimately responsible to itself.[44] It is Plato's merit to have conceived the idea of this absolutely rational science and of it normative function in the development of humanity as an authentic human community. The whole history of European humanity will determine itself by this unique idea, and it will not cease to develop itself in conformity with it, because it acts as the telos which arose in Greek science and culture but proved applicable to the whole history of humanity. What is it indeed if not the history of the desperate struggle for rationality, punctuated by strenuous and incessant battles for an autonomous Reason against all its enslavings and blindings?[45] The history of Europe (and especially of European philosophy) has thus become exemplary among all civilizations in having served as the battleground par excellence between Reason, ultimately conscious and responsible to itself and to humanity – with the care of which it is burdened – and unreason – which closed its eyes to the end prescribed to humanity; but it is further the history of an infinite movement animated by a telos "innate" in European humanity, which will lead it from a Reason latent and buried in the darkness of night toward a luminous and responsible Reason, capable of serving as a norm to an authentically human humanity.[46]

In *Crisis* Husserl ignores both the sequence of empirical events which characterize European history and the writings of historians of philosophy – who are anxious to confirm the objective truths of their historiographic narratives – and strives to disclose the "hidden motives" and the immanent teleology which betray the effective movement of history in order to bring to light "behind the disorderly historical fact" the unity and final harmony of a meaning.[47] Husserl is thus led to two capital conclusions which imply each other: first,

philosophy is nothing but *rationalism*, through and through rationalism . . . it is *Ratio* in its incessant movement to elucidate itself, starting from the first irruption of philosophy

in humanity whose reason, although innate, had till then remained inaccessible to itself, steeped in the darkness of night.[48]

Second, the history of Occidental thought and the whole historical development of European humanity reveal themselves as being marked by the awakening and development of rationalism, which far from limiting itself to the simple domain of positive knowledge of reality, has a universal and normative character: it is only itself if it penetrates and rules all the spheres of human existence. Philosophy, considered as a science, that is, as "the historical movement of the unfolding of universal reason, 'innate' to humanity as such," will thus assume its main function of providing a rational historial foundation for humanity – which now, in contemporary Europe, is on the point of succumbing to a fatal crisis from which it can be saved only by a phenomenological philosophy promoting, amid the torment, a veritable "heroism of reason".[49]

Rationalist, European, and of Greek origin, such is the history of thought which is fundamentally philosophy. One cannot fail to question the privilege accorded to Europe and to the Greek origins of philosophy and will thus accuse Husserl of what has been called his "Europocentrism." Europe and rational and philosophical culture undoubtedly have for him the undeniable privilege of serving as model for all authentic humanity, since they embody the telos of all historicity, namely, universality, radical rationality, and traditionality of infinite tasks. Hence Europe is anything but a simple "anthropological type" comparable with the Chinese or Indian civilizations. The carrier of a "teleological idea," the presence of which history attests right through its failures, its role as an example or model is confirmed and illustrated – admirably and tragically – by the "spectacle of radical Europeanization" of the world and of humanity as a whole – which has led some to ask if it denotes the absolute meaning of the history of the world or the absurdity of a deadly destiny.[50] However, Husserl does not doubt that the history of Europe is exemplary insofar as it is the history of the centuries-old spiritual struggle to recover the meaning of an authentic humanity, by virtue of which it could not have failed to draw in its wake the entirety of humanity. Hence, one should beware of interpreting "Europe" as a simple geographical notion or as designating a region of empirical history to which the hazards and vicissitudes of the development of humanity have conferred an eminent and dominating role. Europe, for Husserl, has at once the meaning of an exemplary period of the history of humanity and of the absolute *Idea*, in which originates the teleology of universal history, at least, of the authentic history of humanity, where absolute historicity and absolute rationality are joined.

III

But what of the historial mission of Heidegger's philosophy? If one turns his attention to the foundations of the thought which uphold Heidegger's vision of history, one cannot fail to see that already the fundamental question of philosophy itself, far from being an inconsequential historical question, is reiterated without ever receiving an answer; yet it is the decisive historial question for our European-Occidental being, one which decides the destiny of Europe and thus consecrates the "act of birth of our own history," not only in its historical origin – which reveals the Greek source of philosophy – but even of the "present epoch of universal history which is called the atomic era".[51] We know of Heidegger's thesis – which has shocked many because no one has wanted, or known how, to understand it: philosophy and "European philosophy" of Greek origin are one. We have here a double affirmation: on the one hand, we have the assertion that philosophy is Greek in its very essence, that is, in its origin, its becoming, and its deepest essence; in other words, philosophy is of such a nature that it was primarily in the Greek cultural universe and only there that it could be born and that it transmitted to Western European posterity the Greek and philosophical character of its origin. This underlines the second aspect of Heidegger's thesis: "The West and Europe are, and they alone are, in their innermost historical movement, originally philosophic."[52] And it is not only the history of philosophy which testifies to the Occidental and Greek character of philosophy; the history of thought, including that of science and technique, demonstrates, according to Heidegger, that the historical path of the European West proceeds from the same source and traces its path in accordance with the same telos which is philosophical. Now the *provenance* of European thought also discloses its *destination*, and expresses its historical mission, by virtue of which it has imprinted itself on the planetary history of man.[53]

Let there be no misunderstanding! Let us not take at face value the apparent "ethnocentrism" of the philosopher. We should not construe the historial preeminence conferred upon Greek and European thought as if Heidegger were asserting the exclusive prerogatives and privileges of an historic people or a geographical region of the world; "Greek" in this case, Heidegger makes clear, does not designate a national, geographical, or anthropological characteristic but something more fundamentally historial; "Greek" indicates, in effect, the belonging to a common *destiny*, common to initial Greek thought and to that of today, and it is this which makes the one and the other Occidental. Similarly, "Occidental" should not be interpreted geographically, in opposition to the Orient; nor does it limit itself to Europe alone; it is, instead, thought in the perspective of world history (*Weltgeschichte*) and of the planetary mission which Occidental thought has fulfilled throughout history. The Occident is that historial scene in which the destiny of the world has been decided and from which its future history will spring.[54]

Let us add that the decisive step for Heidegger, as for Husserl, leading to "philosophy" as *episteme theoretike* par excellence was taken by Plato and completed by Aristotle, to whom befalls already the responsibility for the eminently rationalist destiny of both philosophy and Europe. Who could doubt, asks Heidegger, that philosophy is the "business of reason" and that it is the "true agent of reason," even if one ignores or fails to recognize this to be the case; reason, then, "has become the master of philosophy," because during the history of Europe, it has deployed all its forces to shaping philosophy concurrently with imposing its reign on European humanity in all its spheres of existence.[55] Metaphysics – what is it, if not the very name of philosophy? – Heidegger will not tire of repeating, appears as the "necessary fatality" of the West: from its beginning it has determined the history of Occidental humanity, which it continues to guide in the totality of its relationships to being.[56]

But what exactly is meant when speaking of "metaphysics"? Assuredly, it is not a question of designating some philosophical doctrine which one imputes to this or that philosopher: to Plato, to Aristotle, to Descartes, and to Hegel, and which one might now suspect belongs to the past – and of which one must rid oneself as fast as possible. For Heidegger philosophy has remained what it was at its birth; metaphysics, which was established in its first form by Plato and Aristotle as the understanding of being in its Being, remains ontotheologic in the form in which it "still reigns in the West today, not only within the philosophical doctrines, but in the most mundane activities of daily life".[57]

At all stages of its history metaphysics remains governed by the single question, namely, what is being? and at all stages, from Plato to St. Thomas Aquinas, Descartes, Kant, and Hegel, up to Nietzsche, the answers it will give are directly or indirectly determined by the initially Greek interpretation of the Being of being. As a consequence of that interpretation we are led to think of it as being present in its presence and thereafter to conceive of it in its different modes of "appearing," in the stability of its "essence"; this thought finally conceives it as "object," with the result that the Logos, comprehending itself as Understanding and Reason, "finishes by establishing its reign on Being at the beginning of Greek philosophy," and by engendering all the subsequent modern misinterpretations which will dominate philosophy; for Being is now reduced to a Being-object, and thought is designated as the keystone of human subjectivity. From now on thought, under the historial form of Reason, will have no function other than to apprehend being, to seize it in order to dominate it, to lay hold of it for its own security, to master it so as to better exploit it for man's benefit.[58] The entire history of Western philosophy will thus continue to be the incessant pursuit of being toward its domination by sovereign thought; and it is in this sense that it manifests, from its beginnings with Plato to its first definitive closure with Hegel, a fatal oblivion of Being, to the advantage of the representable being, calculable and exploitable by the various technical modes of domination. If the history of thought thus reveals an ever-growing extension of the reign of Reason – which found its apogee in the modern epoch of a calculating and technical Reason of Science – the oblivion of Being which accompanies it, far from being the expression of an impotence or a weakness of thought, is, on the contrary, "the condition of . . . planetary domination" for the West and the expression of a "necessary fatality": it constitutes, by virtue of this, the fundamental feature of Western European history.[59]

It seems therefore that for Heidegger as well, the history of Western thought is governed by a unique telos, a unique fatality called "metaphysics" which expresses itself in an ineluctable oblivion of the essential foundation of all being: Being. However, if it is true that the history of Europe is that of the foundation and progressive consolidation of the "empire of Reason," it is nonetheless true that Husserl and Heidegger appreciated differently the meaning of this long and confused history marked by the rise of rationalism in Greek thought and its flourishing in modern times. When Husserl thus considers positively the progressive establishment of the reign of universal reason and of a humanity conscious of its power over the totality of being, Heidegger makes a *negative* appreciation; for him the establishment of the planetary reign of a calculating and technical Reason – because technique and technicized science are merely the mask of an accomplished metaphysics of the atomic age[60] – expresses the progressive decadence of the authentic ontological relations of man to Being; in opposition to Husserl, Heidegger does not believe that man's salvation can come to him through this dominating Reason, self-assured and conquering, which he sometimes denounces as the fiercest opponent of true thought.[61]

Would Husserl, who is widely considered "the apostle" and the absolute defender of rationalism, have been naive enough to ignore the failures of historical rationalism and the rise of contemporary irrationalisms? The understanding that he increasingly acquired – and especially following the *Crisis*-period – of the ambiguous, historical significance of the *modern epoch*, is crucial: dominated by the Cartesian ideal of absolute and universal scientificity and the will to complete command of the universe, it tends to bind the entire universe to the endless exploration of the physico-mathematical praxis; and it thereby ushers in an increasing mathematization of nature, which distorts itself into a pernicious *objectivism*, thereby surrendering itself to the exclusive end of enslaving nature (including therein man and his world).[62] The pernicious consequences, which, for humanity, involve domination by a philosophico-scientific objectivism, are considerable although hardly anticipated by that founder of modern reason, Descartes, or by his rationalist successors. The illusion of having discovered an infallible method of scientific control – and hence also a technical one – of the totality of being engenders a sort of fresh naiveté and a rationalism that loses itself in a self-satisfaction which ignores its failures and is oblivious of its true foundation in the prescientific experience of the world of life. It is this that will be the cause of that *crisis* which has arisen and been aggravated in the contemporary epoch, where the unbridled domination of modern science and technique and their incapacity to answer the fundamental questions of man, himself exposed to the nearly fatal upheavals and anguish reflected in the apparent absurdity of his existence, has given rise to a "wave of skepticism": faith in science and philosophy, both expressions of universal and absolute Reason, has crumbled and men have adopted irrationalism and mysticism.[63]

It appears, therefore, that Husserl, although he never disowned his deep faith and hope in the possible reign of Reason, on which in his eyes depended the salvation of humanity, was neither naive nor blind to the gravity of the crisis and of the threats which hung over Europe and over the life of humanity. Even if he convinced himself that humanity could only salvage itself from the barbarism which imperiled it, or from the inertia which lay in wait for it, by a true "heroism of Reason," he nevertheless imputed the responsibility for the crisis to objectivist science and to a misguided rationalism. Science, by representing itself as ultimate and absolute – while favoring a growing technicization of all things, including man, and locking itself into the objectivist oblivion of its own rootedness in the world of life – made possible the breakdown of all theoretical and ethico-cultural values which had served till then as norms for modern humanity. Husserl did not conceal, however, the fact that substitution and distortion effected by science express, in the last analysis, was more a metaphysical aberration than the vigorous implementation of an authentic rationality: this is to say that the modern epoch has missed its ideal and has not known how to realize the idea of scientificity and absolute and universal rationality which had, however, initially inspired it.

Whatever one might say of the bankruptcy of the rationalist ideal – which modern thought has attempted to conceal by triumphantly advertising its brilliant technical achievements – it remains clear that for Husserl, as for Heidegger, science remains the dominant force of our epoch, the reign of which has been in preparation for centuries. But the history of the European sciences has shown itself to be simultaneously the triumph of Reason and the oblivion of the *Lebenswelt*. In compensation, the mortal crisis resulting from it gave a new impulse to philosophy which, in the form of phenomenology, has understood that its essential task is the comprehension of its historical mission; that is, to give a new meaning to philosophy as the foundational science of an authentic humanity. Thereafter, there is but one way to overcome the crisis and the wave of disbelief and irrationalism, namely, to attempt the reconciliation of philosophy and science with the world of life where the very meaning of man's project of scientific rationalization is rooted.[64] For despite what the skeptics may think of the failure of all the historic attempts to realize the teleological meaning of philosophy (which they end in denouncing as an "illusory dream" sought after by generations of philosophers), these failures are in fact themselves comforting, since they prove that the idea has never been abandoned. Thus, in Husserl's eyes, skepticism fulfills a positive function: it gives an edge to our belief in the possibility of realizing the idea of an absolute and universal science which has ever been a main concern of philosophers. The negative balance sheet which Husserl himself draws from the lessons of the history of thought, far from driving him to pessimism and discouragement, leads him instead to pursue his task and to contribute to the edification of the commencement of philosophy, which he recognized, at the end of his life, as being the unending task assigned to humanity by the Greeks.

We thus see taking shape and gaining in clarity – as the father of phenomenology became conscious of his own historical rootedness – the idea of a *universal teleology*. It is for the phenomenologist therefore to reaccept, thoroughly and with complete lucidity, the task inherited from the Greeks, at the very moment when its realization is most threatened – more threatened now than by any of the numerous memorable failures which have punctuated the history of philosophy. For it is indeed history itself – and not some obscure metaphysical speculation – which according to Husserl reveals its teleological structure and the immanent unity which it owes to the common vision of a single end which generations of philosophers pursued with a more or less clear consciousness of their task.[65] But the task will only become fully evident at the end of the historical process of the clarification of its meaning in its ultimate establishment (*Endstiftung*) by transcendental phenomenology, which, as it grasps itself in its integral historicity, simultaneously understands that its realization is a never-ending task devolving upon humanity present and future.[66]

It is contemporary thought confronted by the crisis in which Western humanity risks its death by succumbing to the inertia or to the assaults of a new barbarism provided with the weapons of science, that will accomplish the decisive step toward a new radical commencement; but it must accept with ardor the never-ending task laid down by teleological history. In the light of the teleological idea which the great thinkers of the past, each in his own way, served and which secretly guided their sometimes hesitant steps, the philosophies of the past appear as aborted, but necessary, attempts to realize the task of philosophy; they were steps on the road to the revelation of universal Reason which they obeyed as an "obscure instinct." To the impartial observer the history of the world presents itself under the guise of a progressive development, stage by stage, and culminating in the ultimate rational elucidation of universal Reason under whose aegis a new humanity will take shape, fully in possession of its liberty and of the responsibility of its destiny.[67] The failures and the apparent regressions as well as the final crisis where the teleological idea itself is brought into question, are all due to the "transcendental naiveté" and the objectivist preconceptions of which the thinkers of the past were unable to rid themselves: to those thinkers is due the "fatal historical evolution" which has resulted in that decline or "bankruptcy of European culture born of autonomous science, that is to say, of philosophy"; and far from being a simple historical accident, it expresses the "teleological necessity" to which humanity is committed and which it must assume in order to reconquer its true autonomy founded upon absolute rational science.[68]

IV

The Husserlian conviction that philosophy has not yet truly come to life, that it remains the end and the latent teleological sense of all philosophical activity, was partly echoed by Heidegger, who accepted the same diagnosis apropos thought, which was still, in his view, awaiting to become. The pioneers of thought may well have foreseen the task of thought (*Sache des Denkens*) and its destination, but history soon buried it in oblivion in such manner that it finally constituted the history of *the errantry of thought* and the ever-deepening oblivion of Being, baring itself shamelessly under the reign of technological Reason, itself fascinated by the will to dominate over all being. Hence, the different periods of the history of metaphysics are no more than the expression of the different stages of the errantry of thought starting from its first rise with the Greeks.

However, what again distinguishes the Heideggerian interpretation from that of Husserl is the meaning he gives to historical evolution: for Husserl it is, as we have pointed out, positive, progressive, leading toward a growing self-elucidation of Reason, even if men concealed from themselves, as it were, their own destination in an unsound rationalism or in a no less pernicious irrationalism. For Heidegger, if historical becoming has a meaning, it is not so much teleological as eschatological. It could not be guided by an immanent rational idea which would act as an end tending to realize itself progressively through the hazards of the empirical history of men; the profound history of humanity reveals, rather than progress, an ineluctable decadence or decline, and the end toward which it is moving is truly the realization or completion (*Vollendung*) of philosophy, arrived at its ultimate stage in the form of planetary science and technique, the origin of which is the same, namely, metaphysical and Greek.[69]

Let us understand the Heideggerian interpretation of history as *eschatology of Being*. Let us remember that metaphysics defines itself, from its origins to the period of its completion with Hegel, as ontotheological, that is to say, as the questioning of being on its Being and the ways in which it presents itself to the questioning being that is man. Now, even when it intends to reveal the truth about being, metaphysics finds itself steeped in a singular permanent confusion of Being and being; it almost seems as if the very way of thinking about being dooms metaphysics to construct an obstacle which forbids man from having access to a "primordial relationship to Being".[70] Such is the essential phenomenon of the oblivion of Being in which man, without his knowing it, is trapped. This is not an accident of history but "the fundamental historical event." Its gravity is due not so much to the fact that Being, as such, has fallen into oblivion – as the expression *Seinsvergessenheit* might suggest. We must rid ourselves of the current banal representation of oblivion which regards it as a psychological state. The oblivion of Being is not an event that arises because human thought is forgetful, nor because man has a failing memory; it is, Heidegger asserts quite simply, "consubstantial to Being".[71] It does not have the sense of an anemia, nor of an ontological amnesia by which man would be stricken because, truly speaking, the onus does not lie with man. "The word does not apply," declares Heidegger, "to a sin of man; it designates the essential relationship of man to Being within the relationship of Being to the man's being".[72]

Nevertheless, the philosopher denounces as the first cause of the oblivion of Being from which Western history springs, the fact that man is caught in a "supreme passion for being," is indifferent toward Being, and thus takes to the unrestrained pursuit and to the incessant manipulations of being. The historical situation of decadence of the peoples of Europe, Heidegger wrote in 1935, at the moment when Husserl was denouncing the fatal crisis of European humanity, was due perhaps to

> that which since the beginning is on its march through the provenance of the Occident to an event which all the dupes of all the historians will be unable to perceive, and which yet originates long ago, today, and in the future. What would you say if things were such that man, that peoples, in their greatest concerns and machinations, had indeed a relationship with being, and had yet fallen a long time ago out of the Being without knowing it, and that just this was the innermost and most powerful reason for their downfall?[73]

Was the ignorance of the Western peoples and their irresponsibility in arresting this decline into which they had let themselves be carried judged in a similar manner by Husserl and Heidegger? It would seem here again that one witnesses a strange movement of convergence on the basis, however, of distant positions: although Husserl imputed to the thinkers, even if they be rationalists like Descartes, the responsibility for the historical deviance of the idea of philosophy, he did not conceal from himself that there was a sort of "unfathomable necessity" and an "obscure instinct" which led them to act. Heidegger, who interpreted the history of metaphysics and thence the history of the West as a fatal decline – even if he refused to depict it in the same pessimistic terms as Spengler – considered it unjust to have men bear the responsibility for the historial situation, unless this word be taken in its proper sense, exempt of all ethical reasonance; for it arises from the obscure destiny of Being itself, rather than from the initiative and liberty of men.

What conclusions can one draw from Heidegger's analyses? First, the idea that metaphysics, far from consisting of doctrines succeeding each other in history, is, in its essence, the very historiality of the truth of Being: it is the movement of becoming of truth through the progressive unveiling of being in its beingness. The history of philosophy appears and accomplishes itself with the progressive *veiling* of its origins (in the sense that they designate the originating Greek thought as the scene of its birth and the truth of Being as *aletheia*) and thus also with progressive *unveiling* of its beginnings by what is, from the start, thought of being in its beingness. The history of philosophy is thus marked by a double movement: from the Platonic Idea to the absolute subjectivity of Hegelian metaphysics and to Nietzsche's will to power, all extreme figures of Being; and the growing intelligibility of being as being occurs simultaneously with a movement obscuring Being; in other words, the movement of increasing control of man over being is accompanied both by an increasingly radical enslavement to being and by an accentuated indifference toward Being. And yet, at the very moment when man, pushing to the extreme his oblivion of Being in an "oblivion of oblivion," busies himself in his position as master of the world, reducing all things to the rank of raw material for his technical control, he does not totally escape his "ontological" vocation by which he is summoned before the mystery of Being.[74]

Thus takes shape, little by little, what we, perhaps incorrectly, have called Heidegger's "philosophy of history." It takes more precise forms at the point where there is effected the "reversal" (*Kehre*) of the thought of Being, which initiates just as radical a reversal in the thought of history. For inasmuch as the history of Western metaphysics has shown itself to be the history of the oblivion of Being, so total history (*Geschichte*) undergoes a radical transmutation which expresses itself in a sort of transfer of the subject of history: from man to Being. Since its Greek origins, the truth of Being has known numerous disguises and variations; and through these has arisen a history of thought which has taken the rise of the truth of Being as its destiny. But simultaneously as it institutes itself, the relation of man to Being is an "event" (*Ereignis*) of Being itself, such that its oblivion, through which thought finds itself thrown into errantry, is itself less a fault or shortcoming of man than a fate which befalls Being itself. Such is the enigmatic imbrication of the history of human thought with the history of Being itself. "The history of Being," wrote Heidegger, terminating his course on Nietzsche,

> is neither the history of man and of humanity, nor the history of man's relationship to being and to Being. The history of Being, is Being and nothing but Being itself. However, owing to the fact that Being has recourse to man, in order to found its truth in being, man remains implicated in the history of Being not otherwise than in accordance and conformity with the way in which he assumes his essence on the basis of his relationship to Being – according also to the way in which he loses it, transgresses it, sacrifices it, founds it, or wastes it.[75]

It is a singular conception, it will be objected, which substitutes for the history of thought – indeed, actual history in its Western totality, and clearly and empirically verified development – a mysterious and obscure history of

Being (*Seinsgeschichte*) which, presenting itself in the last analysis as the "Destiny of Being" (*Seinsgeschick*), is itself affected by a somewhat fatalistic resonance. It is a far cry here from the empirical history of the historian and far also from the philosophical manner of conceiving history as "the realization of supratemporal ideas or values" or even of universal Reason such as found in Hegelian or Husserlian teleology.[76] Heidegger recognized this, while alleging that such customary conceptions, contrary to appearances, did not proceed directly from historical experience itself, but were, rather, the transposition to historic becoming of the Platonic metaphysical division of the world into two parts – the one sensible and subject to change, the other suprasensible and immutable. That this philosophical representation of history inscribes itself directly in the history of metaphysical thought which determines it, casts not a shadow of doubt in Heidegger's eyes.

However, the philosopher does not conceal from himself the difficulties thrown up by his conception of history as the destiny of Being, the ultimate basis of all Occidental history. Assuredly, the formula sounds peculiar to one used to the "objectivist" (or, at least, objectifying) thought of the historian. The history of Being has nothing to do with those mythological stories which elders used to tell each other; no more is it something of no concern to man. "The destiny of Being," asserts Heidegger forcefully, "remains in itself the essential history of Western man".[77] Let us not conclude that Being is thought "anthropologically"; it is rather the essence of man remaining installed in the locus of the destiny of Being; and his manner of apprehending his own being reflects the way in which Being will give itself to him. The history of Western thought gives some clues to the meaning of the history of Being, as "destiny of Being," which we could discern if only we discarded the moral or, as it were, fatalistic, interpretation and take the expression in its original and quasi-etymological sense of "destination" (Heidegger derives *Geschichte* from *Geschick* and not from *Geschehen* and thus from the verb *schicken* or *beschicken*). "The destiny of Being is indeed a history of Being, but not thought of from the idea of becoming or from the temporal succession of things."[78]

We are here led to an audacious and enigmatic identification: *Weltgeschichte* rhymes with *Seinsgechichte* or even with *Seinsgeschick*. To understand it, one must be attentive to the history of thought through which Being unfolds its being, which is the destiny assigned to man. Now Being, while no doubt manifesting and revealing itself through being (be it the supreme being), is thus apprehended by thought and reveals itself by erasing, as it were, its own image to the benefit of the multiple figures of being. Heidegger formulated this by observing that "being withdraws in that it discloses itself in being" and thus leads the being into errantry, that is to say, into the oblivion of Being.[79]

Such is the event by which Being adapts itself to being and institutes what Heidegger calls "the scene of the happening of history" to a dimension of Being where being is doomed to errantry and wherein is instituted "the empire of errantry." This phrase does not denote some error of judgment or even the permanent state of one who is mistaken in his appreciations of being; it denotes instead an infinitely more fundamental and original "state" which constitutes the origin of history. The thought of man corresponds each time to a state of occultation and the withdrawal of the opening or the light of Being, which gives rise to "mistakes" of Being – mistakes, however, which are in no way comparable to man's blunders or to deplorable failures or shortcomings on his part: they constitute, on the contrary, the very condition of the unfolding of history. "Without errantry," observes Heidegger in that same text, "there would be no relationship of destiny to destiny, there would be no History".[80]

But History is not only that somewhat inarticulated scene of the coming of the truth of Being; it is articulated in successive stages, a fact explicable only because Being, while giving itself to the thought of being and revealing itself in that same movement, "retains its own truth, not by a definitive holdback," Heidegger argues, "but by a retention such that it is nevertheless accompanied by a given mode of disclosure, of unveiling of being".[81] It follows that each stage of history corresponds to a given retention or reserve of Being destining us to being, which Heidegger qualifies, with a seemingly Husserlian expression, "the *epochè* of Being." In fact, it is not directed in any way at the sovereign attitude of knowing subjectivity suspending its thetic acts. "The *epochè* of Being," we are told, "belongs to Being itself" and points to the epochal essence of its destination wherein originates world history with its successive epochs:

> From the *epochè* of Being arises the epochal essence of its destination wherein lies veritable world history. Each time Being holds itself in its own destiny, there arises suddenly and unforeseeably a world. Every epoch of world history is an epoch of errantry.[82]

The historical epochs are the answer that thought proposes each time to an epochal retention of Being which thereafter manifests itself in a new historial figure, in a different ontologico-historial constellation.

These indications are no doubt insufficient to throw light on the meaning of the history of Being as opposed to and by reference to world history. We might identify the history of Being with the history of the world; the former fuses into the latter with the advent of the errantry of each historical epoch. Thus world history is neither external nor inessential to the history of Being; no more is it purely and simply identical to the history of Being; for insofar as it is an enfolding of destiny in each of its epochs, it is through and through subject to the empire of errantry.

V

There remains one last question whose elucidation will raise again the Husserlian vision of history. What, in the last analysis, is the meaning of the history of Being as world history, which hides it while at the same time revealing it? In other words, what is that *law* which determines the course of history? Here again the philosopher's answer may perhaps appear less illuminating; it may seem instead a mystification of the destiny of Being, implying a globalizing or a totalizing conception of history. And how could it be otherwise? How can one conceive of continuity and unity in historic becoming, ruptured by the discontinuity of the epochs of errantry, were one not to tacitly presuppose the image of one totality in which lie the phases and the events which constitute the concrete reality of history? The law which presides over the history of thought as over universal history is, according to Heidegger, the law of Being itself, and is summed up in the expression "eschatology of Being." This expression, which is somewhat enigmatic, is only partly elucidated by the philological guides provided by Heidegger. History as historial totality characterizes itself in its tendency toward realization and completion (*Vollendung*): by its continuity and its proper logic, it is led toward its end, toward that stage which will constitute the supreme outcome of its tendency to carry itself toward the ultimate oblivion of Being.[83] In evoking the eschatology of Being, Heidegger puts us on guard against all theological reminiscence (from which the inexperienced reader will have difficulties in protecting himself); he refers to the *Phenomenology of Mind*, which is presented as the historical example illustrating the nontheological sense of the eschatology of Being.

One could undoubtedly – and without the slightest problem – uncover epochs in the history of Being; a more difficult task would seem to be the consequence we are asked to draw therefrom: it is difficult to think of the historial process in itself as proceeding toward an ultimate, final stage, one representing the modern and contemporary epoch which began with Descartes and culminated with Nietzsche – a culmination which we suspect will last longer perhaps than the epochs which preceded it. The history of Being is eschatological because it is in its totality that teleological movement, which from its very commencement, is directed toward its own end, and which will realize itself at the instant when Being finds itself, for one last time, summoned in the ultimate "difference" (*Ab-schied*) of itself with being, leaving man utterly adrift from Being.[84] But this is not all; each historical epoch is itself marked by the eschatological law – the grand law of history – and tends toward its own realization in the new suspension of the authenticity of Being: thus the dual character of historical temporality, continuous and discontinuous. There is a sort of continous progression toward an end – which is simultaneously telos and the outcome and ultimate realization of its essence; there is, however, at times also the extreme withholding of Being, a brusque and unforeseeable rupture in historical continuity. The completion (*Vollendung*) which the epoch has reached constitutes therein both its apogee and its end, the point at which the epoch takes, as it were, leave of itself in an ultimate summing-up of all its potentialities.

One thus understands the historial preeminence of the present epoch, which is the eschatological epoch par excellence. It is the herald of both crepuscular time – prefiguring ultimate decline – and future salvation. It is in this present epoch that history, past and future, manifests and opens itself; it has the historical privilege of being, so to speak, the focus of unification of the ineluctable march of history, at the same time as it presents – to one scrutinizing history – the image of the coherence of the historial epochs determined by errantry and the oblivion of Being, unfolding themselves in the successive suspensions of its truth, to which correspond the successive mistakes of humanity. Let us repeat, however, that rather than a "fault" or a weakness of men, it betrays the fatality of the very destiny of Being.

The decline of the Occident finally has no meaning other than that conferred upon it by the end of the history of Being which summarizes in its extreme epoch the consistent unfolding of its origin, the realization, but also the obscurity of what is given at the commencement: the deoccultation of Being as unveiling occultation. Heidegger concludes that the history of the Occident is appropriately named, since it produces itself in the form of a progression of the oblivion of Being and of a fatal decline leading from the commencement to the end, where it will perhaps return to a new commencement. The ideas of the decline of the Occident and of the continuity of its profound history are linked, despite the breaks which intervene and mark the great epochs of its becoming. Western history is eschatological par excellence, explains Heidegger, because Being is in itself eschatological: it tends of itself to summon itself in the ultimate revelation of its differences with being and to escape into an ultimate abandonment in which man finds himself forsaken. This is the phase of the ultimate domination of being as well as of the absolute withdrawal of Being buried in oblivion. The contemporary epoch then spreads itself in its planetary domination; this is also the epoch of the end of philosophy, which it no longer needs; this is the moment when philosophy has become universalized, as prophesied by Marx. At this ultimate stage of its realization philosophy definitively takes on the mask of a planetary science and technique, without being able to conceal its origin in metaphysics and in Greek thought. With this contemporary epoch there ends the historial mission of Europe, from where all has come and which has overrun the whole universe; at least, it is to this end that it seems destined – being nothing other than a "simple cape" for humanity. Echoing Paul Valery, who announced the end of the preeminence of Europe following World War I, Heidegger doubted that it had ceased being "the brain of this vast body of the world, which sets going the calculators of planetary conquest": the Occident appears to be the land of the setting sun, called upon to "become the land of such an evening that from it alone can the world prepare itself for the birth of a new morning?"[85]

To end this imaginary dialogue to which we have invited the two masters of contemporary philosophy, let us try one last time to draw together those features which, despite the apparent distance between the two philosophers, express their profound proximity. To do this we should begin by noting to what extent the place they occupy in the Occidental history of historial thought is common and familiar to them. Indeed, how can one fail to notice that both worked within a double horizon of thought: on the one hand, they were consciously or unconsciously inheritors of the Christian ontotheological tradition. Even if both of them made an effort to put an end to this cumbersome inheritance, they did it in such a way that while contributing to hasten the end of Occidental metaphysics, each in his way was led to assume its destiny. Did Heidegger not say that *Überwindung* meant *Verwindung*?[86] But this made their thought doubly determined: by its origin (the Greek and Christian heritage) and by its destination; in short, their work was shaped by the impact of the "initial" thought and by the "future" thought, which, to their mind, was to end in a reunion which would give rise to new divergences. Nonetheless, these influences would not have shaped their thought were both men not committed, right from the start, to dealing with one main problem: the ineluctable decline of Western thought and humanity. There lies their common point of departure: the experience which they underwent and which they felt with lucidity urged them to examine history from the present, backward and forward. It is the consciousness that they had of the manifestly decadent character of our times that motivated their historical thought to turn, on the one hand, toward the past, the source of our present, which it is necessary to destroy in order to make it appear in its historical-destinal function; and, on the other hand, toward the future, which rests concealed in the present, but whose contours are already outlined and which we and future humanity have the task of realizing.

Thus the historial thought of Husserl, as of Heidegger, is altogether *retrospective* and *prospective*; and it is this that unites the two philosophers, more perhaps than they suspected themselves, to the great tradition of the philosophy of history of the last century. For, in essence, the Husserlian teleology – with its progression and ascension toward universal Reason, conscious of itself in its absolute liberty and of its historial mission – and Heideggerian eschatology – which permits the return to the sources of thought, from where the force of future thought must rise – are not fundamentally opposed. Both historicize the Absolute, the one under the name of Reason, and the other under that of Being, conceiving the movement of history as eschatologically determined by its march toward the end. In other words, both proceed to a reconstruction of history. They start by casting a critical glance at the present epoch, and they imagine the past as the "prehistory" of humanity, which at the same time consecrates, in their eyes, the Occident's privilege – both a distinguished privilege and fatal fortune – and, at the very least, the eminent mark of its historial essence.

For Husserl, as for Heidegger, fundamental history is, in the end, the secret history of the Occident; for the former, it is the history of absolute Reason; with the latter, the destinal history of Being. The one saw in actual history the expression of the weakness of man and of the responsible thinkers who have missed the true beginning and have thus failed in their teleological task. The oblivion of the world of life was the ultimate stage of that failure, and it was imputable to the philosophizing subject no matter how motivated he might have been by the intention of rational radicality. The crisis of European humanity was thus derived; it was not accessory or transitory; the crisis had arisen out of the history of modern times. And if it finally reveals itself as being salutary to us, it is to the extent that it is the condition of radical understanding that humanity must take up its mission, namely, to be the promoter of universal Reason. With Heidegger, on the contrary, the historical destiny of humanity was only the revelation of the destiny of Being in its dialectical movement of unveiling its own truth. It seems as if the historial thought of Heidegger was, in the last analysis, the expression of the unconscious heritage of Christian eschatology which the thinker had drawn from its origins in Western ontotheology and which united ancient Greek thought of Being to the thought of the history of the Judeo-Christian tradition.[87] And as with this, there is, in Heidegger, a reversal of the long thought of meaning of history – which still exists with Husserl – under the sign of progress, into a decadent history which is advancing toward its close where the threat reaches its culminating point; and this point constitutes the contemporary crisis which is the ultimate stage of the eschatology of Being and the path to salvation.

We have thus arrived at the last convergence – the task which each assigns to future thought – where there appears an abyss between Husserl and Heidegger. Against wind and tide, the one remained ultimately rationalist and humanist and did not cease to affirm his faith in man and in his rational destiny, in the humanistic mission of philosophical thought; while the other did not think that salvation could come to man by means of his conquering and domineering Reason, nor from science and technique, invented by Reason to bring to men (even if this be despite themselves) happiness. Heidegger does not, however, deserve the criticism that has often been made against him as the apostle of a radical antihumanism, made because he thought he had to denounce, in the humanistic doctrines of Europe, the metaphysical connivance to which he imputed their failure and their impotence to save humanity from the mortal malady which had enveloped it. The time has come when one must renounce overestimating both philosophy and science. It belongs to future thought to return to what it was at its birth among the Greeks by redescending "into the poverty of its provisional essence" and by understanding at last that one would not again give meaning to the destiny of humanity by sacrificing to the illusionary myth of man as "master of being," which, in spite of appearances, underestimates the humanity of man; instead, philosophers must teach man the simple and humble virtues of the "shepherd of Being," all the dignity of which lies in his vocation of safeguarding Being in all its truth.[88]

NOTES

1 Martin Heidegger, *Was heisst Denken?*, p. 39, fr. Trans. p. 72.

2 E. Husserl, 'Philosophie als strenge Wissenschaft', in *Logos* I, p. 292, French trans. p. 56.

3 " ... geringfügige Nuancen, die philosophische Wege und Abwege entscheidend bestimmen", declared Husserl. 'Cartesianische Meditationen', § 14, in *Husserliana*, vol I, pp. 70/1, fr. trans. p. 27.

4 M. Heidegger, *Vorträge u. Aufsätze*, p. 163 sq.

[5] Op. cit. pp. 325/5 sq., French trans. p. 103 sq.
[6] "We have not become philosophers by means of the philosophies" he wrote. "It is not necessary that the philosophic impulse arises out of the philosophies but of things and of problems", he added, stressing this sentence. Ibid., p. 340, fr. trans. p. 124.
[7] *Ibid.*,
[8] Vol I: 'Kritische Ideengeschichte', in *Husserliana*, vol VII.
[9] On History as the universal horizon of the *Selbstbesinnung* of the philosopher, cf. also 'Die Krisis der europäischen Wissenschaften', in *Husserliana*, vol. VI, passim, especially p. 488, or 'Vom Ursprung der Geometrie', ibid. p. 378, and finally 'Zur Phänomenologie der Intersubjektivität', III, in *Husserliana*, vol. XV, p. 393 sq.
[10] 'Vom Ursprung der Geometrie', loc. cit., pp. 381 & 379/80, fr. trans. pp. 205 & 201 sq.
[11] On the historicity of the cultural forms, cf. op. cit., p. 382, fr. trans. 206/7, also *Krisis*, loc. cit., p. 504 sq.
[12] 'Erste Philosophie', II, in *Husserliana*, vol. VIII, p. 15, our trans. p. 20.
[13] *Ibid.*, p. 196.
[14] "Funktionär der Gemeinschaft", wrote Husserl in 1924 (*Ibid.*, p. 197) "Funktionäre der Menschheit", he specified in 1935 (*Krisis*, loc. cit. pp. 15 & 72).
[15] "Es wird sich offenbaren, dass der *a-historische Husserl nur zeitweise Distanz von* der Historie nehmen musste (die er doch stets im Blick hatte), gerade um in der Methode so weit kommen zu können, an sie wissenschaftliche Fragen zu stellen", one reads, Letter of 27 November 193 quoted by Iso KERN, in *Husserliana*, vol. Xv, p. XXXV.
[16] *Cf.* ex. *Krisis*, loc. cit., p. 72 & p. 488 sq.
[17] 'Erste Philosophie" vol. II, loc. cit., p. 196.
[18] M. Heidegger, 'Die onto-theo-logische Verfassung der Metaphysik', in *Identität und Differenz*, p. 39 sq., fr. trans. in *Questions* I p. 281 sq.
[19] *Ibid.*, p. 44, fr. trans., loc. cit. p. 283.
[20] *Ibid.*, p. 45, fr. trans., loc. cit. p. 284.
[21] Op. cit., § 6, & p. 2, fr. trans. p. 16.
[22] *Ibid.*, p. 21, fr. trans. p. 37.
[23] *Cf. Holzwege*, p. 301, fr. trans. p. 266.
[24] *Sein und Zeit* § 6, especially p. 22, French trans. p. 39.
[25] 'Was ist das – die Philosophie'? p. 33, French trans. in *Questions* II, p. 29.
[26] *Ibid.*
[27] *Ibid.*, p. 34, French trans., loc. cit., p. 29.
[28] 'Hegel und die Griechen', in *Wegmarken*, p. 256, fr. trans. in *Questions* II, p. 46.
[29] 'Was ist das – die Philosophie'? pp. 14/5, fr. trans., loc. cit. p. 16.
[30] 'Hegel und die Griechen', loc. cit. p. 256, fr. trans., loc. cit. p. 46.
[31] 'Was ist das – die Philosophie'? p. 31, fr. trans.; loc. cit. pp. 27/8.
[32] *Über den Humanismus*, bilingual ed., pp. 88/9, and *Einführung in die Metaphysik*, p. 33, fr. trans. p. 54 sq.
[33] *Einführung in die Metaphysik*, pp. 33/4, fr. trans. p. 55.
[34] *Ibid.*
[35] On the meaning of the "archéo-logie", cf 'Vom Wesen und Begriff der physis,' in *Wegmarken*, p. 317 sq., fr. trans. in *Questions* II, p. 190 sq., and 'Was ist das –die Philosophie'?, p. 39, fr. trans., loc. cit., p. 32.
[36] 'Erste Philosophie' II, p. 23, our trans. p. 31.

37 *Ibid.*, vol. I, p. 205.
38 *Ibid.*, p. 14, our trans. p. 19.
39 *Ibid.*, p. 16, our trans. p. 22.
40 *Ibid.*, pp. 205/6.
41 *Ibid.*, p. 15, our trans. p. 20.
42 *Ibid.*, p. 204.
43 *Krisis*, loc cit. p. 15 & p. 72.
44 'Erste Philosophie' I, p. 16, our trans. pp. 21/22.
45 *Ibid.*, p. 207.
46 *Cf. Krisis*, loc. cit., p. 13/4 & pp. 273/4.
47 *Ibid.*, p. 74 & pp. 114/5.
48 *Ibid.*, p. 273, French translation P. Ricoeur, in *Deucalion* III, pp. 123/4.
49 'Heroismus der Vernunft'. Cf. 'Die Krisis des europäischen Menschen tums und die Philosophie' in *Husserliana*, vol. VI, pp. 347/8, fr. trans. P. Ricoeur, in *Revue de metaphysique et de morale*, 1950, pp. 257/8.
50 *Krisis*, loc. cit., p. 14.
51 M. Heidegger, 'Was ist das – die Philosophie'? p. 15 sq., fr. trans., loc. cit., pp. 16–18.
52 *Ibid.*, p. 13, fr. trans., loc. cit., p. 15.
53 *Holzwege*, p. 310, fr. trans., p. 274 sq.
54 *Ibid.*, p. 300, fr. trans. p. 266 sq.
55 "Die Philosophie ... ist nicht nur etwas Rationales, sondern die eigentliche Verwalterin der Ratio". 'Was ist das – die Philosophie'? p. 9 sq., fr. trans., loc. cit., p. 12 sq.
56 *Über den Humanismus*, pp. 172/3 & *Nietzsche* II, p. 343, fr. trans. pp. 275/6.
57 *Einführung in der Metaphysik*, p. 45, fr. trans. p. 70.
58 On this interpretation of the history of the Metaphysics, abridged cf. ex. *Einführung in die Metaphysik*, ch. III & IV.
59 *Vorträge und Aufsätze*, p. 77, fr. trans. pp. 88/9.
60 *Ibid.*, pp. 80/1, fr. trans. p. 92 & *Uber den Humanismus*, pp. 104/5.
61 'Nietzsches Wort "Gott ist tot"', in *Holzwege*, p. 247, fr. trans. p. 219.
62 E. Husserl, *Krisis*, p. 41 sq.
63 *Ibid.* p. 1.
64 *Cf. Krisis*, passim, especially Part II.
65 *Ibid.*, pp. 73 & 172 & also p. 336 sq. & p. 443.
66 *Ibid.*, pp. 72/3.
67 Cf. the original preserved at the Husserl Archives, catalogued under E III 2, p. 66.
68 *Cf.* The original preserved at the Husserl Archives, catalogued under E III 4, p. 57 sq.
69 M. Heidegger, 'Die Überwindung der Metaphysik,' in *Vorträge und Aufsätze.*
70 Cf. 'Was ist Metaphysik'?, p. 10 sq., fr. trans. in *Questions* I p. 29 sq., also *Über den Humanismus*, pp. 100/1 & 50/1 sq.
71 'Zur Seinfrage', in *Wegmarken*, p. 243, fr. trans. in *Questions* I p. 238.
72 *Über den Humanismus*, pp. 81/2.
73 *Einführung in die Metaphysik*, p. 28, fr. trans. p. 48.
74 *Cf.* 'Was ist Metaphysik'? Nachwort, p. 43 sq., fr. trans., loc cit. p. 79 sq., also *ibid*, p. 30 sq.
75 *Nietzsche* II, p. 489, fr. trans. p. 398.
76 *Der Satz vom Grund*, p. 159.
77 *Ibid.*, p. 157.

[78] *Ibid.*, p. 109.
[79] *Holzwege*, p. 310, fr. trans. p. 274.
[80] On the idea of errantry, cf. *Vom Wesen der Wahrheit*, VII, also 'Der Spruch des Anaximander', in *Holzwege*, p. 310, fr. trans. p. 274 sq.
[81] *Ibid.*
[82] *Ibid.*, p. 311, fr. trans. p. 275.
[83] *Ibid.*, pp. 301/2, fr. trans. pp. 266/7.
[84] *Ibid.*
[85] *Cf.* 'Hölderlins Erde und Himmel,' in *Hölderlin-Jahrbuch* XI (1958–60) p. 38.
[86] *Cf.* '*Sur Seinsfrage*', in *Wegmarken*, p. 242 sq., French trans., loc. cit. p. 236, also 'Die Überwindung der Metaphysik', in *Vorträge u. Aufsätze*.
[87] Cf. K. Loewith, *Heidegger. Denker in dürftiger Zeit*, ch. II & III.
[88] "Der Mensch ist nicht der Herr des Seienden. Der Mensch ist der Hirt des Seins", wrote Heidegger in *Letter on humanism*, loc. cit., pp. 108/9.

CLOSURE

PAUL RICOEUR

CONCLUSION AREZZO

Voici comment j'ai compris ma tâche de conclusion. Renonçant d'une part à composer l'exposé synoptique qui rendrait justice á chacun et à tous, j'ai conçu ma tâche comme essentiellement problématique. Par là j'entends l'effort pour faire apparaître comme problème ce qui semble se donner comme solution sous le titre de téléologie. Il me semble en effet que, derrière les usages en apparence dispersés du mot lui-même de téléologie, donc derrière l'unité thématique éventuelle du concept, c'est une même difficulté, une même aporie, un même embarras, qui se creuse, se relance et sans cesse s'aggrave. A vrai dire constituer le problème en problème, c'est constituer l'unité thématique des divers emplois du terme de téléologie chez Husserl.

A première vue, en effet, le concept de téléologie semble avoir deux usages majeurs sans liaison apparente. D'une part, dans la constitution de la chose, d'autre part, dans la théorie de l'histoire. Entre les deux s'étend la région de transition de l'intersubjectivité. Aussi bien, l'ordre des matières de notre congrès a-t-il suivi cette progression. Mais si l'on considère seulement les lieux d'emploi du concept de téléologie, l'unité n'apparait pas du tout. Ces lieux sont multiples. Cette situation initiale du problème est en elle-même surprenante. Dans le passé du concept de téléologie, il n'en a pas été ainsi. Chez Aristote, aussi bien que chez Kant, le problème de la téléologie a un lieu unique: la nature, et, dans la nature, un lieu plus déterminé encore: la nature organique. Même si chez Aristote toute la nature est à penser selon la cause finale, c'est par excellence le vivant, l'être naturel animé, qui présente un *télos*. Sa fin est sa forme, le terme de sa croissance, la configuration en vue de quoi sont ses parties, ses organes, ses stades, etc. Chez Kant, dans *la Critique de la Faculté de juger*, le jugement téléologique s'applique à cette partie de la nature qui, outre la légalité commune conférée par la causalité, présente un ordre d'une autre sorte, parfaitement contingent par rapport à la causalité, celui de la différence des espèces et de leur stabilité. Cet ordre se donne à penser comme subordination de la causalité linéaire à la relation réciproque entre parties et tout. Le concept de téléologie coincide alors avec celui d'organisation. Certes les différences sont énormes entre Kant et Aristote: non seulement la finalité interne n'est pas la cause finale, mais le caractère purement réflexif et heuristique du jugement téléologique kantien s'oppose au réalisme

Tymieniecka (ed.), Analecta Husserliana, Vol. IX, 415–426.

de la cause finale aristotélicienne. Néanmoins le lieu de la téléologie est commun. La première surprise que réserve Husserl c'est de ne pas concevoir de lieu privilégié de la téléologie. Bien plus, le lieu classique, l'organisation de la vie, est pratiquement vide. Il a été montré ici même que Husserl n'a pas réfléchi spécialement sur la biologie, tant il met l'accent sur l'unité de la science galiléenne et post-galiléenne et sur l'unité du processus de mathématisation et d'objectivation de la nature (*Ideen* II, *Philosophie Première*, *Crise*, etc.). Rien donc n'est à tirer, chez Husserl, de la différence entre Darwin et Newton, entre Lamarck et Galilée.

Ce résultat négatif n'est pas sans importance; nulle *Naturphilosophie* ne surgit dans le sillage de la téléologie. Du même coup, la téléologie se déplace vers les deux pôles de la "chose" et de l'"histoire", avec un transit éventuel par l'"intersubjectivité" qui reste à comprendre. D'un côté, le fil directeur de l'unité de la chose dans le divers des perspectives, des profils, des apparitions, – de l'autre, le fil directeur de la raison philosophique, éclairée par la philosophie transcendantale, dans le divers des événements, des conflits d'écoles, des avancées et des reculs. N'y a-t-il pas alors deux concepts de téléologie, l'un appartenant à la constitution de la chose, l'autre à la région des sciences de l'esprit?

M'appuyant sur les travaux de ce colloque, je voudrais faire apparaître l'analogie des deux emplois du terme de téléologie et fonder cette analogie sur l'unité d'une problématique sousjacente à laquelle j'essaierai tout à l'heure de donner un autre nom que téléologie.

Reprenons la *Ding Konstitution*. Pourquoi parler ici de téléologie? Tout simplement parce qu'il est parfaitement concevable qu'il n'y ait pas du tout ce que nous appelons une chose, c'est à dire une chose une. Or c'est précisément la phénoménologie transcendantale qui atteint ce niveau critique du problème. Dans l'attitude naturelle, il va de soi qu'il y a des choses qu'on identifie et réidentifie, auxquelles on donne un nom, sujet logique des prédicats variables. A cet regard, il y a une grande parenté entre la démarche de Husserl et celle de Strawson dans *Individuals*. On peut parler sur les choses et d'abord les reconnaître comme les mêmes. Or l'*épochè* suspend l'unité donnée, naturelle, et fait apparaître comme problème ce qui d'abord va de soi. On l'a dit ici avec beaucoup de force, la constitution commence par la dé-construction et par le dé-ployement des couches sédimentées. Or, ce travail de la phénoménologie génétique en direction de synthèses toujours plus cachées est un travail sans fin qui, à la limite, menace l'idée même de donnée intuitive, laquelle, de dépli en dépli, se recueille dans la fine pointe du présent vivant. Si la téléologie intervient dans la *Ding Konstitution*, c'est parce que l'unité de la chose ne peut être trouvée dans les données elles-mêmes. La situation est paradoxale, – mais nous n'avons pas fini de nous étonner – car l'unité de la chose ne lui est pas surajoutée, mais constitute la chose même. Bien plus, comme Denise Souche l'a montré dans sa thèse, l'intentionalité elle-même n'a son sens fort que quand la visée est visée d'une unité de sens. L'intentionalité serait une trivialité, si elle signifiait seulement conscience de . . . n'importe quoi. L'intentionalité au sens fort, c'est la visée d'un sens un. Le principe de "l'absence de borne de la rationalité" ne signifie pas autre chose que ceci: viser... c'est viser *le même*. C'est pourquoi la phénoménologie ne peut pas commencer avec l'originairement donné: peut-être même celui-ci est-il introuvable. Elle commence en prenant pour fil conducteur l'unité de la chose à constituer. C'est alors que le paradoxe éclate. Car la méthode phénoménologique, en tant qu'elle déconstruit l'objet en profils, en esquisses, en apparences, mine par dessous son propre thème. Elle défait ce qu'elle se donne à faire. La connaissance des transcendances vise une adéquation que par principe elle ne peut atteindre. Le symbole de cette possibilité abysmale est à chercher dans le fameux paragraphe 49 de *Ideen* I: l'hypothèse de la destruction du monde, c'est-à-dire la possibilité de la discordance absolue. Alors, pas de chose, pas de discours, – même pas de perception (au sens de la reconnaissance des même choses). Bref nous découvrons la contigence de l'ordre.

Cette situation n'est pas si différente de celle de la *Critique de la Faculté de Juger*, bien qu'on ne se concentre pas sur la différence entre l'organisation de la vie, dans sa diversité et sa stabilité, et la causalité universelle, monotone et indifférenciée. Chez Kant, en effet, la question de la téléologie procède de la rencontre de la contingence et de l'ordre. Selon la première *Critique*, il était nécessaire qu'il y eût la chute des corps et la gravitation universelle, non qu'il y eût des fleurs et des oiseaux. La téléologie est la rencontre de la question: pourquoi y a-t-il des espèces vivantes plutôt que simplement le mécanisme universel? Avec la *Ding Konstitution* de Husserl, la question est celle-ci: pourquoi y a-t-il des unités de sens, plutôt que la discordance absolue? C'est cette question qui, nous allons le voir, revient analogiquement à l'autre extrémité de la hiérarchie des étants, avec l'histoire de la philosophie dans le milieu de l'histoire des événements, des hommes et des idées. Mais avant de montrer la répétition analogique de la question, cherchons à la creuser comme question, bref à la problématiser à l'extrême.

Husserl, en un sens, l'a fait partiellement. Mais nous allons voir que d'autres facteurs dans notre culture nous contraignent à la problématiser jusqu'au voisinage de la folie.

Comment Husserl l'a-t-il problématisée? En appelant idée le *télos* de la chose, Husserl a ouvert l'espace d'une critique de l'idée même de téléologie. En liant cette idée à l'hypothèse de la destruction du monde, il a préservé la contingence de l'ordre. Enfin en incluant les possibilités de discordance sous le concept d'inadéquation, il a caractérisé l'ordre lui-même comme fragile et menacé. C'est toujours en dépit de quelque catastrophe possible de l'être qu'un ordre se donne à dire. Chez Husserl lui-même l'hypothèse de la destruction du monde donne une idée en quelque sorte fantasmatique à toute une série d'analyses partielles et disparates de portée épistémologique bien délimitée. L'hypothèse joue alors le même rôle que celui du malin génie chez Descartes, qui rassemble dans la forme d'une hyperbole les raisons multiples et diverses de douter.

Parcourons quelques unes de ces analyses que l'hypothèse fantastique regroupe. Selon le perspectivisme bien connu de Husserl, toute perception est unilatérale. L'autre côté de la chose n'est pas donné, mais aperçu comme donné à l'autre, donc absent à la perception actuelle. Ce qu'on appelle percevoir, c'est donc composer de présence et absence. De plus l'acte d'observation, en tant qu'inépuisable, fait apparaître des potentiels d'apparition qui constituent les horizons externes et internes de l'objet; par là il faut entendre les prescriptions de description qui de proche en proche me porteraient, si je les suivais jusqu'au bout, à l'autre extrémité du monde, c'est-à-dire nulle part; ou bien ces prescriptions me porteraient à dévisager en détail la chose, comme on fait dans certaines pratiques yoga ou zen; portées à l'extrême ces prescriptions pourraient me distraire entièrement, me dépréoccuper absolument.

Cette double remarque révèle le caractère interprétatif de la perception. La perception est une *Auslegung*, une lecture infinie du réseau d'intersignifications à quoi l'*époché* m'a livré sans retour.

Enfin si j'intériorise toute l'analyse, comme Husserl le recommande, en passant de l'analyse intentionelle orientée objectivement, vers l'analyse intentionelle transversale, c'est-à-dire intra-temporelle, la situation devient plus étonnante encore: l'hypothèse de la destruction du monde signifie que l'intention, la rétention et la protention pourraient ne pas se recouvrir dans des unités de sens capables d'être reconnues, désignées, nommées, signifiées. Ce que Husserl problématise ainsi, c'est à la fois triple synthèse de Kant: parcourir-retenir-reconnaître, – et le premier chapître de la *Phénoménologie de l'Esprit* de Hegel, qui se demande comment l'unité du dire se produit elle-même dans le disparaître de l'apparaître.

C'est tout cela que l'hypothèse de la destruction du monde dramatise dans la supposition fantastique d'une radicale discordance. Cette supposition fantastique devient ainsi la vision radicalisée – parce qu'étendue à tout objet – de la contingence de l'ordre.

Mais si la téléologie procède de la rencontre de la contingence et de l'ordre, en quoi constitute-t-elle une réponse, donc une thèse philosophique? Ne pourrait-on pas dire que la téléologie est tout simplement l'affirmation tautologique de l'ordre? On vous demande: pourquoi y a-t-il de la concordance plutôt que de la discordance? Vous répondez: c'est un *factum rationis* que les apparences s'accordent en dépit de l'hypothèse inéliminable du désordre.

Je ne considère pas l'accusation de tautologie comme mortelle. Elle est plutôt le signe d'une certaine circularité, essentielle à toute investigation fondamentale. Il n'est pas impossible que le *logos* soit ingénérable et qu'il se précède lui-même. C'est ainsi que l'unité de la chose, prise pour fil conducteur, se construit en se déconstruisant, comme la merveille d'une signifiance qui aurait pu ne pas être. Le mot téléologie serait alors le terme emblématique de cette découverte.

Mais, pour que cet aveu de circularité ne se perde pas dans une nouvelle trivialité – du style: il y a de l'ordre! – il faut le faire travailler critiquement jusqu'au point où le premier sens de la téléologie – la téléologie de la *Ding Konstitution* – se convertira dans le second sens: la téléologie de la raison historique.

J'appelle ici travail critique une bataille serrée avec la prétention de la téléologie de se transformer, par coup de force, de question en réponse. En effet, nous n'en avons pas fini avec la question de savoir quel *logos* articule le *télos* de la chose une. Husserl lui-même nous donne des armes pour continuer au-delà de lui-même l'auto-critique de la raison qui prétend sceller l'unité de la chose. L'arme qu'il nous donne, c'est, on l'a dit ici de multiples façons, l'appellation de ce logos comme une idée au sens kantien, c'est-à-dire au sens de la limite d'une série convergente ou du terme asymptotique d'un procès d'approximation. L'introduction d'un tel conceptlimite dans une philosophie de la constitution ne devrait jamais cesser de nous étonner. Voilà un philosophe qui se met en route pour découvrir l'originaire et qui rencontre en cours de route l'inadéquation comme un trait, non pas accidentel, mais constitutif – constitutionnel, pourrait-on dire – de la connaissance primaire du monde, à savoir la perception.

Mais ce n'est pas tout, car on peut se demander ce que serait l'adéquation. Nous lisons et répétons: ce serait le remplissement intégral. Soit. Mais remplissement de quoi? De quel projet? C'est bien en interrogant l'adéquation elle-même comme ce qui n'est jamais atteint qu'on interroge aussi le *logos* du *télos* dans la téléologie.

Husserl a commencé cette critique interne, non seulement en appelant idée l'adéquation, mais en distinguant toujours plus radicalement la raison des phénomènes de ce que la science moderne tient pour la raison. En ce sens, Husserl continue le combat kantien et hégelien pour distinguer *Vernunft* et *Verstand*. Nos sciences ne se pensent pas elles-mêmes, parce qu'elles confondent rationalité et mathématisation. A cet égard la lutte entre transcendantalisme et objectivisme dans *Philosophie Première* et dans *Crise* portent à l'intérieur de la rationalité moderne la guerre intestine de la raison. La gigantomachie que Husserl déploie dans ces deux grands livres atteste que nous ne savons pas ce qu'est la raison. (Déjà *Ideen* I, IVème partie, proposait un concept de raison beaucoup plus proche de la raison hégelienne concrète que de la raison scientifique, dans l'unité du sens et de la présence). Mais Husserl est-il allé assez loin dans cette critique de la raison? On l'a dit ici, il appelle encore "science" son propre projet. Science des dernières légitimations. Science du commencement responsable de soi. Qu'on se rappelle le *Nachwort* à *Ideen* I.

Une telle prétention à s'auto-fonder soi-même de façon aussi principielle peut-elle rejoindre en quelque façon la question élémentaire et radicale que j'articulais à l'instant: la question de savoir quel *logos* articule le *télos* de la chose une? On peut se demander si Husserl est ici mieux placé que Kant pour répondre à deux difficultés considérables. Première difficulté: si l'on reporte dans le sujet le principe de l'ordre, ne se borne-t-on pas à intérioriser ce qu'on lit sur les choses, à savoir qu'elles sont identifiables, localisées dans l'espace, reliées causalement, etc.? Dire que la conscience transcendantale prescrit l'enchaînement que la conscience naive, naturelle, croit trouver dans les choses, c'est baptiser prescription par la conscience ce qui se lit sur les choses, et, du même coup, faire fonctionner de la façon la moins acceptable la tautologie de l'ordre qui se pose lui-même. Surtout, c'est donner à la conscience un pouvoir exorbitant, Prométhéen, qui, pour moi comme pour un des membres de ce congrès, constitue la plus fondamentale déclaration d'athéisme. Mais ce n'est pourtant pas cette première difficulté qui m'arrêtera mais la seconde. Elle me fournira l'idée centrale de cette méditation et la transition au deuxième concept de téléologie". La deuxième difficulté commune à Kant et à Husserl est celle-ci: si les apparences sont un divers, et si pourtant la chose est une, comment le divers s'articule-t-il dans l'unité? Comme on sait, Kant posait le problème dans le cadre d'une philosophie du jugement: comment, demandait-il, subordonner une intuition sous un concept? Réponse: grâce au schématisme qui participe des deux et qui, peut-être, est leur racine commune. Mais le schématisme lui-même, ou plutôt la double appartenance du schématisme à l'entendement et à la sensibilité, quelle en est l'origine?

Réponse: elle réside dans *l'affinité des phénomènes*, c'est-à-dire dans la disposition même des apparences à l'ordre. Il me semble que cette expression: *l'affinité des phénomènes* est l'index de ce qui se cache sous le concept de "téléologie", je veux dire du problème de la raison concrète qui a été à la fois visé et manqué sous le titre de la cause finale, de la finalité interne, puis de la téléologie transcendantale.

Pourquoi parler d'affinité des phénomènes plutôt que de téléologie? Précisément pour soustraire le problème de la téléologie à la juridiction d'une conscience prétendument souveraine. Ce qui, me semble-t-il, émerge dans la *Ding Konstitution*, mais qui, on le verra, émerge aussi dans la Philosophie de l'Histoire, c'est que nous appartenons à un ensemble de l'être qui a par disposition et par destination le pouvoir d'être dit, de se laisser dire par nous. Je reviendrai sur cette expression pour finir. Mais d'abord je voudrais montrer comment cette critique interne de l'idée de télélogie peut être étendue de la région de la chose à celle de l'histoire et d'abord à la région qui fait transition d'un pôle à l'autre, à savoir la téléologie de l'intersubjectivité.

J'ouvre ici une parenthèse. On pourrait en effet trouver bien d'autres usages de l'idée de téléologie que ceux mis en oeuvre par Husserl entre la *Ding Konstitution* et l'histoire. Je pense en particulier à la psychanalyse, qui repose fondamentalement sur l'hypothèse que la vie profonde des pulsions a une affinité pour le langage, plus précisément, que la sexualité humaine est une demande qui peut être relayée par la parole. L'affinité des phénomènes signifie à ce niveau que la vie s'interprète elle-même comme disait Dilthey. Entre *bios* et *logos* il y a affinité. C'est pourquoi la psychanalyse est possible comme *talk-cure*, comme procès de resymbolisation appliquée à cette part de désymbolisation de nous-mêmes. Je referme ici cette parenthèse. J'ai voulu seulement marquer que la problématique téléologique, approfondie par celle de l'affinité des phénomènes, n'est pas enclose seulement dans les deux extrêmes considérés par Husserl: la chose et l'histoire. Entre les deux, il y a la vie et la psyché. Et l'affinité entre le *pathos* et le *mathos* est une des plus profondes qu'une investigation conduite par l'idée de téléologie puisse concevoir.

Mais la manière dont Husserl fait la transition par l'intersubjectivité ne manque pas d'intérêt. Plusieurs exposés lui ont été ici consacrés, à la charnière précisément de l'épistémologie et de la philosophie de l'histoire. Le problème de l'intersubjectivité est bien, pour Husserl, *le* problème de la transition, en ceci qu'il appartient aux deux problématiques. D'un côté, l'intersubjectivité appartient à la constitution de la chose, comme on voit dans *Ideen* II et dans les *Méditations Cartésiennes*, pour autant que la chose est chose en commun.

L'idée de chose en commun est une idée téléologique, en ce sens que personne ne voit la chose en commun. La chose en commun est le géométral, dirait Leibniz, de toutes les perspectives. Mais ce géométral est simplement présumé, en ce sens que chaque conscience ne peut que "apprésenter" la perspective d'autrui et même doit "apprésenter" sa propre perspective en tant qu'"apprésentée" dans la perspective d'autrui. C'est dans ce jeu d'"apprésentations" croisées, que se constitute la nature commune. Mais d'un autre côté l'intersubjectivité appartient à la constitution de l'histoire en ce sens que l'idée même d'une nature commune est un acquis de la culture dans laquelle la nature est interprétée. L'idée de nature en commun renvoie donc à celle d'une culture en commun. Or celle-ci ne peut être alléguée qu'en un sens tout à fait original, qui inclut des procédures complexes de symbolisation, des traditions sédimentées, dont certaines sont vivantes, d'autres mortes, d'autres en voie d'être réactivées ou renouvelées. Pensez seulement que la nature pour un paysan n'est pas la nature pour un citadin et qu'il n'y avait pas de paysage avant la littérature et la peinture.

C'est ici que la téléologie commence à déployer, d'une manière non plus occulte mais manifeste, les traits problématiques que nous avons à grand peine fait apparaître sur la simple *Ding Konstitution*. Les conditions d'un monde culturel en commun sont d'une incroyable complexité. La discordance n'est plus une possibilité théorique, qui demande pour être conçue le secours d'une hypothèse aussi fantastique que la destruction du monde. Elle est donnée tous les jours comme conflit d'opinions, compétition de points de vue, heurt de traditions. Et quand règne une visée en commun, elle est, pour une part, l'oeuvre d'une tradition vivante qui nous gouverne sans que nous la maîtrisions intellectuellement; mais, pour une part aussi, elle est le précipité et le résidu de préjugés communs. Si on appelle esprit cet être en commun, il faut avouer qu'il conduit les hommes par le rude écolage d'une aliénation qui les rend d'abord étrangers à leur première culture naive. Cela nous l'avons appris au début du fameux chapitre VI de la *Phénoménologie de l'Esprit* de Hegel intitulé précisément *Geist*. C'est donc par la discordance que passe la vérité du Soi. On peut évoquer un autre passage hégelien: seul le scepticisme peut défaire l'identité prématurée de la conscience pensante qui fait dire que le stoique esclave est égal au stoique empereur, qu'Epictète et Marc-Aurèle le même. Voilà bien un exemple troublant de téléologie trompeuse, pour laquelle la pensée bourgeoise a une particulière propension, toutes les fois qu'elle identifie les hommes dans le formel de libertés trop abstraites. On l'a dit ici même, la convergence des consciences passe autant par le conflit que par la sympathie. Mais alors il faut reporter très loin l'idée de paix perpétuelle, l'idée d'une humanité réconciliée. Et cette idée, comme tout à l'heure celle de l'identité de la chose, ne vaut qu'autant que sa puissance critique demeure inentamée, et qu'elle continue de subvertir les totalités et les totalisations qui prétendent exhiber et réaliser l'idée de l'humanité une. Le beau texte de Merleau-Ponty évoqué hier prend tout son relief: "l'histoire est impuissante à rien achever sans des consciences qui la reprennent et par là en décident". Et plus encore l'affirmation de Stephan Strasser, selon laquelle est historique une situation où quelque chose peut réussir ou avorter . . .

C'est sur ce fond d'une téléologie précaire, menacée, que peut être comprise, selon moi, la téléologie husserlienne de l'histoire. On en a souligné les trois traits principaux: premièrement l'histoire est mue du dedans par le *télos* de l'idée d'humanité qui fait de cette histoire une approximation indéfinie de son sens immanent; deuxièmement le *télos* de l'idée d'humanité a son logos dans la raison, en tant qu'unité profonde de la raison théorique et de la raison pratique; troisièmement la raison elle-même a son sens éminent, *archontique*, dans la philosophie européenne issue des Grecs et culminant dans la philosophie transcendantale. Husserl résume tous ces traits dans cette formule d'une extrême densité: "la *ratio* se trouve dans un procès continuel d'éclaircissement pour elle-même à partir de la première irruption de la philosophie dans l'humanité". Cette manière de penser l'idée d'histoire sous celle d'humanité, l'idée d'humanité sous celle de raison, et la raison sous celle de philosophie transcendantale, ne met pas en jeu une problématique différente de celle qui s'est fait jour à l'autre extrémité de l'échelle des êtres. Ce qui fait tenir ensemble les choses est ce qui fait tenir ensemble les hommes. Et ce lien n'est ni plus ni moins étonnant lorsqu'il règne entre les hommes que lorsqu'il règne entre les choses. Sa dialectique interne est seulement devenue manifeste: au lieu de s'écraser sur la chose, elle se déploie sur les hommes. Dire en effet que le *télos* de l'humanité s'éclaire dans la philosophie, et que la philosophie n'est pas autre chose que le rationalisme et le rationalisme pas autre chose que le transcendantalisme, c'est reprendre à un niveau beaucoup plus complexe la même problématique que celle que j'ai essayé d'éclairer par la question de l'affinité des phénomènes.

Pour ce qui concerne specifiquement la notion de téléologie en *histoire*, je me bornerai à trois remarques. Premièrement, tout *télos* que nous pourrions enfermer dans un système dont nous aurions le contrôle, qu'il soit théorique ou pratique, est à refuser. Moins que jamais l'unité qui nous précède et nous meut – "le dialogue que nous sommes" – selon le mot du poète – n'est une structure assignable par l'économique ou la politique, – fût-elle une politique de la culture. Ce n'est pas à vrai dire une unité que nous pouvons produire, dont nous pouvons disposer ou que nous pouvons fomenter. Elle nous instaure et nous mène sans que nous la sachions. C'est en cela qu'elle est idée. Cette unité dans quoi nous sommes peut certes être appelée raison, mais aussi longtemps que la raison ne s'épuise pas dans la science et contient le germe d'une critique de tous les rationalismes connus, y compris celui de la phénoménologie transcendantale. Husserl en dit quelque chose lorsqu'il parle de cette lutte incessante de la raison éveillée pour accéder à la compréhension d'elle-même. L'histoire transcendantale nous manoeuvre plus que nous ne la projettons et ne la régissons.

Deuxièmement, cette histoire compose, plus manifestement que la constitution de la chose, l'ordre et la contingence. S'il appartient à la convergence des apparences de pouvoir se démentir, à combien plus forte raison la convergence des cultures entre elles et dans la figure de la philosophie est fragile. La philosophie est plutôt cette voix à peine audible, dont parle Freud à la fin de *Malaise dans la Civilisation*, mais dont nous attendons que finalement elle l'emporte, parce que déjà elle nous régit et nous met en mouvement. Nulle expérience ne le confirme, nulle vérification empirique ne l'atteste. L'aspiration à la clarté repose seulement sur la certitude socratique qu'une vie non examinée n'est pas digne d'être vécue et que la vie, dès qu'elle est humaine, a de l'affinité pour ce que Socrate appelle examen, *Zètèsis*. Encore une fois, le logos se précède lui-même, mais sans évidence triomphante, comme ce qui se produit historiquement parce qu'il s'anticipe lui-même. Et cette affinité, cette disposition pour l'examen, n'est pas une hétéronomie qui nuirait à notre autonomie, mais l'orient de notre désir. La possibilité du revirement, virtuellement contenue dans l'idée d'une simple chose, devient, à l'échelle de l'histoire, la possibilité de l'échec. Ainsi la problématique n'a pas changé; elle est seulement devenue consciente d'elle-même. Cela veut dire que la contingence de l'ordre, en passant de la chose à l'histoire, se réfléchit désormais dans l'angoisse d'une catastrophe à laquelle la menace atomique donne aujourd'hui un contour plausible. La possibilité que le procès constitutif librement déclenché par la raison soit *en vain* est aujourd'hui une possibilité réelle.

Troisième remarque: affirmer la téléologie, dès lors, n'est pas s'aveugler sur le danger des temps. Ce n'est pas non plus faire fond sur quelque ruse de la raison qui, a-t-on dit ici, nous rappelle trop celle des tyrans modernes. C'est, me semble-t-il, s'attendre à ce que l'humanité ne sera jamais privée d'une voix, d'une parole, d'une sentence, qui aidera encore quelques uns à se comprendre eux-mêmes jusque dans l'ultime déréliction. Je dirai: il y aura toujours un témoin pour dire vrai. Car le fond des choses, à partir de quoi nous sommes, a de l'affinité pour la parole vraie. En cela, l'affirmation de la téléologie au niveau de l'histoire prend la forme d'un pari pour la raison, d'un vouloir pour le sens, en dépit du nihilisme menaçant. C'est ainsi que je comprends que la philosophie puisse se considérer comme sens éminent et archontique.

En conclusion je dirai que le concept husserlien de la téléologie peut être maintenu comme téléologie universelle, c'est-à-dire englobant la constitution de la chose, l'intersubjectivité et l'histoire transcendantale, à condition qu'à son tour ce concept s'approfondisse et se soumette au même genre de critique que l'idée même de raison avec laquelle elle s'identifie. Mais alors la constitution n'est plus un procès dont nous aurions, en tant que sujet transcendantal, le commandement. C'est un procès dont nous avons seulement la garde. Penser cela, c'est penser avec Husserl et contre Husserl.

Je me suis risqué à introduire l'expression kantienne "d'affinité des phénomènes" pour désigner ce que l'idée de téléologie donne à penser et qui en même temps en amorce la critique interne. Le terme d'affinité des phénomènes est à son tour l'index d'une aporie et d'une conviction qui ne cesse de revenir à travers l'histoire de la philosophie. Au début de la philosophie occidentale, nous écoutons que "l'être et le penser sont le même" (ou que "le même est à être et à penser"). Pour pouvoir commencer de parler et même de percevoir, il faut admettre qu'entre l'être, au sens de l'ensemble de ce qu'il y a, et le discours de l'être dans ses catégories, il y a communauté. Même dans son retrait l'être n'a pu rester étranger à notre dire.

C'est ce que j'oserais appeler la merveille du sens, seule réponse au nihilisme porté au jour par Nietzsche. Je dis la merveille du sens, pour ne pas employer le mot de "miracle", dont les prêtres ont trop abusé pendant des millénaires. Par "merveille du sens", nous entendons encore une fois une disposition des choses à se laisser dire, une raison des phénomènes que nous ne pouvons articuler dans nos sciences, même dans la science philosophique.

En ce sens, le concept de téléologie doit lui même se laisser critiquer par celui d'affinité des phénomènes. Le concept de téléologie évoque encore trop la causalité d'un projet dont l'homme serait le maître ou dont il aurait le secret. Je n'entreprendrai pas ici cette critique de la téléologie par elle-même sous le signe de l'affinité des phénomènes. Je laisse seulement entrevoir que cette critique aurait beaucoup à apprendre de Nietzsche, aussi bien que de la dialectique négative d'un Horkheimer ou d'un Adorno. Toutes nos idées d'unité, d'ordre, sont trop courtes pour mesurer la disposition de l'être au logos. Dans les idées d'unité et d'ordre, qui sont les nôtres, il se cache encore trop de ressentiment contre les différences, trop de prétention à la domination, trop d'accoutumance à la loi d'équivalence qui règle l'échange des sociétés marchandes. Pour accéder à l'authentique disposition de l'être pour le discours, il faut peut-être oser affronter jusqu'à la notion d'une "anti-téléologie", pour autant que téléologie exprime *notre* idée de l'ordre, de l'unité, de l'identité. Par "anti-téléologie", j'entends ceci: qui l'être dans quoi nous pensons et parlons se tient en retrait par rapport à tout ce que nous désignons et maîtrisons par l'idée d'un but, d'une fin, d'un projet, – et nous fait signe dans des expériences qui contredisent à nos idées parcimonieuses d'ordre, d'économie, d'équivalence, je veux dire dans des expériences telles que le jeu, le gaspillage et la surabondance.

Husserl n'aurait certainement pas voulu aller jusque là. Mais penser dans cette direction n'est pas encore dire adieu à Husserl. Je pense plutôt que seule la critique interne de l'idée de téléologie, pour autant qu'elle est réglée par cela même que cette idée donne à penser, fait tenir ensemble les deux usages majeurs de la finalité chez Husserl, dans la *Ding Konstitution* et dans l'histoire transcendantale. Ce que ces deux usages ont en commun c'est finalement la problématisation de l'idée de téléologie par elle-même.

University of Paris (X) and Chicago

COMPLEMENTARY SECTION

PHENOMENOLOGY IN ITALY

ANTONIO BRANCAFORTE

A HISTORICAL NOTE ON THE PRESENCE OF BRENTANO IN SICILY AND ON THE FIRST LINKS OF ITALIAN CULTURE WITH THE PHENOMENOLOGY OF HUSSERL

I. INTRODUCTION

A history of late nineteenth- and early twentieth-century Italian philosophical thought is presently unavailable. Yet one is certainly needed, if only to correct the distorted view that has arisen as a consequence of the emphasis on the work of Croce and Gentile – before which all other philosophical studies have been made to appear secondary, or even negligible.

In order to redress this neglect, and thus provide a more complete account of Italian philosophy, and, in particular, of the place of Sicilian culture, I will begin with a few facts about G. Amato Pojero and the Philosophy Library in Palermo. I will then discuss the reception of Brentano's thought in Sicily in the early years of this century and how that philosophy, under the influence of the philosopher of Marienberg, led to the consideration of problems very similar to those from which Husserlian philosophy started. Finally we shall see that the fact that the decision to begin an Italian edition of the periodical *Logos* was formed in the Philosophy Library in Palermo, was not fortuitous.

II. NOTES ON GIUSEPPE AMATO POJERO AND ON THE PHILOSOPHY LIBRARY IN PALERMO

Giuseppe Amato Pojero was born in Palermo in 1863 of an upper-middle-class family and died there in 1940.[1] A contemporary of Husserl, he knew Brentano, to whom he was attached by strong bonds of friendship, though critical of certain aspects of his philosophy. Having gained his high school diploma in classical studies in 1880, Pojero enrolled for a degree course in mathematics and physics and graduated at Pisa in 1884. In 1886 he received a degree in jurisprudence from the University of Naples, and followed this, in 1888, with the study of medicine, which he continued from 1889 in Berlin. It was during this stay in Germany that he met Brentano.

An able business man, Pojero and his brother Michele had founded the Banca Siciliana, and Pojero remained on the board of directors until 1890. Heart trouble, however, forced him to interrupt his medical studies and to

Tymieniecka (ed.), Analecta Husserliana, Vol. IX, 429–439.

leave Berlin to return to Palermo, where he spent the remainder of his life. Illness profoundly affected the life and personality of "Dr. Amato," as his friends and acquaintances called him with affectionate veneration, and it forced him to turn in upon himself, to face decisively such philosophical problems as the problem of the meaning of existence, and to find compensation and balance in the search for something which could satisfy his religious, moral, and ascetic aspirations.

With adundant financial means at his disposal he began to gather round him some of the most famous men of his time, both Italians and foreigners, entering into correspondence with them and offering them hospitality, sometimes for long periods, at his home. Here, in his large library of more than ten thousand books, later donated to the University of Palermo, daily friendly conversations and animated discussions were held on the latest, most immediate problems of literary criticism, mathematics, physics, astronomy, political economy, and law seen in relation to one another and in relation to philosophy. Soon Dr. Amato became the moving spirit and central point of reference of an ever-growing group of thinkers and scholars who, in this atmosphere of sympathy and human solidarity, spontaneously formed a remarkable intellectual coterie.

Among the members of this coterie was formed something which the growing scientific and technological specialization was already beginning to make extremely difficult, namely, interdisciplinary contact not at a generic and superficial meeting, but specific and in depth, characterized by a common search for a fundamental and unifying principle of knowledge. Because of this search scientists, men of letters and philosophers came into contact with one another, integrating reciprocally the most diverse attitudes to problems which were established from time to time. Pojero managed to include within his coterie men like William James, the philosopher; J. Wagner, the discoverer of the therapy against malaria, and his rival Weigandt; the biotypologists Voronoff and Pende; the physicist Weil of Zürich; and P. Barth, G. Itelson, L. Rougier, and V. Lutosławski, among others. Naturally the most regular members were Italians, philosophers like Gentile, Croce, Guastella, Orestano, Aliotta, Carabellese, and Vailati; scientists like Lombroso, Bagnera, Corbino, and Giardina; lawyers such as V. E. Orlando, Impallomeni, Di Marzo, and Riccobono; and men of letters like Borgese, Cesareo, Zingarelli, Romagnoli, and Bignone.

One of the most important events for the coterie happened when Franz Brentano, who had moved to Florence in 1895, accepted an invitation to visit Sicily and pass the spring months of 1899 and 1900 at the Villa Amato.

These stays must have been particularly pleasant and evocative; for a time the Austro-German philosopher was even tempted to remain permanently on the island, together with his wife and his son Giovanni. It is certain, however, that the active participation of Brentano contributed to raise further the already high tone of the cultural meetings. The discussions became keener and more rigorous, but above all there was a more precise selection of the philosophical problems in the direction to which the two thinkers, whose views were undoubtedly different and in a certain sense complementary, but both equally disciplined by scientific habit, already leaned spontaneously.

As the coterie acquired a particular aspect, characterized by a certain methodological unity and a common concept of philosophy considered in itself and in its relationship with the sciences as well as in the variety of the systematic positions, its transformation into the Philosophy Library of Palermo began to mature and was officially constituted in 1910 "with the aim of promoting philosophical culture, in its widest sense, and with the purchase of books and periodicals of a philosophical nature to be kept at the disposal of the members, discussions, lectures, lessons, special courses, and publications."[2] In 1912 the *Annuario della Biblioteca Filosofica* became the group's official organ; in 1914 this journal merged with the periodical *Logos,* which had originated in Germany in the same year in which the Philosophy Library of Palermo was founded, and which was to have, at least by intention, an international character. Each issue of *Logos* included two or three German articles selected to be translated into other languages, while the remainder of the number was composed of other articles published in the language of the particular foreign edition concerned. In fact, French, English, Russian, and Italian editions were projected. It was, as can be seen, an international enterprise (although German contributions enjoyed a favored position: Germany could export two or three chosen articles to all the other nations while the other nations exported only to Germany).

The ideology implicit in this operation, apparently simple and disinterested, can be summed up in this proposition: the communion of peoples as a concrete realization of universality proceeds from Germany and is brought about in Germany. It is easy to see that it is only a short step from the idea of Germany as promoter and carrier of universality to pan Germanism, as Husserl was soon to realize, and Heidegger a little later. But in 1914 the periodical *Logos* seemed to correspond perfectly to the political-universalistic ideal which, in Husserl's mind, was destined to be expressed in a great Prussian Germany: that Germany to which Husserl still remained subjugated, if only for a short time, and because of which he had already found himself in disagreement with his master, Brentano, who had tried in vain, several times, to put him on his guard.[3] None of the editorial staff of *Logos* and none of the Italian and Sicilian philosophers, however, suspected that which is so easy to see now. Thus the cultural task was brought to a successful conclusion and the Philosophy Library of Palermo printed an Italian edition of *Logos*, following the examples of France, Russia, and England.

An Italian directive committee formed of B. Varisco and A. Bonucci was added to the German editorial committee composed of Mehlis, Eucken, Gierke, Meinecke, Rieckert, Simmel, Troeltsch, Weber, Windelband, and Husserl. This generous international collaboration lasted, however, only up to 1915, the year in which Italy entered the war, first against Austria, then against Germany.

In 1920 (following World War I), with the resumption of the Italian edition of *Logos* at the instigation of A. Aliotta (who became its director), the international character had become a simple question of collaboration between various Italian scholars and an American, an Englishman, and a Frenchman, and even this was gradually fading out. To this was to be added the fact that the National Socialist movements – which were about to debouch into Nazism and Fascism – were beginning to condition cultural life ever more heavily, suffocating any yearning for universality and restricting the horizons of knowledge. Husserl, in any case, had already founded his magazine of phenomenology in 1913, well before *Logos* died in Germany and was reborn, substantially nationalized, in Italy.

We must wait until 1923 before the contact between a certain Italian culture and the original center of Husserl's phenomenology, begun in Sicily but immediately interrupted, was reactivated. The credit for this reactivation goes to Antonio Banfi, a philosopher open to European cultural movements, who had known Husserl personally but who had also followed carefully (the fact is worth stressing) the ups and downs of the periodical *Logos*.[4] To be precise, however, it must be noted that Giovanni Calo had already begun to study Brentano and some themes of phenomenology in 1908 and that in the same year, 1923, Francesco De Sarlo published an essay about similar problems with many references to Brentano and Husserl.[5]

After the first writings of Banfi on Husserl there followed immediately studies by Grassi, Alliney, Filiasi Carcano, Paci, Preti, Bobbio, Stefanini, Vanni Rovighi, and Lazzarini, all of which contributed greatly toward breaking the dreariness of Italian culture between the 1920s and World War II. After the war interest in Husserl's phenomenology in Italy became widely diffused. But what it gained in widespread interest it may have lost in intensity: that intensity which comes from direct contact with the sources. I think, therefore, that we must wait for a "turning point" in Italian studies of Husserl: a "turning point" toward a more phenomenological interpretation of phenomenology.

III. PRESENCE OF BRENTANO AND TRACES OF HUSSERL IN SICILY

In 1899 Brentano arrived in Sicily, but his philosophy had preceded him and had spread with surprising rapidity – so much so that the Palermitano, G. Amato Pojero, during a visit to Germany in 1889, had wanted to meet the famous philosopher: a clear sign that Brentano not only represented a *Weltanschauung* or a method but was also a reference point for the many intellectuals who found themselves at that time, worried and bewildered, at the hub of events laden with religious and political significance; in particular, the disagreement within the Catholic world between traditionalists and modernists and the tension in the political world between conservative and extreme-progressive forces.

In these circumstances Brentano had shown a free, strong personality, intolerant of compromise. The fact that he represented in religion a lay-theism and in politics a liberalism of the center, as well as the prestige in which he was held because of his well-known personal vicissitudes had contributed to make his philosophical position more worthy of consideration in spite of the fact that it was so rich in nuances. In the face of idealism, positivism, and neo-Kantianism Brentano had, paradoxically, put forward again a new form of Aristotelianism, impregnated with comparative reflection and criticism of science and, above all, above suspicion of collusion with the ideological valencies of the emerging neo-Thomism. In complete naturalism he had claimed the preeminence of psychology over the natural sciences, basing it on the principle that that which is a phenomenon is never really real and that which is real is never really a phenomenon. Physical phenomena are never really real because they are always *signs* of something real and therefore produce mediated certainties. On the contrary, psychic facts, which are never completely objectified and therefore never really become phenomena, are real because they are signs of themselves and produce immediate certainties.

The knowledge of psychic acts in their completion as acts and in their showing themselves to themselves in this completion is, for Brentano, the model of certain knowledge. Brentano advocated, therefore, psychologism, but by coupling it with intentionality of consciousness thought to avoid subjectivism. The principle of "intentionality" would open and extend cognition:

the acts are directed toward the objects, the consciousness is always consciousness of something, or, better still, consciousness of tending toward something. Brentano thus supported Descartes but turned back to Aristotle, with the secret aim of neutralizing Hegel.

Fundamentally Husserl never questioned either the theoretical premises or the historical coordinates of the master, but he remained concerned with justifying in the most correct manner possible the autonomy of logic as regards psychology, while avoiding, Platonic objective idealism and Fichtian-type subjective idealism, empirical realism, and metaphysical realism through the phenomenological attitude. Only recently, moreover, has it become evident that Brentano's thought was much vaster and more articulate than is apparent in his published works, as little by little the project of editing the *Opera omnia di Brentano* reaches its conclusion.[6] It is surprising to see, therefore, that G. Amato Pojero was fully aware of this when he wrote in a letter addressed to Giovanni Vailati on 18 August 1899:

> In the first months of this year at least I enjoyed the company of Brentano, who acted like an antidote upon me, strengthening my resistance to boredom and worries.... Brentano set against my monadological views a system of his own (largely to be found in his unpublished works) to which he attributes much greater importance than to the psychological works to which he owes his fame.... Brentano rejects all monadology, and intends to construct a theory of matter which gives to atomism the philosophical bases it needs, especially in its present phase."[7]

This letter seems to me very important, since it signals the historic moment of the influence of the Pierceian pragmatist G. Vailati on Brentano's thought.

As Pojero not only knew Brentano personally and critically appraised his work, but also took considerable trouble to diffuse it among those who seemed to him most capable of understanding it, Vailati was one of the best qualified to receive this propagation of Brentanian studies. The fact that this choice was a most fortunate one is substantiated by the wealth of Vailati-Brentano papers. Brentano and Vailati, in fact, having met at Dr. Amato's house in Palermo, soon formed a close friendship and began to exchange ideas and opinions in writing.

While on this subject it is a great pleasure for me to be able to announce that the Vailati-Brentano papers have been increased by a letter, hitherto unpublished, given to me by Professor Maria Concetta Amato Comparato, Dr. Amato's widow: it is a particularly precious letter, because in it Brentano sums up in clear terms his ideas concerning the limits of analogy between the field of theoretical judgments and that of practical options.[8]

The results of these investigations, in effect, lead us to think that G.

Amato Pojero was undoubtedly among the most profound connoisseurs of the thought of Brentano in the early years of the twentieth century in Italy; that his knowledge of Brentano was highly critical; that this did not, however, impede him in his work of diffusion of Brentanian doctrine – as we can see, for example, from the letter of 20 October 1899 from Dr. Amato to G. Vailati:

I have read Marty's pamphlet ... which closely follows the ideas of my friend Brentano which are somewhat different from my own. This has given rise to interminable discussions between him and me ... I hope that you will be able to meet Brentano here as it is by no means impossible that he could come back ... Vailati can also be included among the most careful readers and critics of Brentano in Italy: he was one of the first to write about him.[9]

These conclusions concerning Pojero's study of Brentano can be largely confirmed by reading the Vailati-Brentano and Amato Pojero-Brentano papers as well as that part of the Vailati-Amato Pojero papers which deal with Brentano; but the same impression is obtained from evidence furnished first by members of the coterie and later by associates of the Library, who were present at debates and conversations and who spoke and sometimes even wrote of them.[10]

As this point I should like to satisfy the curiosity of those who wish to know something of the points of divergence between, first, Amato Pojero and Brentano and, second, between Brentano and Vailati. For Vailati this is relatively easy since in addition to the evidence in the papers we have Vailati's *Scritti.* For Dr. Amato there is little available to document his relationship with Brentano, apart from the letters written to Brentano, now apparently included in the Brentano Archive. There are of course the three notebooks of *Pensieri* by Pojero published in 1968; but these are often fragmentary thoughts which do not permit us to make a complete and organic reconstruction of his philosophical vision and especially of his methodology and of his manner of demonstrating his thesis. We may assume, however, that Brentano must have found an extraordinarily interesting interlocutor in Dr. Amato, judging by the length of Brentano's letters (now in the Amato Pojero Archives).

Another scholar and critic of Brentano in Sicily was Leonardo Grassi of Catania, a lecturer at the university, and author of several works on science, literature, and philosophy, among which *L'unita dei fatti psichici fondamentali* is particularly noteworthy.[11] In this work Grassi provided a critical examination and exposition of Brentanian psychology, and observed that:

Brentano who, even though he starts from an empirical point of view, recognized a certain ideal intuition (*eine gewisse ideale Anschauung*) as perfectly reconcilable with this and believed that the principal source of psychology had to be internal perception (*innere Wahrnehmung*), to be carefully distinguished from internal observation (*innere Beobachtung*) which is absolutely useless because observation could only apply itself to exterior objects, states as an essential fundamental character of psychic phenomena that they are representative. It is universally admitted that every intellectual act supposes an object; but this character is not attributed, especially from Hamilton onward, to feeling and to sensibility in general. Now this for Brentano is a mistake: joy, hate, love are always given as relating themselves to something, the quality of intentional existence is the foundation of psychic activity: every psychic phenomenon refers to an object . . . .[12]

This passage gives an idea of the clarity and fidelity of Grassi's exposition of Brentano's thought.

Among other Brentanian scholar's in Sicily may be noted Mario Puglisi of Catania, who published in the *Annuario della Biblioteca Filosofica* (1912) "Teologia di Aristotele secondo F. Brentano" and subsequently translated Brentano's *Von der Klassification der psychischen Phänomene.*[13] References to Brentano occur in the works of other Sicilian authors, most of whom taught philosophy in high schools between 1900 and 1920. It is probably unnecessary to add that Grassi and Puglisi regularly attended meetings at the Philosophy Library of Palermo and remained affectionate friends of Dr. Amato.[14]

It would seem true to say that the secret of the Philosophy Library of Palermo was to be found, at least in part, in the intensity of the dialogue between Amato Pojero and Brentano and in the phenomenological structure of the problems then debated. This dialogue gave an identity and a specific character to a small intellectual circle which could have lived a wretched, self-enclosed life and which instead opened itself to the most significant cultural movements of Europe, choosing to debate quite different problems from those chosen by Gentile and Croce (both of whom – and especially Gentile – were heavily indebted to the Library).

Of the problems considered perhaps the main one concerned the discovery of a point of union and communication among the sciences, in short, a global foundation of all the sciences through a new concept of philosophy open to ontology: an ontology reduced to an essential principle that is something less positive than Aristotelian metaphysics, but also less negative than the pure and simple metaphysical requirement, which remained even after the discrediting of Kantianism. More precisely, the main objective was the discovery of a connective tissue between philosophy and science and between philosophy and ontology, the latter understood no longer in an apodictic but in a problematic sense: that connective tissue that presented particular connections with the phenomenological attitude of Husserl, consisting above all of an intermediary point of view which establishes the sciences and includes the possibility of an implicit ontology. This inquiry may possibly account for the discontinuation, in 1914, of *Annuario della Biblioteca Filosofica* as an organ of the association. In its place was substituted *Logos,* whose editorial committee included Husserl among its members. In the first issue of the Italian edition of *Logos* we find the following reference to Husserl:

> . . . the investigation of nature has yielded conscientiousness, clarity, and exactness of method that must serve as a model for all "arranging" even of the purely philosophical type and above all for the culturally philosophical type; a model, I say, for an arrangement in general, not for an arrangement of a particular form. Husserl when he requires philosophy to be like exact science did right to think of this methodological pattern of the natural sciences, in the narrowest sense, as a doctrine of the *laws.* (p. 84)

Nevertheless, during the next several years Italian contributors to *Logos* evidenced little concern with Husserl and continued nominally to ignore him even thought they debated fundamental philosophical themes in the light of their dialogue with Brentano.

Yet Husserl's phenomenology had an indirect effect, by providing a point of mediation between the antimetaphysical, pragmatical-phenomenological line represented by Vailati and Cosmo Guastella[15] on one side and the metaphysical Leibnizian-Aristotelian line of Amato Pojero and Brentano on the other side: these being the two fundamental poles of the theoretical debate of the Philosophy Library of Palermo.

We can ask why Husserl was practically ignored at the Philosophy Library of Palermo when the members were so well informed about German culture. Indeed, why is the name of the founder of phenomenology absent from the letters of Brentano to Amato Pojero while the names of Meinong, for example, and of other pupils, are mentioned? To tell the truth, there are various reasons for Brentano's reticience, as well as of Brentano's friends; and these reasons can be linked with the famous declaration of Chapter 2 of the third book of *Ideen:* " . . . Brentano has always remained a long way from a phenomenology in the present-day sense, and this is true up to today – to the point that he refuses to recognize progress in our movement."[16] In fact, Husserl did not realize that he was asking too much of a man like Brentano – about whom he had confessed that "he had earned enormous merit simply by making phenomenology possible" – when he expected him to recognize in phenomenology a more progressive movement than Brentanism. Whatever the case may be, even if the question is interesting, there is no space for a more specific study of the subject within the limits of this brief paper. I will only mention, therefore, that to clarify this point it is necessary to possess an in-depth knowledge of the personality and thought of Husserl and of the personality and thought of Brentano, along with many other apparently insignificant external circumstances. And, in addition, it would be useful to try to understand why two disciples of Husserl, Edith Stein and Roman Ingarden, in continuing their master's work have found themselves arriving at the neo-Aristotelianism of Brentano and have there met some of the most important themes of Brentanism.

University of Catania *Translated by Margaret Mason Sanfilippo*

NOTES

1 To reconstruct the biography of G. Amato Pojero and the history of the 'Biblioteca Filosofica di Palermo' I have relied on the direct evidence of friends and relatives of Dr. Amato, especially his widow and the sources cited in *Atti del V Congresso Regionale di Filosofia*, vol. 2, ed. SPES Milazzo 1974 and in G. Amato Pojero, *Pensieri* ed. Concetta Maria Amato Comparato, vol. 3, SPES Milazzo, 1968.

2 'Statuto e Regolameno della Biblioteca Filosofica di Palermo,' *Annuario della Biblioteca Filosofica* 1 (Palermo, 1912), 13.

3 For the biographical notes on Husserl consult L. Kelkel and R. Schérer, *Husserl*, Milan, 1966.

4 A. Banfi collaborated also on *Logos* with the article 'Il Naturalismo romantico del Guyau e le correnti del pensiero contemporaneo'; cf. *Logos* 7, no. 3 (July–September, 1924), 179–96. On A. Banfi and Husserl cf. in this volume: Roselina Salemi, 'The first Italian interpreter of phenomenology.'

5 Cf. Giovanni Calo, 'Concezione tetica e concezione sintetica del giudizio', *Cultura filosofica* 1 (1908), 178–221. Cf. Francesco De Sarlo, 'Lineamenti di una fenomenologia dello spirito', *Memorie dell 'Accademia dei Lincei*, ser. 5, 17, (1923), fasc. 6.

6 For a general exposition of the thought of Brentano we have as yet in Italy no good monograph but only an occasional essay, such as the notable article by A. Bausola, *Conoscenza e moralità in F. Brentano*, Milan, 1968. It is necessary therefore to draw from E. Gilson, O. Kraus, and A. Kastil: the last two edited *Gesammelte Philosophische Schriften*, vol. 10, Leipzig, 1922–1930. The 'Brentanogesellschaft' has been published in Prague since 1931.

7 The letters of G. Amato Pojero to Vailati are collected in the 'Fondazione Vailati' which is part of the Institute of Philosophy of Milan University.

8 The photostat of this letter was kindly given to me by Professor Maria Concetta Amato Comparato, currently in charge of the Pojero Archives. I shall limit myself to quoting this passage: "Villeicht wäre hier ein längeres Verweilen erwünscht, doch der Augenblick gestattet es mir nicht. So füge ich nur noch eine ähnlich kurze Entgegnung auf Ihre 2te Bemerkung bei: Sie sagen, es fehle das Analogon zum "Summum bonum" auf dem Gebiet der Wahrheit. Diesem stimme ich Ihnen bei. Auch habe ich in meinen Abhandlung über den "Ursprung Sittlicher Erkenntniss" den Grund dafür namhaft

gemacht. Wir haben nämlich auf dem Urtheilsgebiet zwar ein Analogon des Gegensatzes von "Lieben" und "Hassen" ("Bejahen" und "Verneinen") aber kein Analogon des Unterschieds zwischen einfachen Lieben und Vorziehn. So besteht denn auch auf dem Urtheilsgebiet zwar ein Analogon der Richtigheit der Liebe und ihres Correlatives (des Guten) aber kein Analogon der Richtigheit des Vorziehns und ihres Correlatives (des Besseren). So fehlt denn auch ein Analogon zu dem, was besser als alles andere Gute ist, zum Summum bonum."

[9] G. Vailati wrote *Sulla portata logica della classificazione dei fatti mentali proposta dal prof. F. Brentano,* a communication presented at the Third International Congress of Psychology in Paris in 1900 and published in *Rivista di Filosofia* 2 (1901), fasc. 1 and later in *Scritti,* Florence-Leipzig, 1911, pp. 336–40.

[10] For the Vailati-Brentano papers consult G. Vailati, *Epistolario (1891–1909),* Einaudi, Turin, 1971; for the Amato Pojero-Brentano papers consult the Amato Pojero and the Brentano archives separately; for the Vailati-Amato Pojero papers see as well as *Epistolario* and 'Lettere di G. Vailati ad Amato Pojero,' ed. Antonio Brancaforte, *Rivista critica di Storia della Filosofia* 32 (1977), fasc. 1, pp. 50–71. For the direct evidence cf. *Pensieri,* (1st quarter), 85–145. I have other letters of G. Vailati that have yet to be published.

[11] Cf. Leonardo Grassi, *L'unità dei fatti psichici fondamentali,* Muglia, Messina, 1905, pp.164–66.

[12] Leonardo Grassi, Ibid., p. 164–65.

[13] M. Puglisi also wrote 'F. Brentano: notizie e ricordi,' *Bilychnis* (1921), fasc. 7, pp. 1–12.

[14] For this the proof lies in the letters of Grassi and Puglisi to Dr. Amato, available in the Amato Pojero Archives.

[15] Cosmo Guastella (1854–1922), a member of Biblioteca Filosofica di Palermo and affectionate friend of Dr. Amato, wrote: *Saggi sulla teoria della conoscenza,* Palermo, 1897; *Filosofia della metafisica,* Palermo, 1905; and *Le ragioni del fenomenismo,* Palermo, 1921–1922. He is especially indebted to John Stuart Mill.

[16] E. Husserl, *Ideen zu einer reinen Phänomenologie und phänomenologischen Philosophie,* trans. Alliney and Enrico Filippini, Einaudi, Turin, 1965, p. 837.

ROSELINA SALEMI

ANTONIO BANFI, THE FIRST ITALIAN INTERPRETER OF PHENOMENOLOGY

I. INTRODUCTION: THE ORIGINS OF THE BANFI-HUSSERL CONNECTION

It is generally agreed that Antonio Banfi first introduced the ideas of Husserl in Italy[1] in two essays ('La tendenza logistica nella filosofia tedesca e le Ricerche logiche di Husserl'[2] and 'La fenomenologia di E. Husserl e l'autonomia ideale della sfera teoretica')[3] published in 1923, the same year in which Husserl, on a journey to Italy, met Banfi. The two men were to remain in close contact from that time. Husserl's influence is visible thereafter in Banfi's philosophy: the work prepared just before the latter's death and published posthumously in 1958 deals, in fact, with Husserl's *Crisis.*

Banfi's interest in Husserl can be traced, on the one hand, to his deep dissatisfaction with Italian culture, mainly provincial and exclusively dominated by a Gentile-Croce oriented idealism, and, on the other, to the need for cultural renewal after the positivist crisis with its arbitrary absolutism claiming scientific criteria for philosophical validity.

The eighteenth-century cultural inheritance, characterized by an extreme effort to unite the finite and the infinite, the concrete and the universal, expressed at its best by Hegel, crumbles, in fact, at the end of the nineteenth century into a series of contrasting views. The process of the destruction of the primacy of speculative reason and the fragmentation of experience which has its epicenter in the Vienna of the end of the last century begins in this period.[4]

In Italy Banfi is perhaps alone in realizing the abstract nature of these idealistic disputes, with their vain ambition to reform the Hegelian dialectic; in reaction, he turns his attention to the intellectual world beyond the Alps (e.g., Renouvier, Bergson, Avenarius, Mach, Hegel), particularly to Germany where he studied at the Friedrich Wilhelms Universität in 1910–1911. In Berlin Banfi finds himselt in the midst of a rich philosophical experience in which the main streams of neo-Kantianism and neo-Hegelianism mix. He is inspired by Simmel's conception of life as continuously transcending all determinate form, but feels the need to reconcile this view with reason, understood in the critical positive sense of the transcendentalism of Marburgh.

Tymieniecka (ed.), Analecta Husserliana, Vol. IX, 441–460. *All Rights Reserved.*

It is here on the level of the relationship between reason and life that Husserlian investigation of the authenticity of experience and the transcendentalism of reason meet.

Two fundamental roles of interest link Banfi to the phenomenology of Husserl: first, the problem of intuition and the possibility of its analysis and description as a means of reaching the essence of life and reality; second, the problem of a critical and antidogmatic system of knowledge. Both thinkers share the refusal of the metaphysics of being, the accentuation of the value of the philosophical method, and the reaffirmation of its role in contemporary civilization.

Husserl's concern, like Banfi's, is to discover how philosophy can formulate a rational system of knowledge. To do so, it has to free itself from the false preconception that where reason is called upon to exercise its requirements to the full and to unify its spontaneous and divergent directions there is no need for philosophy, which is then confined to dealing with the irrational and to expressing itself arbitrarily through inconsistent points of view. Sharing this fundamental concern of Husserl, Banfi is not only the first Italian interpreter of Husserl but also the first lucid exponent and at the same time shrewd critic of phenomenology. Here we shall try not only to examine Banfi's interpretation of Husserl's thought, but also to evaluate his own originality and his limitations when compared with Husserl.

II. HUSSERL'S THOUGHT AS INTERPRETED BY ANTONIO BANFI

1. The General Outline of the Banfi-Husserl Connection

Banfi discerns in Husserl an emphasis upon intuition and the eidetic viewpoint; and he finds in Husserl's work an answer to the demand for knowledge which draws upon experience in all its complexity and on various intersecting levels. The "*epoché*" suggests to him the problem of the historic continuity of reason.

In the attempt to avoid relativism which admits an order of logical truth independent of any factual truth that can form the basis of a thought system, Husserl revalues pure logic, the object of which becomes the ideal structure of the logical scientific system. For Banfi, this revaluation stresses the principle of the autonomy of reason as the basis of scientific knowledge, independently of any consideration of content or point of view, empirical or anthropological.

Husserl's work to which Banfi chiefly refers (and in general his interpretation remains always the same) is the *Logische Untersuchungen* of the *Philosophie als strenge Wissenschaft,* and the first volume of *Ideen.* Banfi expresses clearly his attitude toward Husserl's thought in certain important pages of *Principi di una teoria della ragione;* this approach to Husserl is never much modified except when he later clarifies in his *Husserl e la crisi della civilta europea*[7] the place intuition must have in a theory of reason.

In fact, Banfi agrees with Husserl that intuition and reason are not two contrasting forms of knowledge, but two moments (the one intuitive and the other rational) in the cognitive process. The intuitive moment, which represents the first encounter of the ego and the world, must simply avoid the claim to be absolute.

For Banfi, Husserl's "intuition" should not be conceived as static: it is the movement of experience toward rationality. Its validity consists in the possibility of reference to other intuitions, and for this motive, to be precise, we should speak of an "intuitive level" (rather than of a single intuition) from which knowledge starts.

For Banfi the syllogism "All men are mortal, *X* is a man, *X* is mortal" has meaning only if the concept of "man" is given a precise intuitive meaning.[8] He also clearly maintains, against the critics of phenomenology, that it is not the immediacy, but the negation of the universality of the immediacy of empirical knowledge, which fixes and abstracts fact from experience that lies at the basis of the boundless right of rationality.

On this foundation Banfi takes profound interest in certain particular features of Husserl's thought, especially in its antidogmatic method, its systematic expression linked to its concept of philosophy, and its search for the connecting link between reflection and praxis.

2. *Phenomenology as an Antidogmatic Philosophy*

From 1926 onward Banfi is concerned with the antidogmatic method of phenomenology. Husserl's researches reveal beneath the dynamic complexity of reality the presence of a pure dialectic system as a basis of scientific thought. But, one could object, does not the concept of the autonomy of essence hide the more or less explicit assumption of an ontological dogmatic rationalism? For Banfi the independence of the "eidos" does not mean its existence outside of empirical reality in another sphere of being, but means situating empirical reality within a purely rational order. It is the methodic principle of an analysis of experience.

Inasmuch as Husserl is aware of the infinity of the theoretical task, this excludes any dogmatic attitude.[9] In fact, Banfi interprets the position of phenomenological thought as precisely the opposite of a dogmatic approach.

The latter concerns the transcendent object, whereas phenomenological intuition neutralizes the theory of natural objectivity, not in the sense of casting doubt on its existence but in the sense that this latter ceases to be of interest as the essential moment in the thought process.

The existential neutralization of the object, in phenomenology, refers as much to real objects as to ideas, as much to logical forms as to the transcendent essences of material ontologies: their transcendent status must be explicated in terms of phenomenological immanence. Phenomenological cognition becomes thus completely independent of any other form of knowledge; it is considered as the first principle, without presuppositions, in which other cognitive procedures find their basic methods.[10] It is on this basis that, with the assessment of the immance, the transcendental foundation of the problem of cognition is recaptured, and phenomenology emerges as the science establishing the objectivity of transcendental knowledge.

In the phenomenological approach the ontological element has its realization and its importance: for Banfi Leibniz's ontological rationalism and Kant's transcendentalism seem to unite in Husserl's philosophy. Banfi also holds that any dogmatization of this philosophy, in the sense of attributing an absolute priority to the essence over the transcendental constitution, is to be attributed not to Husserl himself but to the various thinkers who took part in the development of the school of phenomenology, as represented in particular by Conrad-Martius, Hering, and Ingarden.

3. The Phenomenological Method

In examining the characteristics of the phenomenological method Banfi reveals that the analysis of ontological relationships which determines the various spheres of reality is not an end in itself, but represents a moment in a rational system which is referring to the same universal form of experience. In other to explore and bring to light the scheme of ontological reality, Husserl applies the method of transcendental phenomenology where the structure of each cultural level is assessed in its relationship with other structures. Phenomenology thus becomes, according to Banfi, a method of rational integration, a new *universalis sapientia* which embraces all the fields of scholarship in a systematic scheme from the very foundations of their method. It unites the problem of logic to that of reason in general and offers a methodological basis for epistemology.

"Philosophy," writes Banfi,

"is not the awareness of the presence of an ideal order in life, but it is part of the internal dynamic structure of life itself, in the sense that it develops within itself its own problems, states its own ideals and renews them in touch with reality."[11]

Banfi agrees with Husserl in affirming the possibility and validity of philosophy, understood as a rigorous science ("Philosophy *als strenge Wissenschaft*"). He sees it as an endeavor aiming to get away from the rigidity of the interpretative schemes of experience and to offer an answer to the crisis of the loss of an adequate unity among the disparate fields of research, a result of the specialization of science.

This very desire for reconstruction is for Banfi the deepest theme of phenomenology which calls philosophy back to an awareness of its rational universality, liberating it from an arbitrary selection of principles and of methods with their limitations. Banfi is convinced that Husserlian philosophy aims to resolve, not only the crisis of rational consciousness, and with it the crisis of contemporary culture, and to assure in the fashion of a "*philosophia perennis*" the further development of knowledge. It helps, he maintains, to provide a method capable of fulfilling a multiple task. First, it defines the autonomy of cultural unities with reference to a law which helps us to understand the process of their inception and development through the variety of subject matter. Second, the phenomenological approach is capable of grasping the diversity of feelings and of their relationships and bringing together the laws in various fields by coordinating them under the general idea of the spirit – and thus enables us to recognize the lines and strata of reality in their reciprocal connections.

Philosophy thus appears not as a theory of absolute reality from which we would deduce particular forms, but as the organization of human experience in all its complexity. It corresponds then to the nature of its method in the light of which systems are no more than a provisional synthesis of problems and solutions. What constitutes the truth of a philosophy as such, Banfi believes, are not its results when confronted with a subjective thought process but the rational system in which these results are enclosed. Clearly, to anyone who considers dogmatically the pure and simple conclusions of philosophical thought, philosophy can easily appear to be a field of contradictions when, in fact, it is a rich dialectical movement which takes place and corrects itself in history.

For Banfi, in reconstructing a line of continuity of speculative thought and in reaffirming its internal dialectical unity as well as harmonious discordance, Husserl has explained the rise of skepticism, of irrationalism, and of empiricism as a reaction against the dogmatization of rationalism which has been identified exclusively with the cognition of an absolute object. From a theoretical point of view the history of philosophy is seen as a gradual development, through the dialectical opposition of the different systems, of a free system of knowledge, the fruit of a critical consciousness which has become constructive. Husserlian philosophy assumes for Banfi the specific value of a philosophy of culture insofar as it stresses in its criticism and in its method the different levels on which culture tends to work out the antithesis between the individual and the world; a philosophy which intends, on the one hand, to distinguish, and on the other hand, to examine critically, the various directions of knowledge in their dialectical movement. There corresponds to every form of the life of the mind a philosophical science which makes contact with the particular disciplines, purifying them of every pragmatic limit. This does not imply on behalf of philosophy the assumption of a metaphysical approach. On the contrary, it emphasizes its role of a co-ordinator in the movement of ideas in the various sectors of experience.

A theory of culture, therefore, is not a theory about what leads men to the supreme ideals of life, but it is the active search for living values, constant and never accomplished, that is expressed through the incessant *emendatio* of the intellect; it is the way to free the mind from ever recurring prejudices. In this perspective it is clear how, for Banfi, Husserl does not postulate a metaphysical reality but intends to proceed to the analysis of the formative moments of the cultural world and thence to recognize the unity between the world and man, maintaining the subject as a point of reference, because "if there is a center from which speculation takes its starting point and to which it always returns, it is man."[12]

4. Praxis and Reflection

Banfi maintains that in seeking the link between praxis and reflection, Husserl raises the problem of history in an unusual fashion, and that embracing in this way the totality, phenomenology points out the meaning of the world.[13] According to Banfi, Husserl would like to reunite the extremes severed in contemporary thought and reconnect the particular action to the universal, not in order to annul it, but, on the contrary, to give it meaning. This line of thought raises several problems which Banfi feels keenly. The conciliation of the state of scission and of crisis that permeates man's individual life must happen in a politico-historical sphere and not only in representation and in thought: it is in the light of this profound necessity to use cognitive results in praxis that the meeting of reason and history is seen. In fact, the separation of these two levels, of cognition and praxis, a recurrent risk in all speculative philosophy, could, according to Banfi, lead to an incomprehension of reality as long as the level of reason is identified with that of a neutral theoretical structure, distorted by cultural abstractions, and the level of history with that of a pragmatic action that emerges from irrational impulses, not clearly identified.

Abstract formalism and pragmatic irrationalism would on such an understanding become the two insoluble extremes of the conflict. And, for that matter, a purely theoretical reflection detached from history and removed from the same reality which produced it would be useless. For Husserl the problem could be expressed in terms coined later in Enzo Paci's essay:[14] "How is it possible that an action be concrete, that it is linked to particular problems which require particular solutions and at the same time that it represents an expression of the universality of reason?" And, more clearly: "How can the particular be known in such a way that the praxis be free of prejudices and yet universally rational?" It is at this point that the theme of "pragmatic reflection" appears in Banfi, a key notion, not opposed to, but complementing that of "theoretical reflection," a notion which is the fruit of Banfi's intense meditation on the problems of Husserlian phenomenology. "Pragmatic reflection" appears with all its problems in the work which Banfi dedicates to Husserl for his seventy-fifth birthday: *Sui principi di una filosofia della morale.*[15] For Banfi, in every field of the experience and thus of life experience, a particular type of reflection arises and is enacted; it can be termed "pragmatic" because it is not governed by purely theoretical concerns but has as its end the solution of problems for the benefit of praxis. The affirmation of the pragmatic roots of thought is offset with that of the necessity of the purely theoretical connection of the same thought, and pragmatic reflection thus takes on the meaning of an invitation to change the world.

In this way Banfi expresses the requirement already met with in Husserl of a phenomenology as praxis and historicity but he moves noticeably away from it. The commitment to revitalize the theory putting into it all the richness of the concrete will take Banfi a long way from Husserl, even if it is in Husserl's concept of "the *life-world*" that Banfi is to recognize the possibility of that "pragmatic transcendentalism of reason" which is to be the fundamental theme of his ethical realism. Banfi thus sees phenomenology as the synthesis (or at least the possibility of a synthesis) between rationalism and historicism, but the reason for this conclusion lead his thought to a totally different level from that of phenomenology. Thus Banfi begins to deviate from Husserl's line of thought.

III. BANFI'S CRITICISM OF PHENOMENOLOGY

We have previously analyzed the underlying motives which link Banfi to Husserl and the most important points of contact between the two philsophers. Banfi makes a careful selection of Husserlian themes which he then examines from a carefully defined viewpoint. In examining his criticism we shall try to recognize its originality and to discover its motivation, bringing to light also certain interesting anticipations by Banfi of the Husserlian thought yet to come, present in Banfi's writing, an anticipation which is probably the fruit of frequent meetings between the two philosophers.

1. Does Banfi Anticipate Husserl?

According to Enzo Paci, Banfi blamed Husserl for many things, though never on a petty level, because of his interpretation of the phenomenological position (for example, certain reserves toward logical neo-positivism expressed by Banfi and later taken up again by Giulio Preti are those which Husserl himself would have made).[16] A singular proof of this never static, never passive identification of Banfi with Husserlian thought are the anticipations easily found in *Principi di una teoria della ragione* of positions eventually adopted by Husserl and later by philosophers involved in the rebirth of phenomenological studies in Italy. Banfi is among the first to realize that the "eide" of Husserl are not ideas in the sense of Platonic realism.[17] Husserl's object is not real, but intentional. The object he speaks of has sense only after its neutralization, the "epoché." This leads to the "cogito" and allows us to understand the objectivity; the object is the "noema," that is the content of the "noesis," the intentional act. In *Principi di una teoria della ragione* Banfi writes that reason must be understood in the phenomenological perspective. This statement is important because this becomes clear only in the later Husserl.

Banfi writes:

> Thus phenomenology appears here as the science of the general formation of the objectivity of transcendental consciousness, of, in other words, the synthetic-transcendental structure of reality. It refers to all types of objects of superior order, theoretical, valuative, and practical, to cultural objectivity, law, state, church, customs, etc. The various levels of the constitution of reality unfold in this way.[18]

It is remarkable that Banfi wrote this in 1926 while the same views which Husserl maintained since 1912, remained unpublished (the second volume of *Ideen* was published only in 1952).

Another Husserlian conclusion fairly clearly anticipated in *Principi di una teoria della ragione* is that the authenticity of life is reached not by accepting life as such, but by experiencing it as a new possibility; the "epoché" in this way becomes the refusal of the conclusivity of pure fact and the establishment of the sense of the *life-world* in the transcendental form; the negation of factual life, but also its "recovery" in the sense of Kierkegaard (its *"Behaltung der Zeit"*). The use of the "epoché" helps us to rediscover the freedom of reason and at the same time to capture the authenticity of reason and of life.

Thus Banfi enters up to a certain point in the sphere of Husserlian philosophy developing some of its themes. He applies his phenomenologically influenced reflection in particular to morality and aesthetics. The fruitfulness of his researches lies principally in the fact that they have suggested valuable lines of approach for the understanding of the inception of experience in categorical form, opening Italian culture to the suggestion of a phenomenlogical vision of reality. In spite of Banfi's partial division and separation from Husserl, there remains a deep understanding and dialogue between them, indeed, a dialectical relationship, because a critical and antidogmatic theory of reason which he aims at, must take into account, as a starting point, a phenomenological method that does not spring from one-sided, prejudiced assumptions but leaves the concepts to emerge from the real experience of phenomena. Banfi's criticism of Husserl begins as a criticism of phenomenology from the inside and it is directed toward emphasizing the problems in need of resolution.

It is in this way that Banfi arrives at what I called above "anticipations" of Husserl's philosophy. Banfi understands the risk of ambiguity present in certain themes of phenomenology and propelled by the need to clarify, treats them in his own way, raising queries and suggesting solutions, to test its possible "dogmatic limits." In fact, Banfi analyzes, on the one hand, the capacity of phenomenology to answer exhaustively the questions which arise from within it, and on the other hand, its ability to treat constructively the problems raised by the crisis of contemporary civilization. The first concern is expressed early, that is, in the *Principi di una teoria della ragione;* the second, in his later thought in the article already quoted on Husserl's *Crisis.*

Contrary to the opinion of those who maintain that Banfi's criticism of Husserl is already present in its entirety in 1926, it is useful to point out the importance of the change in his attitude which occurred between 1926 and 1957. In fact, while in *Principi di una teoria della ragione* Banfi corrects what he sees as the mistakes of his phenomenology with phenomenology itself, rigorously developing, that is, transcendental rationalism and seeing the difficulties from the inside, in contrast, in his article on the *Crisis* he tries to correct phenomenology with insights stemming from outside the phenomenological approach. Having realized that the phenomenological attitude (as, for that matter every particular attitude) is insufficient, Banfi looks elsewhere for completing it and a more integral view.

2. *The "Dogmatic Limit" of Phenomenology*

As already mentioned, for Banfi the fundamental characteristic of phenomenology is its tendency to construct a critical, antidogmatic method of cognition which would be correlated with experience in all its complexity. The first problem confronting this task is then to test whether phenomenological practice corresponds in fact to its stated idea of a purely rational method. Above all, he is concerned with the question whether the concept of "intuition" is sufficient to establish the direction of the process by which the autonomy of ratiocination is recognized and enacted, or if it does not rather imply a certain equivocation of meanings. Banfi points out, first, that the concept of "intuition" seems to refer equally to the tangible fact and to the rational essence. Consequently, it does not seem to him suitable for expressing the rational moment in its specific quality. For Banfi the intuitive element represents within the context of the concept the moment of its empiric, pre-theoretical determination. This moment, according to him, cannot be recognized in all its purity unless its essential independence of that intuitive determination through which it is grasped, but in which it is not comprised completely, is not also recognized.

The fundamental error of "dogmatic rationalism," as he calls it, lies precisely in blurring this relationship so that while on the one hand it transforms the intuitive richness of the conception into an abstract form (which conflicts somewhat with the concrete experience), on the other hand, it reveals itself without any unitary principle except an empty form in which the abstract persists in its own identity. A pure universal rational method does not aim at annulling intuitive concrete determination, but only at defining its meaning. This is expressed by the fundamental phenomenological principle of Husserl: what in intuition is originally offered is to appear self-explanatory as to how it is offered and within what limits it is offered.

For Banfi, nevertheless, intuition only expresses the first moment of the methodological principle of phenomenology (that is, the ideal meaning of every real or ideal datum) and it is not necessary to explain the second, that of the rational method, in which the intuitive datum must be presented, to escape an abstract-dogmatic comprehension, in which even the essences or the ideas given intuitively must find their theoretical unitary connection. Husserl maintains that this second moment takes place in the phenomenological intuition of the pure consciousness in which intuition and descriptive analysis must be integrated. But to this Banfi objects: if the facts of intuition all exist on the same level, how can we single out the dominating element of meaning, that is, the first moment, and justify its priority? It is true that Husserl introduces a distinction between the transcendental essence and the immanent essence of pure consciousness. It is from the latter alone that perfect pure intuition is possible. However, if we do not wish to fall into the psychologism, how else can we think of immanent essence if not as being the objective of the systematic meditation never explicitly stated but implicitly assumed by the transcendental analysis? And how can we relate a particular intention to the structure of pure consciousness when the description of its component parts can be understood only as the result of a long and complex rational elaboration of speculative concepts?

Finally, when Husserl defines the idea of reason as intuitive, the differences between the views of both thinkers become emphasized and Banfi's queries multiply: how can the idea of reason be a "given fact" if it is the law for the explanation of facts? Again, how can an intuitive content lie behind the infinite process of conceptual progress even if this is made up of intuitive concepts?

The concept of "consciousness" entails similar problems, according to Banfi. "Consciousness," unless it is to be understood simply as a metaphysical projection, represents only an aspect of the cognitive process; while the character of a central factor that Husserl attributes to it stems from the crystallizing tendency of the knowing itself. The dogmatic conception of consciousness as a sort of a central agent is the usurpation of the prerogative of the much richer concept of "reason" which comprises the transcendental ego. In effect, for Banfi, the two moments of the reasoning process, the ontological and the phenomenological, instead of finding their unity in Husserlian philosophy, remain divided. The second seems, indeed, a negation of the dogmatic meaning of the first, but it is simply juxtaposed to it. No explanation of its priority is given and a unitary law of their relationship is missing. Their coincidence, which never comes about in Husserl's thought, appears on a metaphysical level, with a strong religious flavor, in the thought of such representatives of phenomenology as Reinach and Mahnke. This demonstrates for Banfi that Husserl has left unsatisfied the deep necessity of unity in the reasoning process and left undefined the significance and the value of the relationship between the various levels of knowledge. While recognizing the validity of the Husserlian standpoint, Banfi maintains that there is an ontological point of view involved in the difficulties of Husserl's "rational dogmatism"; the distinction of science of essences and science of facts brings us, notwithstanding, to a dogmatic attitude. It is responsible for a distinction between rationality and intuition which remain two antithetical functions, even though they are indivisible in knowing. Reason, according to Banfi, aims at incorporating the data of intuition in an ever-widening system of relationships, but these relationships emerge only in a higher intuitive insight which permits a new mediation. As a result, every concept exists only in the form of its intuitive content, more or less complex, and has rational value only insofar as it expresses that content and in confronting it with other ones brings it to its ideal realization as the moment of higher synthesis.

The rational truth of a concept lies then for Banfi in this dialectic, through which alone its dogmatic determination may be overcome and the concept may express itself as a moment in a free speculative system. Husserl has not sufficiently stressed this dialectic quality which remains as an intrinsic postulate of his thought.

As already mentioned, the problem which inspires Husserlian research is, for Banfi, the possibility of a universal system in which not only the complex variety of experience would find a unitary interpretation, but also the diverse directions of knowledge would meet in a harmonious rational unity.

According to Banfi, the phenomenological connections, themselves apprehended intuitively, are often defined as concepts which have an essentially empirical meaning. This involves the danger for phenomenology of degenerating into psychologism. Banfi therefore thinks that although the problem of a pure system of reason free from dogmatic attitudes has its radical formulation in phenomenology, yet it does not find in it a definitive, universal solution.

In spite of the partial validity of the theoretical attitudes, there reappears a residue of dogma, which hampers the Husserlian need to express the life of the spirit in its freedom; the theory of the transcendental consciousness understood as a system which determines at the same time the relationship of the subject and the object may be considered still open to further factors to be integrated. Husserl reaffirms the principles of the theoretical unity of knowledge, leaving to philosophy the task of outlining the system of the sciences without imposing abstract schemes from the outside, but he fails to clarify in this context the differences of method to be taken into account as well as failing to treat the question whether the relationship which is established between the single science and philosophy may justify the variety and mobility of the relationships between philosophy and the particular sciences. Paradoxically, it seems to Banfi that the phenomenological conception of the system of knowledge does not succeed in taking into account the dynamic and dialectical qualities within knowledge itself. Furthermore, it seems that it serves to coordinate the various directions of knowledge provisionally rather than to penetrate their permanent structural laws.[19] As far as experience itself is concerned, phenomenology, while taking into account its primordial importance, is seen as giving it an abstract interpretation. Banfi objects that in the development of his researches Husserl gives priority to a genetic and simplifying description of the appearance of experiential data rather than to the search after, and definition of, that system of laws which *constitute the structure of experience.* For Banfi every type of experience exhibits necessary structures. The problem then is not to describe the occurring experiential instances but to understand and point out, under the seemingly arbitrary act, the formation of rationality. Furthermore, the phenomenological interpretation turns in spite of itself to the transcendental essences while neglecting the existent, and, in the end, prefers the abstract to the concrete. Thereby, Husserl does not respect, in the last resort, the critical and antidogmatic premises which he proposed as a basis of the phenomenological method. We may say that Husserl leads to the evaluation of the Cartesian "cogito," to philosophy as reflection upon the original structure of experience. Banfi, in contrast, attributes significance to the transcendental system of cognition only in the sense that it constitutes the formal structure of the specific character of knowledge. Husserl, according to him, tries to supply cognition with a basis that allows the analysis of the origins of the consciousness which investigates the world; Banfi, in contrast, raises the problem of the intuitive transposition of fact. Husserl is looking for the original structures of the subject which account for the subject's cognitive relationships with the world; Banfi holds that all knowledge is the fruit of rational mediation.

The road covered by Husserl leads to the constitutive analysis of objectivity, to reflection on the constituent structures of a consciousness understood in an ever more realistic sense, and therefore to the problem of originary experience. The investigation of the origins is the focal point of the Husserlian thought, while for Banfi every search for origins is already mediated in the nonoriginal. Finally, Banfi interprets Husserl's philosophy as a specific type of reflection which aims at normative results; for Banfi himself, philosophy, as already stated, is a method of the rational integration of facts and laws, a regulative idea of the process of culture. As can be seen, Banfi takes the most important themes of Husserlian philosophy, examines them amply, and looks at them from a personal point of view. At a certain point of his critico-constructive analysis there arises the need of an historico-practical approach which Husserl leaves partly unsatisfied. In particular, the problems raised by the connection between the "epoché" and history are direct incentives for Banfi's reservations expressed in the essay *Husserl e la crisi delle scienze europee.* He feels very strongly the problem of commitment to concrete reality and the need to give a historical correlative to speculation. Banfi's writings on the *Crisis* of Husserl should be interpreted in this sense.

3. Banfi and the Crisis

The theme of the crisis of occidental culture occurs frequently in the literature of the early twentieth century. There is talk of "theology of the crisis" and of "philosophy of the crisis" with greater or lesser irrational subtleties. Barth, Tillich, Klages, Nietzsche, and Simmel are among those who express an awareness of the imminent crisis of European culture. Husserl joins in, therefore, on a theme already widely discussed and analyzed, on which, it seems, it would be difficult to say anything new that might also be true. Banfi, in the period of his collaboration on a neo-Calvinist periodical *Conscientia,* devotes some studies to the state of diffused crisis characteristic of the first years after World War I, and his already vivid interest in the subject makes him feel a particular affinity with the same theme as treated by Husserl. The novelty of Husserl's approach strikes him. Indeed, Husserl's interpretation of the crisis differs radically from the studies of Barth and Tillich, of the irrationalists of the Nietzschean mold, and of vitalists like Simmel. In the first place, his interpretation is not metaphysical but historical, that is, consists of an attempt at explaining the development starting by its very origins. In the second place, the characteristic orientation of the other philosophers is that of a metaphysical evasion; they attempt to explain the crisis without seeking for a solution but restricting their scope to the contemplation of the catastrophe. Husserl, in contrast, feels the obligation to analyze rigorously the reasons of such development. As a result of this analysis, Husserl defines the crisis of the occidental culture as a crisis of the consciousness that man has of himself and of the world, and ultimately, as the crisis of philosophy seen as a structural system of the consciousness of the contemporary man. The present system of knowledge, which may be otherwise seen as a triumph of organization and classification of the human *life-world*, is the most striking example of this crisis which can and must be resolved historically through the renewal of philosophy. This renewal is seen in the continuing process of the rational self-enlightenment or clarification of consciousness. "So ist Philosophie nichts anderes als Rationalismus durch und durch, aber nach den verschiedenen Stufen der Bewegung von Intention und Erfüllung in sich under-schieden Rationalismus, die Ratio in der ständigen Bewegung der Selbsterhellung."[20]

Banfi notes that the continuous movement of reason toward its own clarification through concrete self-recognition in the stages of historical development comports a critical consciousness in whose perspective rationalism and historicism can really be integrated. The first problem, therefore, is presented by the critical function of reason. For Banfi, an interpretation of the critical function of reason as self-reflection, which puts itself before itself as an object and constitutes itself in the universal being of the spirit, resuscitates the ghost of an idealistic form of the metaphysical dogmatism emphasized by analogy with personal self-consciousness. A similar formulation has been already present in post-Kantian idealism and has brought about some rather remarkable consequences. In fact, on one side, he sees in it the hardening of the methodologico-systematic level and the immobilizing, freezing of the *life-world* rather than a free unfolding of its vitality; on the other, the turning of the history of philosophy into a final and definite world view (*Weltanschauung*), that in which rationalism and transcendental idealism identify themselves. All the rest of Husserl's genetic interpretation of Western cultural progress is in his view only a prehistory, interpreted with reference to the opposition between objectivism and subjectivism, envisaged as two extreme poles separated without any possibility of mediation.

Thus interpretative principle appears to Banfi extremely damaging, because once the principles of approach mentioned above are adopted, history will appear as moving within strongly defined limits and the concepts and principles produced by the development of thought will become ontologically static. Abstract analysis of structure is thus substituted for the living dialectic, freezing systems of values. In fact, Banfi insists that in a similar context the values are not seen as emerging from historic reality, but as dominating it, and are present in it only as ideas. Thus in practice Husserl, while affirming the historicity of the crisis and the necessity of resolving it on the concrete basis of praxis, does not succeed in pointing to a solution that would not be speculative. It seems as if the demiurgic function of the philosopher, while committing him to a great theoretical responsibility, frees him from practical responsibility, consecrating him to a life "au dessus de la melee." For Banfi the transcendental idea is not born of a theoretical operation like "epoché," but is the result of a historico-practical relationship; the transcendental quality of the idea is expressed in historical existence as its political value.

There is in Banfi a clear consciousness of the historicity of reason. Its concept can be thought of only as an ideal in a process whose genesis can not be explored by reflecting on the structures which constitute experience, but by a historical examination. One can speak of "pure philosophy" only in a formal sense. Therefore any argument which refuses the formal rules of cognitive objectivity becomes a metaphysical-type argument in which the dogmatic use of concepts is parallel to its ahistoricity. As a further consequence Banfi challenges the a priori of the *life-world* (a pregeometric spatiality, preobjective temporality, a premathematical causal link) towards which the meditation of Husserl in his later period moves. Banfi does not aim at developing a philosophy conceived as a "dogmatic science" but sees nevertheless rationality as the ideal of historical knowledge. The contemporary crisis thus becomes for Banfi a crisis of growth, a productive crisis which must be solved with the help of a free method of cognition which dissolves all mythico-metaphysical residue and in which the Husserlian "epoché" is translated into "revolutionary will." It is here that Banfi's thought is put into a Marxist perspective. Seeking to be essentially concerned with praxis, Banfi moves away from Husserl, even though he keeps his thought as a constant point of reference, because, as is affirmed in an article published in the periodical *Telos*, "Thus the call of phenomenology for a return to the subject and the call of Marxism for the overthrowing of the capitalist order are two complementary answers to the same problems."[21]

We have seen that in his writing on the *Crisis* Banfi corrects phenomenology with "something else" than phenomenology. This "something else" can be, in fact, identified with Marxism which fills an otherwise abstract theoretical structure of the Husserlian enterprise with its historical concreteness. Banfi holds indeed that the theory of critical reason finds its expression in Marxism both as doctrine and as social praxis. This attempt of Banfi to "correct" in this way Husserl raises a series of problems which make it necessary to consider whether the Banfian interpretation of Husserl can be considered adequate. Has Banfi kept in mind all the main concerns expressed by Husserl or has he in some way a limiting vision of them?

IV. LIMITS AND SIGNIFICANCE OF THE BANFIAN INTERPRETATION OF HUSSERL

Our analysis of Banfi's philosophical attitude toward Husserl's thought shows that he has considered Husserl not only as his master but also as a thinker whose philosophical experience coincides with his own. For Banfi, Husserl is the philosopher of critical reason and the philosopher of culture who tends toward a theoretical universal organization of experience. Banfi seems not to notice the fact that this ideal of rational unification springs in Husserl from a descriptive analysis of the structures in which the object is given and inspectable ("inspectable" means that which can be given in direct experience to a consciousness). Without taking into consideration this fundamental aspect of phenomenology, its complete program remains unclear. In neglecting this crucial point, Banfi has instead treated Husserlian thought with arguments appropriate for the criticism of the neo-Kantian position. In fact the Kantian and neo-Kantian conception of experience (the references to Cohen, Natorp, and Cassirer are constant) like that of Hegel and of modern historicism permeate deeply the style and thought of Banfi. This influence made it difficult for him to see the originality of Husserl's views.

It is not difficult to demonstrate the Kantian-Hegelian roots of some of Banfi's disagreements with phenomenology. It may appear clearly in this example: in Husserl's phenomenology the criticism is exercised with reference to the transcendental analysis which reaches the pre-predicative structures of the subject which investigates the world; Banfi attributes the critical function to the transcendental idea of reason which shows the pragmatic nature of cultural and scientific reflection. The same neo-Kantian blend of Banfi may be seen in Banfi's concern with the "integration of experience." Husserl is not concerned with a "rational integration of experience" but with reflection upon the intentional structures of transcendental subjectivity as the constituents of experience as well as of the objectivity of the *life-world*. Banfi's concern with the "integration of experience" means nothing else but his recognition of "positivity" in a Kantian sense, the fact that it is a direction to take in the investigation of reality. The Banfian philosophical horizon comprises also the Hegelian theme of mediation and dialectic which fuses with the Simmelian need to understand life in all its complexity. Banfi proposes to phenomenology problems which need to be solved, and, paradoxically, reaches the conclusions not only that it is life and not philosophy which teaches us to live and that it is life which teaches us to think. We see here in Banfi a more optimistic vision than in Husserl of the possible overcoming of the crisis of modern science. In fact, Husserl's main concern is to give again its full, valid meaning to the direct experience of the world which surrounds us, and to establish its continuity with scientific experience. It presupposes that true philosophy and science coincide, because both of them refer, albeit differently, to the same cognitive process oriented toward truth.

Banfi, on the contrary, expresses certain reservations about the concept of "truth" which seems to him to involve an element of "dogmatic" risk and prefers the concept of "reality." In this way all reference to knowledge as a "being" is excluded; the reality of which Banfi speaks is identified with the boundaries of the process of knowing. Here the difficulties of his own position appear. Indeed, once the meaning of science and philosophy as true theory is lost, no criterion which is not exclusively pragmatic remains to make a philosophical criticism of knowledge work. Truth is seen by him as founded exclusively in the vital manifestations of the subject, in an experience (*Erlebnis*) which is conceived in a way situated between the views of Dilthey and Simmel. Thus Banfi oscillates, in spite of himself, between pragmatism and historicism: the fact is assumed and explained by its historical genesis but deprived of the possibility of any other type of explanation.

Moreover, it is only right to remark that, in contrast to Husserl, who in his continuous attempt to arrive at the foundation of knowledge, has reached in his analysis truly radical conclusions with an exceptional wealth of research, Banfi in even his major work previously discussed (*Principi di una teoria della ragione*) has determined only a few fundamental points. But Banfi never identifies gnoseological philosophy with philosophy *tout court* and, in fact, his ideal of philosophy as the arrangement of culture leads him to the analysis of other sectors of the life of the mind.

Finally the question should be raised whether the Banfian assumption of the Marxist attitude does not imply precisely an option for an evaluating choice of the point of view with the consequent risk of crystallizing the systematic structure of knowledge? In this case it is Banfi who, instead of blaming Husserl, could blame himself for having taken a rigid, one-sided position. Banfi avoids falling into a "dogmatic attitude" but at the price of reducing phenomenology to some particular themes which are more in tune with his interests.

V. CONCLUDING NOTE

The Banfian version or brand of phenomenology which I have attempted to present, albeit in an outline, appears to be restricted when compared with

Husserl, leaving out several important requirements. To conclude our study we cannot refrain from asking: "What did it mean for Banfi to introduce the Husserlian reflection to a culture such as Italian culture, so slow to open toward the progress of European thought?" We may answer this question by surmising that it means, in the first place, bringing into the limelight a systematic and efficacious attempt to delineate the infinite development of knowledge independently of any presuppositions. Then, notwithstanding the divergencies between the two philosophers, the Banfian interpretation of Husserl is to be seen as a desire to weld together the two souls, the theoretical and the practical, of European thought. But above all, it means – putting aside the mediocrity of the idealistic disputes – to affirm the need to "begin again at the beginning." We owe to Banfi the development of some of the directions in which the thought of Husserl has been dominant both abroad (in 1946 Banfi was asked to join the editorial staff of the periodical *Philosophy and Phenomenological Research*) and as the Banfian school shows in Italy. Enzo Paci maintained that Banfi is one of the few philosophers who, with or against Husserl, has followed the road (which a large part of contemporary philosophy still has to travel) which leads to the understanding of phenomenology. He has understood the "heroism" of the Husserlian reason and has properly appreciated in radically positive character.[22]

Banfi wrote in 1939:

> With Husserl the greatest contemporary philosopher dies. For it was he who, in the crisis of thought at the turn of the century, amid the conflicts of traditions, the uncertainty of directions, the extreme limitations of speculative vision, rediscovered, we might say, the same idea of reason, affirmed its absolute validity as a criterion of philosophy and of knowledge in general, translated it into a system open to all experience, and relived for himself and illuminated for others the meaning and the spiritual value of this new wisdom.[23]

For this reason "the words of Husserl still find an echo which is perhaps even more alive in us and which only the words of human truth awaken." [24]

University of Catania

Translated by M. Mason Sanfilippo

NOTES

[1] Cf. I. Bona in *Omaggio a Husserl*, Il Saggiatore, Milan, 1960 p. 284; Aldo Masullo, *Logica, psicologia e filosofia*, Il Tripode, Naples, 1962, p. 40; and Carlo Sini, *La fenomenologia*, Garzanti, Milan, 1965, pp. 18, 28, 149, 153.

[2] Cf. A. Banfi, 'La tendenza logistica nella filosofia tedesca e le Ricerche logiche di Husserl,' *Rivista di folosofia*, no. 3 (1923).

[3] Cf. A. Banfi, 'La fenomenologia pura di E. Husserl e l'autonomia ideale della sfera teoretica,' *Rivista di filosofia*, no. 3 (1923).
[4] Cf. A. Janik and S. Toulmin, *La grande Vienna*, Garzanti, Milan, 1975.
[5] Cf. A. Banfi, *Principi di una teoria della ragione*, Editori Riuniti, Rome, 1967, p.X.
[6] Cf. *Ibid.*, pp. 433–449.
[7] Cf. A. Banfi, 'Husserl e la crisi della civiltà europea,' *Aut-Aut*, no. 43–44 (1958), –17.
[8] Cf. A Banfi, *Principi di una teoria della ragione, op. cit*, pp. 51–60; and 'Del conoscere intuitivo,' *Aut-Aut*, no. 54 (1959), 363–73.
[9] *Ibid.*, pp. 436–37.
[10] *Ibid.*, pp. 439–40.
[11] *Ibid.*, p.45.
[12] A. Banfi, *La crisi*, preface by Carlo Bo, All'insegna del pesce d'oro, Milan, 1967, p. 11.
[13] A. Banfi, *Principi di una teoria della ragione*, pp. 441–42.
[14] Cf. E. Paci, 'Antonio Banfi vivente, in *Antonio Banfi e il pensiero contemporaneo.* Atti del Convegno di studi banfiani Reggio Emilia, 13–14 March, 1967. La Nuova Italia, Firenze, 1969, p. 40.
[15] Cf. A Banfi, *Sui principi di una filosofia della morale*, in *La ricerca della realtà* II, Sansoni, Firenze, 1959. On the theme of pragmatic reflexion in Banfi cf. also Guido Davide Neri 'La riflessione pragmatica nella fenomenologia di Banfi,' *Aut-Aut*, pp. 75–80, and 'Riflessione pragmatica e riflessione teoretica,' in *Antonio Banfi e il pensiero contemporaneo*, pp. 18–33.
[16] Enzo Paci, 'Antonio Banfi vivente,' p. 37.
[17] *Ibid.*, p. 39.
[18] A. Banfi, *Principi di una teoria della ragione*, p. 441.
[19] Cf. A Banfi, 'La fenomenologia e il compito del pensiero contemporaneo,' in the anthology *Omaggio a Husserl*, p. 41.
[20] E. Husserl, *Die Krisis der europäischen Wissenschaften und die transzendentale Phänomenologie*,' in *Husserliana*, vol. 6, Den Haag, 1954, p. 273.
[21] P. Piccone, Review of *Funzione delle scienze e significato dell'uomo, Telos* 1 no. 1 (Spring, 1968), 52–53.
[22] A. Banfi, *Filosofi contemporanei*, Parenti, Firenze, 1961, p. 125.
[23] A. Banfi 'Husserl e la crisi della civiltà europea,' p. 10.
[24] *Ibid.*

ROSELINA SALEMI

BIBLIOGRAPHY OF HUSSERLIAN STUDIES IN ITALY

With an introduction by Angela Ales Bello

ANGELA ALES BELLO

INTRODUCTION

The following bibliography, arranged chronologically, permits the reader to follow the development of phenomenological studies in Italy in parallel with other, contemporary, cultural currents. From this list it can be seen that knowledge of Husserl's work begins in 1923 with the studies of A. Banfi. Phenomenology, however, did not immediately receive a warm welcome. It contrasted with the then dominant neo-idealism (as has been made clear by G. De Ruggiero), but for this very reason it also found adherents among the opponents of idealism. These were either distant heirs of positivism, who accepted Husserl on account of his scientific approach and rigor, or Christian-oriented thinkers, who, following an initial period of diffidence toward the antimetaphysical attitude of phenomenological analysis, gradually began to use this method as an antiidealist instrument – even though the problem remained of Husserl's own transcendental idealism and the value to be attributed to it.

Despite the difficulties encountered on the way, the numerous studies carried out in Italy prior to World War II make it clear that the better known philosophers who have left a mark on Italian culture already had begun to take a discreet interest in phenomenology.

The fact that only five essays were published between 1940 and 1945 (namely, those of A. Banfi, E. Paci, G. Preti, A. Aliotta, and L. Geymonat) is readily explained by the difficulties that the serene pursuit of intellectual goals encountered in this time of conflict; on the other hand, these essays represent the link between the prewar period and the resumption of phenomenological studies following the return of peaceful conditions, and they therefore assume particular importance. .

An outstanding position among the aforementioned thinkers was occupied by E. Paci, not only on account of the large number of his publications devoted to phenomenology, but because he was one of the very few to try to apply the phenomenological method in a concrete manner to the various fields of knowledge. Moreover, to this day he remains the only Italian phenomenologist who can be said to have founded a school of his own; his numerous disciples dedicated themselves to phenomenology, though many of them later came to turn their attention to other philosophical experiences.

Tymieniecka (ed.), Analecta Husserliana, Vol. IX, 463–467. *All Rights Reserved.*

Evidence of this can be found in the review *Aut-Aut*, a publication directed by Paci, which in recent years (and especially after Paci's death) has gradually lost the phenomenological hallmark that constituted its original characteristic.

Following an early rejection, and notwithstanding a continued diffidence among Christian-oriented philosophers who are usually bound to neo-Thomist metaphysics one can now discern a deepening interest in phenomenology among these philosophers that is indicative of their confidence in being able to employ it in a constructive manner.

Though a series of discussions documented in the proceedings of the 1955 Congress of the Gallarate Study Center and in the volume *Bilancio della fenomenologia e dell'esistenzialismo* (*Balance Sheet of Phenomenology and Existentialism*), and as a result of the interest of S. Vanni Rovighi, thinkers such as P. Valori and M. Sancipriano have exerted a considerable influence in the development of Italian phenomenology in Italy.

Some perplexities arose in connection with the problems caused in the gnoseological field by Husserl's realism or idealism, and in the metaphysical field by the reduction of phenomenology to a pure method or its metaphysical aperture. Interpreted as a methodology, Husserl's views were either rejected or accepted by the very fact of being such: metaphysicians considered his phenomenology to be reductive, while those who believed in the identity of methodology and philosophy believed that they had found a valid solution to their problems.

On the other hand, it was as a method that phenomenology seemed applicable to two fields of enquiry that, in the 1950s, had received much attention in Italy, namely, science and history.

The group of thinkers working with and around Paci proved to be particularly sensitive as regards these two types of research. Although he did not accept the absolutization of science in the neo-positivist sense, Paci considered the development of scientific research as something fundamental; he wanted to go to "the heart of the matter" by leading research back to its "precategorial matrix" – which he considered to be authentically "rational" – and permitting it to regain its philosophical dimension. Moreover, it is the merit of Paci's school to have stimulated phenomenological research in a number of special fields, as is evident from the contributions that he and his disciples made to *Aut-Aut* dealing with the various expressions of modern (contemporary) art or the relationships between phenomenology and such philosophical and cultural currents as structuralism and anthropology (that were gradually replacing historicism).

As far as the enquiry into history is concerned, one can note a clash or an encounter with Marxism that depends on whether the individual concerned is or is not ready to consider Marxism reconcilable with phenomenology in the project for constructing a humane world (e.g., Paci, G. Semerari, F. Melandri). The reasons underlying the possibility of an encounter, at least for those who held this opinion, were to be found in the fact that Marxist historicism, even though it was put forward as an alternative to Croce's historicism, inherited the "humanist" aspect; moreover, it was believed that a scientific reading of history could respect the need for radicality that is peculiar to phenomenology.

Concerning the interest in science, the separation between science and scientificity did not always emerge very clearly, indeed, the separation was perhaps problematical at times even for Husserl himself, even though it was clearly expressed in other parts of his analysis. For this reason one finds that the works that endeavour to highlight Husserl's epistemology are rather few and far between (G. Scrimieri, E. Melandri, F. Voltaggio), while those that exalt a generic ideal of science are far more numerous.

If we concentrate our attention on the more recent publications, in an attempt to discern the orientations of the various contributors, we note that by about 1970 there no longer existed any organized groups of phenomenology scholars; all research in this field was now carried out not only by individuals, but by individuals working in isolation. This led some people to feel the need for a research centre, not so much as a means of forming a "school," but rather to permit a confrontation of the various interpretations of phenomenology and its theoretical development.

The *Centro Italiano di Ricerche Fenomenologiche* (Italian Center for Phenomenological Research) was founded in Rome in 1974, following the initiative of A.-T. Tymieniecka, by a group composed of Mario Sancipriano, Angela Ales Bello, Paolo Valori, F. Liverziani, Elio Costantini, Aurelio Rizzacasa and others. Affiliated at first with the I.H.P.R.S. (*International Husserl and Phenomenological Research Society*), it later extended its affiliation to *the World Institute for Advanced Phenomenological Research and Learning, Belmont, Ma.* It came to embrace thinkers who had undergone different formation processes, including (1) phenomenologists who accepted Husserl's method only after a confrontation with other philosophical experiences and who had then applied it to various fields of research and (2) young scholars who had followed the phenomenological orientation from the very beginning. Some of the members of the Italian Center (e.g., Sancipriano and Valori) were among the first to start phenomenology in Italy. In fact, Sancipriano was one of the first scholars to attempt a detailed analysis of Husserl's writings, including such unpublished texts as the one about the

Urkind; he then proceeded, via an examination of Husserl's logic and the dynamics of intentionality, to investigate the constitution of the spiritual acts, the results being given in his lectures at the Italian Center of Phenomenology. Sancipriano, indeed, has dedicated himself with continuity and coherence to the development of his interpretation and to the gaining of greater theoretical insight into Husserl's phenomenology.

Valori, on the other hand, highlighted the ultimate significance of the phenomenological method and, more particularly, applied it to the field of ethics. In fact, his two fundamental works are concerned with the possibilities of the method itself and its coherent utilization in ethical research. Nevertheless, even though he accepts the need for radicality and sincerity, the anti-dogmatism and the antisystematicness of the method proposed by Husserl, he noted an insufficiency on the ontological level and holds that this should be the subject of further research.

The further study of of life-world (*Lebenswelt*) – a main concern of phenomenological literature at the international level during the last two decades and already studied by both Paci and Melandri – is now undergoing an interesting development among the members of the Italian Centre. A. Rigobello has highlighted its ethical implications by showing that it represents the fundamental moment that has to be complied with by a moral law.

A. Ales Bello has pointed out its fundamental importance in connection with the theme of history, which may thus be made independent of any formulation of a historicist type. B. M. D'Ippolito has utilized it as the ground for the ultimate confrontation of historicity and science in their dynamic tension. S. Zecchi, who remains substantially faithful to the position assumed by Paci, seeks the roots of ideal objectivity, and therefore of the objective sciences, in the originary evidence of the life-world. D. Conci considers the life-world as a determinant moment for an authentic phenomenological enquiry dealing with the precategorical. One has to rediscover the validity of Husserl's method in order to purify it from the residues of (1) objectivism that can still be found in the eidetic intuition and (2) subjectivism connected with the cogito; it is then necessary to pay particular attention to the development of the analysis of the precategorial processes (*Erlebnisse*) as being beyond the grasp of all scientific and metaphysical construction.

Considered as a new reflection about method, one can justify this interest in the phenomenological movement in all its various ramifications; in this connection one cannot forget thinkers like a R. Ingarden and E. Stein, in order to sample the significance of their relationship to Husserl.

Nevertheless, an examination of the recent literature shows one that there still remains the ambiguity between the attempt to subtract the method from each and every contamination (in order to rediscover it in its authenticity) and the desire to utilize it in terms of other formulations or other philosophies, something that is brought out, for example, by P. Rovatti in his Marxist and phenomenological researches into history.

A further example may be offered by the applications of phenomenology to psychiatry; although these are producing very noteworthy results, one often encounters a contamination by and through existentialism (e.g., V. Callieri), even though some scholars (e.g., M. De Negri) express the need for remaining rigorously anchored to the method developed by Husserl.

For this reason it seems necessary to revise the method, a revision that could be a useful instrument of comprehension in the various fields of knowledge. To this end it would be highly desirable to have an Italian translation of all of Husserl's published works, a task that has not yet been completed and, indeed, has made no progress at all during the last twenty-odd years.

Italian Center for Phenomenological Research,
Rome

ROSELINA SALEMI

BIBLIOGRAPHY OF HUSSERLIAN STUDIES IN ITALY

1923

Banfi, A. 'La tendenza logistica della filosofia tedesca contemporanea e le *Richerche logiche* di E. Husserl,' *Rivista di filosofia,* no. 2 (1923).

Banfi, A. 'La fenomenologia pura di E. Husserl e l'autonomia ideale della sfera teoretica,' *Rivista di filosofia,* no. 3 (1923).

1926

Banfi, A. *Principi di una teoria della ragione,* Milan: Isis, 1926, pp. 557–77.

1928

Grasselli, G. 'La fenomenologia di E. Husserl e l'ontologia di M. Heidegger,' *Rivista di filosofia,* no. 4 (1928).

1929

Grassi, E. 'Sviluppo e significato della scuola fenomenologica,' *Rivista di filosofia,* no.2 (1929).

1931

Banfi, A. 'Filosofia fenomenologica,' *Cultura,* no. 6 (1931).

Mazzantini, C. 'L'undicesimo volume dell'annuario della scuola fenomenologica di Husserl,' *Rivista di filosofia, neoscolastica,* no. 3 (1931).

1932

Pastore, A. 'Contributo all'interpretazione dell'ontologia eidetica di Husserl,' *Rivista di filosofia,* no. 4 (1932).

1933

Della Volpe, G. 'Fenomenologia,' in *Enciclopedia italiana di scienze, lettere e arti:* Treccani, vol. 14, Milano, 1933, pp. 1106–7.

Tymieniecka (ed.), Analecta Husserliana, Vol. IX, 469–484.

De Ruggiero, G. Husserl e la fenomenologia' in G. de Ruggiero, *Filosofi del novecento,* Bari: Laterza, 1933: 2nd ed., 1942, pp.
Drago, P.C. *La genesi del problema fenomenologico,* Messina: Principato, 1933.
Grassi, E. *Dell'apparire e dell'essere,* Florence: La Nuova, Italia, 1933.
Grassi, E. 'Edmund Husserl, in *Encicilopedia italiana di scienze, lettere e arti:* Trenccani, vol. 18, Milan, 1933, p. 613.
Pastore, A. 'Husserl, Heidegger, Chestov,' *Archivio di filosofia,* no. 2 (1933).
Vanni Rovighi, S. 'Il valore della fenomenologia, a proposito di una discussione promossa dalla Société Thomiste,' *Rivista di filosofia neoscolastica,* no. 3 (1933).

1934

Bobbio, N. *L'indirizzo fenomenologico della filosofia sociale e giuridica,* Torino: Istituto Giuridico, 1934.
Preti, G. 'Filosofia e saggezza nel pensiero husserliano,' *Archivio di filosofia,* no. 1 (1934).

1935

Alliney, G. 'Rassegna di opere sulla fenomenologia di Husserl: Levina, Driesch, Kraft, Kreis,' *Sophia,* no. 1 (1935).
Bobbio, N. 'La filosofia di Husserl e la tendenza fenomenologica,' *Rivista di filosofia,* no. 1 (1935).
Preti, G. 'I fondamenti della logica formale nella *Wissenschaftslehre* di B. Bolzano e nelle *Logische Untersuchungen* di E. Husserl,' *Sophia,* no. 2–4 (1935).

1936

Filiasi Carcano. P. 'Da Cartesio a Husserl,' *Richerche filosofiche,* no. 1 (1936).
Pastore, A. *Logica del potenziamento,* Naples: A Rondinella, 1936.

1937

Vanni Rovighi, S. 'Il 'cogito' di Cartesio e il 'cogito di Husserl', *Cartesio,* Milan, 1937.

1938

Martinetti, P. 'La teoria del valore di F. Brentano,' *Rivista di filosofia* (1938).
Paci, E. *Il significato del Parmenide nella filosofia di Platone,* Milano: Principato, p. 230 ff.
Pelloux, L. 'E. Husserl,' *Studium* (1938).
Stefanini, L. 'Esistenzialismo eidetico' in L. Stefanini: *Il momento dell'educatione,* Padua: CEDAM, 1938, pp. 15–61.

Vanni Rovighi, S. 'Edmund Husserl,' *Rivista di filosofia neoscolastica,* no. 3 (1938).

1939

Banfi, A. 'Edmund Husserl,' *Civiltà Moderna,* no. 1 (1939) 52–63.
Banfi, A. 'Edmund Husserl e il razionalismo umanistico,' *Rêvue Internationale de philosophie,* no. 1 (1939).
Massolo, A. 'Husserl e il cartesianesimo,' *Giornale critico della filosofia,* no. 6 (1939).
Vanni Rovighi, S. *La filosofia di Edmund Husserl,* Milan: Vita e Pensiero, 1939.

1940

Banfi, A. 'Husserl postumo,' *Rivista di filosofia,* no. 1 (1940).
Paci, E. 'Temi fondamentali del pensiero di Husserl' in E. Paci: *Pensiero, esistenza e valore,* Messina, 1940.
Preti, G. 'E. Husserl,' *Nuova Italia,* no. 2 (1940).

1941

Aliotta, A. 'La fenomenologia, il circolo di Vienna e il mio sperimentalismo,' *Logos,* no. 1 (1941).

1945

Geymonat, L. 'Husserl e il problema dell'evidenza' in L. Geymonat: *Studi per un nuovo razionalismo,* Turin, 1945.

1946

Vanni Rovighi, S. 'La fenomenologia di E. Husserl,' *Humanitas,* no. 2 (1946).

1947

Vanni Rovighi, S. *Husserl,* Brescia: La Scuola, 1947.

1948

Castelli, E. *Introduzione ad una fenomenologia della nostra epoca,* Florence: Fussi, 1948.
Pfister, F. *Il metodo della scienza,* Florence: Sansoni, 1948.

1949

Sancipriano, M. 'Note sulla teoria della conoscenza e sull'ontologia di Sofia Vanni-Rovighi (Recensione),' *Humanitas*, Brescia, 4, no. 7 (1949) 702–7.

1950

Abbagnano, N. *Storia della filosofia*, Turin, 1950, pp. 627–42.
Alliney, G. 'Il posto delle *Ideen* nella fenomenologia di Husserl,' pref. alla trad. italiana di E. Husserl: *Idee per una fenomenologia pura*, Turin, 1950, pp. 13–31.
Castelli, E. 'Il solipsismo e l'analisi fenomenologica,' *Archivio di filosofia* (1950).
Paci, E. *Il pensiero scientifico contemporaneo*, Florence: Sansoni, 1950.

1951

B. N. 'Recensione di E. Husserl: Die Idee der Phänomenologie,' *Rivista di filosofia*, (April–June, 1952).
Ballanti, G. 'Psicologismo ed esistenzialismo,' *Rivista di filosofia neoscolastica*, 43 (1951).
Lazzarini, L. 'L'intenzionalità nella conoscenza metafisica,' *Giornale di metafisica*, 6 (1951).
Spirito, U. 'Significato della fenomenologia; nella raccolta: Fenomenologia e Sociologia,' *Archivio di Filosofia* (1951).
Vanni-Rovighi, S. 'Il colloquio internazionale di fenomenologia,' *Rivista di filosofia neoscolastica*, 3 (1951).

1952

Castelli, E. 'L'indagine filosofica e la filosofia dell' alienazione,' *Filosofia e psicopatologia, Archivio di filosofia* (1952).
Lazzarini, L. 'Etica negativa e intenzionalita,' *Giornale di metafisica*, 7 (1952).

1953

Formaggio, D. 'L'arte come comunicazione,' *Fenomenologia delle tecnice artistice*, Milan, 1953.
Valori, P. 'Husserl e Kierkegaard,' *Archivio di filosofia*, no. 2 (1953).
Valori, P. 'Essenza e significato della fenomenologia husserliana,' *Giornale di metafisica*, 8 (1953).

1954

Diano, C. 'Fenomenologia dell'arte,' *Delta*, no. 6 (1954).

Fabro, C. *Storia della filosofia,* Rome: Coletti, 1954.

Valori, P. 'Inediti husserliani sulla teleologia della storia,' *Archivio di filosofia,* 1 (1954).

1955

La Fenomenologia. Atti dell'XI Convegno di Gallarate, Brescia, 1955.

Giacon, C. 'La fenomenologia,' *Giornale di metafisica,* 11 (1955).

Giacon, C. 'Husserl al Convegno di Gallarate,' *L'Italia che scrive,* no. 12 (1955).

Lazzarini, R. *Intenzionalità e istanza metafisica,* Rome: Bocca, 1955, Chap. 8.

1956

Sancipriano, M. 'Intuizione e possesso nella fenomenologia di Edmund Husserl,' In *La fenomenologia,* Atti dell'XI Convegno del Centro di Studi Filosofici tra professori universitari, Gallarate, 1955, Brescia: Morcelliana, 1956, pp. 151–60.

Valori, P. 'Storia della fenomenologia husserliana,' In *La fenomenologia,* Atti dell'XI Convegno di Gallarate, Brescia: Morcelliana, 1956, pp. 68–84.

1957

Faggin, G. 'Fenomenologia,' in *Dizionario di filosofia,* Milan, 1957, pp. 296–98.

Faggin, G. 'Il compito della fenomenologia', *Archivio di filosofia,* Padua, 1957.

Filiasi Carcano, P. *La metodologia nel rinnovarsi del pensiero contemporaneo,* Naples, 1957, (Chap. 7, 'La funzione metodologica della filosofia.')

Paci, E. 'Per la logica di Husserl,' *Aut-Aut,* no. 42 (1957).

Paci, E. *La filosofia contemporanea,* Milan: Garzanti, 1957, Chap. 7.

Paci, E. *Dall'esistenzialismo al relazionismo,* Messina: Firenze, 1957, Chaps. 13–14.

Sancipriano, M. *L'evoluzione ideale–Fenomenologia pura e teoria dell'evoluzione,* Brescia: Morcelliana, 1957.

Valori, P. 'Il punto di partenza della filosofia di Husserl e in Zamboni,' in *Studi sul pensiero* di G. Zamboni, Milan, 1957.

Valori, P. 'Fenomenologia,' In *Enciclopedia filosofica,* Centro di Studi Filosofici di Gallarate, Venezia-Rome, 1957, Vol. 2, pp. 334–40.

Vanni Rovighi, S. 'Husserliana,' *Giornale di filosofia neoscolastica,* 49 (1957).

Vanni Rovighi, S. 'Fenomenologica–Scuola,' In *Enciclopedia filosofica,* Centro di Studi Filosofici di Gallarate, Venezia-Rome, 1957, Vol. 2, pp. 340–41.

Vanni Rovighi, S. 'E. Husserl,' in *Enciclopedia filosofica,* Centro di Studi Filosofici di Gallarate, Venezia-Rome, 1957, Vol. 2, pp. 1146–52.

Vanni Rovighi, S. 'Il colloquio sulla fenomenologia,' *Rivista di filosofia neoscolastica,* 49 (1957).

1958

Banfi, A. 'Husserl e la crisi della civiltà europea,' *Aut-Aut,* nos. 43–44 (1958), 1–17.

Bartolini, P. *Fenomenologia e pedagogia,* Bologna: Malipiero, 1958.
Paci, E. 'Meditazioni fenomenologiche,' *Aut-Aut,* no. 47 (1958).
Paci, E. 'Introduzione,' a: M. Merleau-Ponty, *Elogio della filosofia,* Turin, 1958.
Paci, E. 'Tempo e percezione,' *Archivio di filosofia,* 1 (1958).
Pedroli, G. *La fenomenologia di Husserl,* Turin: Taylor, 1958.
Petruzzellis, N. 'Logica e fenomenologia,' *Rassegna di scienze filosofiche,* 2 1958).
Valori, P. 'Evidenza e verità nella fenomenologia husserliana,' In *Aquinas,* Rome, 1958, pp. 224–40.
Vanni Rovighi, S. 'La teoria dell'intenzionalità nella filosofia di Husserl,' *Rivista di filosofia neoscolastica* (1958).

1959

Becchi, E. 'Fenomenologia e Gestalt,' *Aut-Aut,* no. 50 (1959).
Costa, F. *Introduzione à filosofia come scienza rigorosa,* Turin: Paravia, 1959.
Fanizza, F. 'Un'idea della fenomenologia,' *Aut-Aut,* 54 (1959), 423–29.
Filippini, E. 'Sull'idea di una storiografia nella *Krisis* di Husserl,' *Aut-Aut,* 54 (1959) 413–16.
Lugarini, L. 'L'idea della filosofia prima di Husserl,' Aut-Aut, 54 (1959) 377–93.
Neri, G. 'Nota su Banfi e Husserl,' *Aut-Aut,* 54 (1959) 373–77.
Paci, E. *Il nulla e il problema dell'uomo,* 2nd ed. Turin, 1959.
Paci, E. 'Tempo e riduzione in Husserl,' *Rivista di filosofia,* (1959).
Paci, E. 'Sul significato del platonismo di Husserl,' *Acme* 1–3 (1959).
Paci, E. 'Pagine da un Diario,' *Archivio di filosofia* (1959).
Paci, E. 'Introduzione alla fenomenologia,' *Aut-Aut,* 50 (1959).
Paci, E. 'Sulla relazione Io-Tu,' *Aut-Aut,* 52 (1959).
Paci, E. 'Esercizio sull'evidenza fenomenologica,' *Aut-Aut,* 54 (1959).
Paci, E. 'Tempo e dialettica in Husserl,' *Pensiero,* 2 (1959).
Paci, E. 'Wright e lo spazio vissuto,' *Casabella,* Milan, 1959.
Paci, E. 'La crisi della cultura e la fenomenologia dell' architettura contemporanea,' *La casa,* (Rome) 6 (1959).
Paci, E. 'Ungaretti e l'esperienza della poesia,' *Letteratura,* nos. 35–36 (1959).
Paci, E. 'Sul significato dello spirito in Husserl,' *Aut-Aut,* 54 (1959) 345–63.
Piana, G. 'Sul problema dell'intersoggettività,' *Aut-Aut,* 54 (1959) 416–18.
Sancipriano, M. 'Henri Bergson e Edmund Husserl,' *Humanitas,* 14, no. 11 (1959).
Santucci, G. *L'esistenzialismo e la filosofia italiana,* Bologna: Il Mulino, 1959.
Scrimieri, G. *Problemi di logica. Studio sui 'Prolegomena' con ausilio di manoscritti inediti,* Bari: Milello, 1959.
Semerari, G. 'L'introduzione alle *Meditazioni Cartesiane,*' *Aut-Aut,* 54 (1959) 393–412.
Valori, P. *Il metodo fenomenologico e la fondazione della filosofia,* Rome: Desclée, 1959.

1960

Bilancio della fenomenologia e dell'esistenzialismo, Relazioni di E. Paci, E. Garin, P. Prini, Padua: Liviana, 1960.

Omaggio a Husserl, Saggi di A. Banfi, E. Filippini, G. Guzzoni, L. Lugarini, E. Malandri, G.D. Neri, E. Paci, G. Pedroni, R. Pucci, G. Semerari, S. Vanni Rovighi. A cura di E. Paci. Milan: Il Saggiatore, 1960.

Tempo e intenzionalità husserliana. Scritti di Husserl, E. Paci, P. Caruso, E. Renzi, S. Bosio, V. Fagone, G. Piana, W. Biemel, L. Lugarini, A. Plebe, F. Bianco, J. Wyrsch. *Archivio di filosofia,* 1 (1960).

Anceschi, L. '*Krisis*, Beilage III zur pfg. 9a,' *Pensiero,* 5 (1960).

Cantoni, R. 'E.H. e la critica dell'atteggiamento naturale,' *Il pensiero critico* (July–September 1960) 1–46.

De Crescenzio, G. 'I valori nella husserliana filosofia della storia,' *Sapienza,* 13 (1960) 435–48.

Lugarini, L. 'La problematica husserliana della filosofia prima,' *Pensiero,* 5 (1960) 248–60.

Melandri, E. *Logica ed esperienza in Husserl,* Bologna: Il Mulino 1960.

Paci, E. 'Doxa e individuazione nella fenomenologia di Husserl,' *Rivista di filosofia* (1960) 144–61.

Paci, E. *I problemi del tempo nella fenomenologia di Husserl,* Milan: La Goliardica, 1960.

Paci, E. 'Coscienza fenomenologica e coscienza idealistica,' *Il Verri,* no. 4 (1960) 131–38.

Paci, E. 'Sul problema dell'intersoggettivita,' *Pensiero,* 5 (1960) 291–325.

Riverso, E. 'Il valore della logica in E. Husserl,' *Sapienza* 13 (1960) 251–56.

Sancipriano, M. 'Spunti ontologici e appunti di metodo fenomenologico,' *Theorein,* 1 (1960) 127–35.

Sancipriano, M. 'Il mondo, l'io e il tempo nella fenomenologia,' *Filosofia e vita* (Turin) 1, no. 4 (1960) 73–80.

Sciacca, G.M. *Esistenza e realtà in Husserl,* Palermo: Palumbo, 1960.

Scrimieri, G. 'Dalle *Logische Untersuchungen* di E. Husserl,' In *Annali della facoltà di lettere e filosofia,* Bari, 1960, pp. 169–204.

1961

Bosio, F. 'La genesi della logica formale dall'esperienza antepredicativa, in *Erfahrung und Urteil* di E. Husserl,' *Pensiero,* 6 (1961) 334–56.

Chiodi, P. 'Husserl e Heidegger,' *Rivista di filosofia,* 3 (1961) 192–211.

Garulli, E. 'Due fasi del pensiero husserliano: le *Richerche Logiche* e la *Krisis,*' *Studi urbinati di storia, filosofia e letteratura,* 35 (1961) 192–235.

Minozzi, B. *L'idealismo fenomenologico di E. Husserl,* Bologna: Il Mulino, 1961.

Paci, E. Tempo e verità nella fenomenologia di Husserl, Bari: Laterza, 1961.

Paci, E. 'Ueber einige Verwandtschaften zwischen der Philosophie Whiteheads und der Phenomenologie Husserls,' *Revue Internationale de philosophie,* 15, nos. 56–57 (1961) 237–50.

Paci, E. 'Fenomenologia e obiettivazione,' *Giornale critico della filosofia italiana,* 40 (1961) 143–52.

Paci, E. 'Fenomenologia, psicologia e unità delle scienze,' *Aut-Aut,* no. 63 (1961) 214–34.

Paci, E. 'Relazionismo e significato fenomenologico del mondo,' *Pensiero* (1961) 28–51.

Sancipriano, M. 'Mondo, io e tempo nei manoscritti di Husserl di Gerd Brand' (Recensione), *Humanitas,* (Brescia) 16 no. 1 (1961) 85–86.

Sancipriano, M. 'Nel centenario di E. Husserl, 1859–1959: verita e libertà nella fenomenologia,' In Atti del XVIII Congresso nazionale della società filosofica italiana *(Verità e libertà).* Palermo: Palumbo 1961, Vol. 2 pp.657–64.

Sancipriano, M. 'Per una fenomenologia della conversione' in *Il problema dell'esperienza religiosa.* Brescia: Morcelliana, 1961, pp. 138–41.

Sancipriano, M. *L'evoluzione ideale. Fenomenologia pura e teoria dell'evoluzione,* 2nd ed. Brescia: Morcelliana, 1961.

Scrimieri, G. 'I sensi della *Krisis der europaischen Wissenschaften* und *die transzendentale Phänomenologie* di E. Husserl e la mia prospettiva filosofica,' *Annali della facolta di lettere e filosofia,* (Bari) no. 7 (1961) 271–84.

Valori, P. 'Recenti studi sulla fenomenologia husserliana,' *Civiltà cattolica* 1, no. 112 (1961) 278–85.

1962

Bosio, F. 'Psicologia e logica nella fenomenologia di Husserl,' *Aut-Aut,* no. 68 (1962) 131–54.

Bosio, F. 'La lotta contro lo psicologismo e l'idea della logica pura in Husserl,' *Aut-Aut,* no. 71 (1962) 373–82.

Bosio, F. 'L'intenzionalità e il concetto di coscienza nelle *Logische Untersuchungen,*' *Aut-Aut,* no. 72 (1962) 479–504.

Bosio, F. 'Gli inizi della fenomenologia – la filosofia dell' aritmetica,' *Aut-Aut,* no. 70 (1962) 294–308.

Fànizza, F. 'Tempo e verità nella fenomenologia di Husserl,' *Giornale critico della filosofia italiana,* 41 (1962) 405–17.

Garulli, E. 'Realtà e giudizio nella logica husserliana,' *Studi urbinati di storia, filosofia e letteratura,* 35 (1962) 282–98.

Garulli, E. 'La phénoménologie de Husserl vue par E. Paci,' *Rèvue de Mètaphysique et de morale,* 67 (1962).

Gatta, E.A. 'I temi della fenomenologia e l'origine della geometria,' *Annali della facoltà di lettere e filosofia* (Bari) 8 (1962) 5–26.

Masullo, A. *Logica, psicologia, filosofia,* Naples: Il Tripode, 1962.

Negri, A. 'Husserl filosofo senza storia?' *Giornale critico della filosofia italiana,* 41 (1962) 177–93.

Paci, E. *Fenomenologia e antropologia. Corso di filosofia teoretica, anno acc. 1961–62.* Milan: La Goliardica, 1962. Appendix: C. Sini, 'Lezioni introduttive al pensiero di Husserl.'

Piana, G. 'Accomunamento, storicità e tradizione nello Husserl inedito,' *Aut-Aut,* no. 68 (1962) 113–30.

Pozzan, A.M. 'La Krisis nell'ambito delle correnti fenomenologiche europee e americane,' *De homine,* 1, nos. 2–3 (1962) 205–11.

Sancipriano, M. *Il logos di Husserl – genealogia della logica e dinamica intenzionale,* Turin: Bottega d'Erasmo, 1962.

Scrimieri, G. 'Introduzione allo studio sulla *Dingvorlesung* del 1907 di E. Husserl,' *Annali della facoltà di lettere e filosofia* (Bari) 8 (1962).

Valori, P. 'La filosofia di fronte alle scienze nella fenomenologia husserliana,' In *La filosofia di fronte alla scienze*, Bari: Milello, 1962, pp. 2–6.

1963

Bona, I. 'Su alcuni manoscritti di Husserl,' *Aut-Aut*, no. 75 (1963) 7–15.

Bosio, F. 'Problematica della riduzione e costituzione trascendentale,' *Aut-Aut*, no. 74 (1963) 62–83.

Bosio, F. 'La riduzione al mondo della vita,' *Aut-Aut*, no. 75 (1963) 62–83.

Bosio, F. 'Il paradosso della soggettività,' *Aut-Aut*, no. 76 (1963) 17–35.

Cammarota, P. 'Revisione odierna del problema gnoseologico. Le conclusioni fenomenologiche dello Husseril,' *Studia Patavina*, 10 (1963) 450–57.

Chiodi, P. *Esistenzialismo e fenomenologia*, Milan: Comunità, 1963.

Costa, F. *Che cos'è la fenomenologia? Husserl e la corrente filosofica più attuale e rigorosa*, Milan: Silva, 1963.

Franchi, A. 'Realtà e mediocrità della filosofia di Husserl,' *Giornale di metafisica*, 18 (1963) 561–75.

Iacobelli Isoldi, A.M. *Il tempo in Kant e i suoi sviluppi in Husserl e Heidegger*, Rome: E. de Santis, 1963.

Oggioni, E. *La fenomenologia e il pensiero contemporaneo*, Bologna: Patron, 1963.

Pucci, R. 'Io transcendentale e mondo della vita nella fenomenologia di Husserl,' *Rassegna di scienze filosofiche*, 16 (1963) 67–89.

Pucci, R. 'Il mondo della vita e la storia,' *Rassegna di scienze filosofiche*, 16 (1963) 327–56.

Pucci, R. *La fenomenologia contemporanea ed il problema dell'uomo*, Naples: Libreria Scientifica Editrice, 1963.

Rovatti, R. And S. Veca 'Per un discorso fenomenologico sul teatro,' *Aut-Aut*, no. 81 (1963) 15–54.

Tommasi, A. 'Il logos di Husserl,' *Rivista rosminiana di filosofia e cultura*, 57 (1963) 282–85.

Valori, P. 'Fenomenologia e filosofia italiana oggi,' *Humanitas*, (Brescia) (July, 1963).

Vanni Rovighi, S. 'Studi husserliani,' *Rivista di filosofia neoscolastica*, 55 (1963) 522–36.

1964

Agosti, V. 'Galilei visto da Husserl,' *Giornale di metafisica*, 19 (1964) 779-96.

Garulli, E. 'Husserl, Kant e i neo-kantiani nell'interpretazione di Iso Kern,' *Pensiero*, 9 (1964) 125–46.

Garulli, E. 'Problemi della filosofia husserliana I e II, *Studi urbinati di storia, filosofia e letteratura*, 38 and 39 (1964) 450–72.

Garulli, E. *Coscienza e storia in Husserl*, Urbino: Stev. 1964.

Giganti, M.A. *Dialettica della coscienza Morale nel mondo contemporaneo,* Palermo, 1964.
Lugarini, L. *Esperienza e verità,* Urbino: Argalia, 1964.
Masullo, A. *Lezioni sull'intersoggettività I: Fichte e Husserl,* Naples: Libreria Scientifica Editrice, 1964.
Paci, E. 'Annotazioni per una fenomenologia della musica,' *Aut-Aut,* nos. 79–80 (1964) 54–56.
Paci, E. 'Fenomenologia e cibernetica,' *Aut-Aut,* no. 84 (1964) 25–32.
Pietra, L. *Fenomenologia e marxismo,* Cologne: Monsese, 1964.
Pucci, R. 'Edmund Husserl' in *Les grands courants de la pensée mondiale contemporaine,* Milan: Marzorati, 1964.
Sini, C. 'Per una propedeutica fenomenologica,' *Aut-Aut,* no. 84 (1964) 19–51.

1965

Altamore, G. 'La filosofia come scienza rigorosa e la riduzione fenomenologica in E. Husserl,' *Teoresi,* 15 (1965) 157–94.
Masullo, A. *La comunità come fondamento: Fichte, Husserl, Sartre.* Naples: Libreria Scientifica Editrice, 1965.
Paci, E. 'Whitehead e Husserl,' *Aut-Aut,* no. 84 (1965) 7–8.
Paci, E. 'Struttura temporale e orizzonte storico,' *Aut-Aut,* no. 87 (1965).
Paci, E. 'Antropologia strutturale e fenomenologia,' *Aut-Aut,* no. 88 (1965).
Paci, E. *Tempo e relazione,* Milan: Il Saggiatore, 1965.
Piana, G. *Esistenza e storia negli inediti di Husserl,* Milan: Lampugnani Nigri, 1965.
Sancipriano, M. 'L'Urkind di Husserl,' *Aut-Aut,* no. 86 (1965) 7–26.
Scrimieri, G. *La matematica nel pensiero giovanile di Edmund Husserl,* Bari: F. Cacucci, 1965.
Sini, C. *La fenomenologia,* Milan: Garzanti, 1965.
Sini, C. *Introduzione alla fenomenologia come scienza,* Milan: Lampugnani Nigri, 1965.
Uccelli, A. 'Problematica husserliana logico-algebrica,' *Aut-Aut,* no. 85 (1965) 31–45.
Voltaggio, F. *Fondamento della logica di Husserl,* Milan: Edizioni di Comunità, 1965.

1966

Bosio, F. *Fondazione della logica in Husserl,* Milan: Lampugnani Nigri, 1966.
Paci, E. *Relazioni e significati III, critica e dialettica,* Milan: Lampugnani Nigri, 1966.
Paci, E. 'Per lo studio della logica di Husserl,' *Aut-Aut,* no. 94, (1966) 7–25.
Paci, E. 'Tema e svolgimento in Husserl,' *Aut-Aut,* No. 95 (1966).
Paci, E. 'Il senso delle parole. Fenomenologia della prassi e realta obiettiva,' *Aut-Aut,* nos. 96–97 (1966–1967)
Piana, G. 'Un'analisi del colore,' *Aut-Aut,* no. 92 (1966) 21–30.
Rovatti, P. and S. Veca 'Fenomenologia, psicologia e teatro,' *Aut-Aut,* no. 91 (1966) 49–75; no. 92 (1966) 47–66.
Valori, P. 'La fenomenologia,' *Studio e insegnamento della filosofia,* (Rome) (July 1966): AVE, 3–11.

1967

Agosti, V. 'Brevi note husserliane,' *Giornale di metafisica,* no. 22 (1967).
Biral, A. *L'unità del sapere in Husserl,* Padua: Cedam, 1967.
Bosio, F. 'La costituzione fenomenologica del mondo dello spirito nelle *Ideen* di Husserl,' *Pensiero,* 12 (1967) 56–67.
Cerri, S. 'La nozione di intenzionalità in Husserl e in San Tommaso: appunti per un esame comparativo,' *Rivista di filosofia neoscolastica* (December 1967).
Conci, P. *La conclusione della filosofia categoriale,* Rome: Abete, 1967.
Forni, G. *Il sogno finito – Saggi sulla storicità della fenomenologia,* Bologna: Il Mulino, 1967.
Paci, E. 'Fondazione e chiarificazione in Husserl,' *Aut-Aut,* no. 99 (1967) 153–54.
Paci, E. 'Per una interpretazione della natura materiale in Husserl,' *Aut-Aut,* no. 100 (1967) 47–73.
Paci, E. 'Natura animale, uomo concreto e comportamento reale in Husserl,' *Aut-Aut,* no. 101 (1967) 26–47.
Paci, E. *La formazione del pensiero di Husserl e il problema della natura materiale e della natura animale: Corso di filosofia teoretica,* Milan: La Goliardica, 1967.
Paci, E. 'Per una semplificazione dei temi husserliani fino al I volume delle *Ideen,*' In *Studi in onore di A. Massolo,* Urbino: Argalia, 1967, Vol. 2, pp. 67–87.
Pantaleo, P. *La direzione della coscienza-intenzione nella filosofia di Kant e Husserl,* Bari: Laterza, 1967.
Raggiunti, R. *Husserl: dalla logica alla fenomenologia.* Florence: Le Monnier, 1967.
Sancipriano, M. *L'ethos di Husserl. Communicazione intersoggettiva e etica sociale,* Turin: Giappichelli, 1967.
Zecchi, S. 'Note di estetica fenomenologica,' *Aut-Aut,* no. 102 (1967) 63–81.

1968

Bacchini, G.R. 'Una interpretazione di Husserl,' *Giornale di metafisica,* 23 (1968) 582–85.
Conci, P. 'A proposito della *Filosofia come scienza rigorosa,* di E. Husserl,' *Giornale di metafisica,* 23 (1968) 186–207.
Daghini, C. 'Motivazione, irrealtà e il tema della praxis,' *Aut-Aut,* nos. 105–6 (968) 7–43.
De Negri, M. "Esperienza e giudizio" analisi husserliana rapportata al problema genetico della conoscenza infantile: 'Infanzia anormale' *Quaderno,* no. 12 (1968).
De Vidovich, 'Husserl in URSS,' *Aut-Aut,* no. 107 (1968) 104–6.
D'Ippolito, B.M. *Ontologia e storia in Husserl,* Salerno: Rumma Edit. 1968.
Ferrari, M.V. *Dalla fenomenologia pura alla trascendenza assoluta,* Rome: Universita Lateranense, 1968.
Forni, G. 'Fenomenologia della storia e idealismo intenzionale in Husserl,' *Giornale di metafisica,* 23 (1968) 408–28.
Paci, E. Motivazione, ragione, enciclopedia fenomenologica,' *Aut-Aut,* nos. 105–6 (1968) 100–28.

Paci, E. and P.A. Rovatti 'Persona, mondo circostante, motivazione,' *Aut-Aut,* nos. 105–6 (1968) 142–71.
Rovatti, P.A. 'Whitehead e Husserl: una relazione,' *Man World* 1 (1968) 587–603.
Sancipriano, M. 'Fenomenologia e rivelazione,' *Humanitas,* (Brescia) 23, no. 12 (1968) 1212–16.
Sini, C. 'Genesi e costituzione in riferimento al secondo libro della *Ideen,*' *Aut-Aut,* no. 105
Tavrizian, G.M. 'La fenomenologia di Husserl e l'esistenzialismo contemporaneo,' *Voprosi filosofii,* 1 (1968).
Tomassini, R. 'Il paragrafo 64 di *Ideen* II,' *Aut-Aut,* nos. 105–6 (1968) 183–91.
Veca, S. 'Implicazioni filosofiche nella nozione di ambiente,' *Aut-Aut,* nos. 105–6 (1968) 172–82.
Zecchi, S. 'Causalità e percezione,' *Aut-Aut,* nos. 105–6 (1968) 66–90.

1969

Ales Bello, A. 'Teologia e teleologia nella fenomenologia de Husserl,' *Vita Sociale,* 26 (1969) 107–61.
Ales Bello, A. 'L'uomo e Dio nella fenomenologia de Husserl,' In *Ristrutturazione antropologica dell'insegnamento filosofico,* Atti del II Convegno nazionale dei docenti italiani di filosofia nelle Facolta, Seminari e Studentati d'Italia, *Rivista Sapienza* (1969) pp. 556–59.
Altamore, G. *Dalla fenomenologia all'ontologia. Saggio interpretativo su Edmund Husserl (Il pensiero moderno S III,* Padua: Cedam, 1969.
Baratta, G. *L'idealismo fenomenologico di Husserl,* Urbino: Argalia, 1969.
Costa, C. 'Sull'evoluzione del problema del tempo in E. Husserl,' *Filosofia,* 20 (1969) 473–502.
Gatta, G. and C. Forni *Sguardi sulla filosofia contemporanea,* Turin: Ed. di Filosofia, 1969.
Marini, A. 'Considerazioni critiche su *Husserl e Kant* di Kern,' *Aut-Aut,* no. 111 (1969) 73–90.
Prezioso, F.A. 'L'intenzionalità della coscienza nella fenomenologia di E. Husserl,' *Rassegna di Scienze Filosofiche,* 22 (1969) 187–220.
Vircillo, D. 'I problemi della psicologia e l'unità del Sapere in Edmund Husserl,' *Teoresi,* 24 (1969) 42–70.
Zecchi, S. 'Diallettica: possibilità e realità,' *Aut-Aut,* no. 111 (1969) 40–72.

1970

Ales Bello, A. 'Coscienza e soggettività nella fenomenologia di Husserl,' *Incontri culturali,* 3 (1970) 69–78.
Ales Bello, A. 'Neutralizzazione e filosofia nella fenomenologia di Husserl,' In *Aporie della filosofia contemporanea,* Parma: Studium Parmense, 1970, pp. 169–97.
Conci, P. *Prolegomeni ad una fenomenologia del profondo,* Rome, 1970.

Paci, E. 'Due temi fenomenologici: I. Fenomenologia e dialetica. II. La fenomenologia e la fondazione dell'economia politica,' *Aut-Aut*, no. 116 (1970) 7–37.

Ponsetto, A. 'Edmund Husserl: dalla critica della società alla domanda intorno a Dio,' In *Civiltà Cattolica*, no. 4 (1970) pp. 69–78.

Prezioso, F.A. 'Nuove prospettive ontologiche della fenomenologia,' *Rassegna di scienze filosofiche*, 23 (1970) 22–53.

Raggiunti, E. *Introduzione a Husserl*, Bari: Laterza, 1970.

Rovatti, P. 'Per un'analisi fenomenologica del marxismo: ritorno al soggetto e dialettica della totalizzazione,' *Aut-Aut*, no. 116 (1970) 60–81.

Semerari, G. 'L'alternativa fenomenologica,' *Aut-Aut*, nos. 119–20 (1970) 110–21.

Vircillo, D. 'Le scienze umane e la fenomenologia da Husserl a Merleau-Ponty,' *Teoresi* 25 (1970) 235–73.

1971

Ales Bello, A. 'La fenomenologia di Husserl,' In *Il tamburo di latta*, Parma (1971) pp. 173–76.

Ales Bello, A. 'Ragione e utopia nella prospettiva fenomenologica,' *Incontri culturali*, 4 (197) 190–98.

Faggiotto, P. *L'interpretazione husserliana del pensiero moderno. Guida alla lettura della parte I e II della* Crisi, Padua: Gregoriana, 1971.

Forni, G. 'Commento alla *Crisi* di E. Husserl,' *Giornale di metafisica*, 26 (1971) 137–70.

Forni, G. 'Commento alla *Crisi* di Husserl II,' *Giornale di metafisica*, 26 (1971) 453–76.

Gessoni, A. 'Il significato di *"L'idea della fenomenologia"* nello sviluppo del pensiero di E. Husserl,' *Proteus*, 2, no. 6 (1971) 95–112.

Masi, G. *Disperazione e speranza: Saggio delle categorie kierkegaardiane*, Padua: Gregoriana, 1971.

Piana, G. 'Husserl, Schlick e Wittgenstein – Sulle cosidette 'proposizioni sintetiche a priori',' *Aut-Aut*, no. 122 (1971) 19–41.

Valori, P. 'Il problema ermeneutico in Husserl e Heidegger,' In *Dimensione antropologica della teologia*, Milan: Àncora, 1971, pp. 321–24.

1972

Ales Bello, A. *Edmund Husserl e la storia*, Parma: La Nazionale, Nuovi, Quaderni, 1972.

Ales Bello, A. 'Valore e significato della filosofia di Anna-Teresa Tymieniecka,' *Incontri culturali*, nos. 2–3 (1972) 319–22.

Brugnaro, F.G. 'Jacques Maritain e la fenomenologia husserliana,' *Humanitas*, 27, nos. 8–9 (1972) 694–704.

Callieri, B. *Lineamenti di una psicopatologia fenomenologica*, Rome: Il pensiero scientifico, 1972.

Conte, G. 'In margine a un manoscritto husserliano sull' estetica,' *Aut-Aut*, nos. 131–32 (1972) 102–7.

Forni, G. *Il soggetto e la storia,* Bologna: Il Mulino, 1972.

Paci, E. 'The *Lebenswelt* as ground as *Leib,*' In *Husserl: somatology, psychology, sociology in patterns of the life-world.* Edited by Lester Embree. Evanston: Northwestern University Press, 1972, pp. 123–38.

Paci, E. '*Life-world*, time and liberty in Husserl,' in *ibid.*

Paci, E. 'A proposito di fenomenologia e marxismo. Considerazioni sul 'dialogo' di Vajdie,' *Aut-Aut,* No. 127 (1972) 44–57.

Penati, G. *Alienazione e verità. Husserl, Hartmann, Heidegger e l'ontologia come liberazione. Studi filosofici,* 7, Brescia: Paideia, 1972.

Rovatti, P. 'Teoria critica e fenomenologia,' *Aut-Aut,* nos. 131 -32 (1972).

Ruggerini, M. *Verità e soggettività. L'idealismo fenomenologico di E. Husserl,* Verona: Fiorini, 1972.

Sancipriano, M. 'Fenomenologia del sacro e figurazione del divino,' *Incontri culturali,* (Rome) no. 1 (1972) 35–46.

Zecchi, S. *Fenomenologia dell'esperienza: Saggio su Husserl,* Published by the Facoltà di Lettere e Filosofia of the Università di Milano, 66, Florence: La Nuova Italia, 1972.

Zecchi, S. 'Un manoscritto husserliano sull'estetica,' *Aut-Aut,* nos. 131–32 (1972) 80-94.

1973

Forni, G. *Fenomenologia – Brentano, Husserl, Schel, Hartmann, Fink, Landgrebe, Merleau-Ponty, Ricoeur*, Milano: Marzorati, 1973.

Marini, A. 'Edmund Husserl,' In *Belfagor. Rassegna di Varia Umanita.* Florence 1973, Vol. 28, pp. 557–91.

Pacchiani, C. *L'idea della scienza in Husserl.* Pubblicazione della Scuola di Perfezionamento in Filosofia dell' Università di Padova. *Quaderni di storia della filosofia,* Padua: Cedam, 1973.

Poggi, S. *Husserl e la fenomenologia: scuola aperta – scienze umane*, Florence: Sansoni, 1973.

Rovatti, P. *Critica e scientificita in Marx.* Milan: Feltrinelli, 1973.

Toldo, S. *Ontologia trascendentale e antropologia in E. Husserl.* Collana di studi filosofici, no. 25, Padua: Gregoriana, 1973.

Vanni-Rovighi, S. *La fenomenologia di Edmund Husserl: Appunti delle lezioni,* Dispense, no. 26, Milano: Celue, 1973.

Vanni-Rovighi, S. 'La fenomenologia di Husserl,' *Verifiche,* 2, no. 1 (1973) 3–17.

1974

Ales Bello, A. 'Un approccio fenomenologico al problema della società,' *Vita sociale,* no. 1 (1974) 58–61.

Ales Bello, A. 'A proposito della 'Philosophia perennis', Tommaso d'Aquino e E. Husserl, nell'interpretazione di E. Stein' in *San Tommaso d'Aquino – Saggi per il VII centenario. Sapienza,* 1974, vol. 27, nos. 3–4. Naples: *Rivista Sapienza*, 1974, pp. 441–51.

De Negri, M. 'A Propos du dévélopement cognitif: quelques remarques sur le 'délire' chez l'enfant,' *Révue de neuropsychiatrie infantile*, 22 (1974).

Ruggerini, M. *Verità e soggettività. L'idealismo fenomenologico di Edmund Husserl.* Quaderni veronesi di varia letteratura, Verona: Fiorini, 1974.

Sancipriano, M. 'The Activity of Consciousness; Husserl and Bergson,' *Analecta Husserliana*, vol. 3 (1974) 161–67.

Tymieniecka, A.-T. 'Hope and the present instant,' in *San Tommaso d'Aquino – Saggi per il VII centenario, Sapienza*, 27, no. 7, Naples: Rivista Sapienza, 1974, 541–545.

Valori, P. 'A che punto è la fenomenologia?' *Gregorianum* (Rome) 55, no. 2 (1974) 393–403.

1975

Ales Bello, A. 'Fenomenologia e creativita,' *Sapienza*, no. 2 (1975) 229–33.

Ales Bello, A. 'Indicazioni fenomenologiche per la fondazione della morale,' *Sapienza*, no. 3 (1975) 338–41.

Ales Bello, A. 'Nota su due recenti congressi dedicati alla fenomenologia,' *Ethos* (1975) 242–5.

Del Carlo, G. and A. Giannini *Approccio fenomenologico nelle psicosi infantili precoci, "Dalla parte dei bambini,"* Pisa, 1975.

De Negri, M. and L. Saccomenni 'Psychodynamique de la disfonction cérebrale a minima,' *Révue de neuropsychiatrie infantile*, 23 (1975).

Prezioso, A. F. 'A proposito della fenomenologia di Husserl,' *Sapienza*, 28 (1975) 219–28.

1976

Ales Bello, A. 'Husserl filosofo borghese?' In *Filosofia e impegno politico*, edited by Comunità di Ricerca-Massimo, Milan, 1976, pp. 235–47.

Ales Bello, A. 'Culture and Utopia in the Phenomenological Perspective,' In *Analecta Husserliana*, (1975) Vol. 5, pp. 305–42.

Ales Bello, A. 'Il VI congresso internazionale di fenomenologia,' *Sapienza*, no. 3 (1976) 284–88.

Ales Bello, A. 'Il VI congresso internazionale di fenomenologia,' *Bollettino filosofico*, (Padua) November 1976, p. 170.

Ales Bello, A. 'Rinascita della fenomenologia in Italia?' *Vita sociale*, no. 6 (1976) 399–402.

Ales Bello, A. 'Empathy, A Return to Reason,' In *Analecta Husserliana*, (1976) Vol. 6, pp. 143–9.

Ales Bello, A. 'Libertà e liberazione nella prospettiva fenomenologica,' *Incontri culturali*, nos. 1–2 (1976) 145–51.

Ales Bello, A. 'Edith Stein: Da Edmund Husserl a San Tommaso d'Aquino,' *Tomismo e Antitomismo, Memorie domenicane*, no. 7 (1976) 265–76.

Ales Bello, A. 'Traduzione italiana di E. Stein. La fenomenologia di Husserl e la filosofia di S. Tommaso d'Aquino. Tentativo di confronto,' *Tomismo e Antitomismo. Memorie domenicane*, no. 7 (1976) 277–93.

Sancipriano, M. 'Homo – Deus, considerazioni di fenomenologia della religione,' In *Il Christo dei filosofi*, Atti del Convegno del Centro di Studi Filosofici tra Professori Universitari, Gallarate (Varese) 1976, pp. 217–221.
Tymieniecka, A.-T. 'Liberta e Liberazione Creatrice,' in *Filosofia e impegno politico*, edited by Communità di Ricerca, Massimo, Milan, 1976, pp. 247–355.
Valori, P. 'Fenomenologia ed esistenzialismo', In *Alla ricerca dell'uomo*, Rome: Ucüim, 1976, pp. 125–48.

1977

Valori, P. 'Phenomenology of Personalistic Morality,' *Analecta Husserliana*, (1977) vol. 6, pp. 81–87.
Valori, P. 'Voce fenomenologia,' In *Dizionario teologico interdisciplinare*, Turin: Marietti, 1977, pp. 192–97.

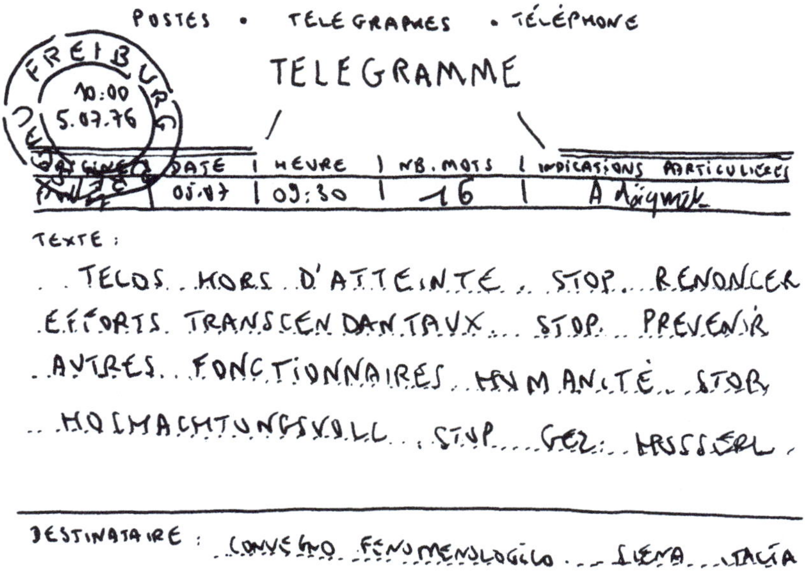

POSTES • TELEGRAPHES • TÉLÉPHONE

TELEGRAMME

FREIBURG 10:00 5.07.76

ORIGINE	DATE	HEURE	NB. MOTS	INDICATIONS PARTICULIERES
	05.07	09:30	16	

TEXTE: TELOS HORS D'ATTEINTE STOP RENONCER EFFORTS TRANSCENDANTAUX STOP PREVENIR AUTRES FONCTIONNAIRES HUMANITÉ STOP HOCHACHTUNGSVOLL STOP GEZ. HUSSERL

DESTINATAIRE: CONVEGNO FENOMENOLOGICO SIENA ITALIA

Telegram received by the participants of the Arezzo/Siena Conference after the closing session of the Conference.

INDEX

ANALECTA HUSSERLIANA

The Yearbook of Phenomenological Research

Editor

ANNA-TERESA TYMIENIECKA

1. *Analecta Husserliana.* 1971, ix + 207 pp.
2. *The Later Husserl and the Idea of Phenomenology.* 1972, vii + 374 pp.
3. *The Phenomenological Realism of the Possible Worlds.* 1974, vii + 386 pp.
4. *Ingardeniana.* 1976, x + 438 pp.

*5. *The Crisis of Culture.* 1976, vii + 383 pp.

*6. *The Self and the Other.* 1977, xi + 186 pp.

*7. *The Human Being in Action.* 1978, xvii + 261 pp.

8. *Japanese Phenomenology.* 1979, xi + 291 pp.

*9. *The Teleologies in Husserlian Phenomenology.* 1979, xvi + 495 pp.

*10. *The Acting Person.* 1979, xxiii + 367 pp.

*Indicates the volumes to which Karol Wojtyla has contributed.